INTERIOR HOME IMPROVEMENT COSTS

Updated Eighth Edition

The practical pricing guide
for Homeowners
& Contractors

RSMeans

INTERIOR HOME IMPROVEMENT COSTS

Updated
Eighth Edition

*The practical pricing guide
for Homeowners
& Contractors*

RSMeans

Copyright 2002
Construction Publishers & Consultants
63 Smiths Lane
Kingston, MA 02364-0800
(781) 422-5000

The editors for this book were Barbara Balboni and Andrea Keenan. The managing editor was Mary Greene. The production manager was Michael Kokernak. The production coordinator was Marion Schofield. The electronic publishing specialist was Paula Reale-Camelio. The proofreader was Robin Richardson. The book and cover were designed by Norman R. Forgit. Cover photographs by Norman R. Forgit. Illustrations by Carl Linde, Richard W. Lowrey, R.A, Robert Megerdichian & Associates, and Barbara Balboni.

Printed in the United States of America

10 9 8 7 6 5 4 3

Library of Congress Catalog Number 2002511399

ISBN 0-87629-656-8

 Reed Construction Data

TABLE OF CONTENTS

[Handwritten notes: "Doors 1.3 man hour per side per coat", "Trim 40¢ per linear ft.", "Paper 60 sq ft per hour", "based on 30 per hour", "RS Means.com"]

TABLE OF CONTENTS *Continued*

TABLE OF CONTENTS *Continued*

Acknowledgments

The editors would like to express appreciation to Scot Simpson for allowing us to reprint selected terms from the glossary of his book, *Builder's Essentials: Framing and Rough Carpentry*.

The "Questions to Cover with the Building Department" and "Contractor Evaluation Sheet" forms in this book were provided by Richard Connolly, President of Cornerstone Consulting, Inc., Weymouth, Massachusetts. Cornerstone Consulting assists consumers in identifying and hiring designers, contractors and subcontractors. The company also helps homeowners develop specifications and negotiate bids, and in many cases, coordinates the overall project, whether new construction or remodeling.

INTRODUCTION

This book is designed to take some of the mystery out of interior projects, both by helping you understand what is involved – in terms of work, materials, and expertise – and by giving you an idea of what it all costs.

This revised eighth edition includes 66 projects—with current cost information and building practices.

Project-by-project, material and labor costs are listed, along with the hours required for installation. The discussion that accompanies each project will help you determine which parts of the job are appropriate for your level of expertise, and which are better left to a contractor. If you are thinking about doing the work yourself in order to save money, consider how much you will really save. If your work has to be corrected later by a professional contractor, you may actually spend more in the long run.

In each project in this book, material quantities are those required in a standard renovation. The contractor's charge includes an estimate of overhead and profit. Overhead includes both direct and indirect costs of doing the work. This includes contractor's site visits, planning and estimating time, and the miscellaneous costs associated with doing business.

A section called "Adjusting Project Costs to Your Location," found at the back of the book, provides cost factors for over 900 cities and towns, organized alphabetically and by zip code. Multiply the project cost by the factor for your location to get the most accurate estimate.

All or parts of the projects in this guide can be done by homeowners, either alone or with advice or assistance. (Keep in mind that electrical, plumbing, structural or foundation work must meet local building code requirements and is best performed by licensed professionals.)

Doing your own work can save you money and give you the satisfaction that comes from creating something of value. But, not everyone can or should undertake every task. Honestly assess your own skills and experience before tackling any home improvement project. Also ask yourself if you really *want* to do the work yourself. If you are doing it just to save money and don't enjoy the work, you probably won't give it your best effort.

Throughout the book, we have referred to three levels of do-it-yourselfer: beginner, intermediate, and expert. A beginner might be attempting a home improvement project for the first time. Generally, a beginner will be in the process of acquiring the tools and equipment necessary for small construction projects. An intermediate is acquainted with tools and has limited experience in building. An expert has extensive experience – having completed many home improvement projects – and has collected a variety of tools. Also, remember that most renovation projects require strength and a good back. If you don't have these attributes, it's probably best to leave major projects to a professional.

Interior projects require accurate layout of the project, and precise measuring, cutting, and assembly of materials. When in doubt, call in a contractor. Although the cost of the project will increase with professional help, the long-term benefits of fine workmanship far outweigh a reasonable initial expense.

Preparation of the site can be a major part of interior remodeling. Old fixtures may have to be removed, and floors, walls, and ceilings reconditioned and prepared to receive new coverings. Experienced do-it-yourselfers can handle many of these basic tasks, but beginners may have to spend some extra money for professional help. Correction of structural problems, particularly, should be left to qualified tradespeople and experts.

Prior to undertaking a project that will require structural changes (that is, the removal or alteration of a supporting wall that will affect the strength of the structure), consult an architect/engineer or local building officials. The proposed changes may require a permit and plans approved or certified by a licensed engineer or architect.

A permit is essentially a license to do the work. Depending on the scope of the project, a permit application may require a description of the property, a drawing of the proposed changes, and a site plan drawing. Most jobs that require the assistance of licensed professionals also require a permit.

A permit generally has a fee attached. Regulatory agencies are established to enforce minimum code requirements and to promote public safety, and permit fees are a small investment compared to the penalties and problems that arise if this part of the construction process is ignored.

When you obtain a permit, ask about the inspection schedule. An inspection is generally required to ensure the work was done in accordance with code requirements. Contractors generally take care of the permit and arrange for the inspection.

Heating, air-conditioning, plumbing, and electrical work can be tricky in interior remodeling, especially in older houses where the heating system, piping, or wiring is in questionable condition or inconveniently arranged. The projects in this book include the cost of basic plumbing and electrical work. Heating system modification costs have not been included. Should your project require changes to the heating system, refer to Part Three, "Details" for help in determining these costs.

How long the project will take is a major consideration, especially in kitchen and bathroom remodeling projects, which can be very disruptive to the household. If you intend to do the work yourself,

determine whether you really have the time to focus on the project and get it done quickly. If the project will take a total of eight days, and you have only weekends to devote to the work, it will be four weeks before it is complete. Again, while hiring a contractor increases your project cost, it may be worth it when compared with weeks of major inconvenience.

Another consideration is delivery and storage of materials. Plan how you will transport the materials to your house and to the appropriate room in the house. Take into account the size of entryways when moving large appliances or building components, and arrange a location for storing materials, especially those that are not weather-resistant.

Finally, consider how the completed work will affect the resale of your home. If you think you'll sell your house within the next few years, weigh any remodeling investment carefully. Avoid spending more than you can get back at sale time. Also, remember that lending institutions expect homes they finance to meet code requirements. In your desire for a beautiful addition to your home, don't neglect environmental and safety issues.

To ensure that your improvements will appeal to future buyers, use fairly neutral colors particularly on items that are costly to change, such as bathroom fixtures, tile walls and floors, and countertops. Use quality materials that look good and wear well, and plan the improvement to blend with the style and character of your house.

The following suggestions are from professional contractors, who have learned by experience the most efficient ways to approach remodeling projects. Those who take the time to read this section improve the odds for starting their projects in a more organized manner, enjoying the experience, and winding up with a result they can be proud of.

Before getting started, establish a time frame for the project. Clearly, the season of year and your own needs and convenience have a great deal to do with the time period chosen for certain projects. Major projects can often be done in phases, which can help to break up the time of disruption; but this approach may also require some rearranging of your living space.

When working on home improvements, try to plan your day so you are finished at a reasonable hour. Painting, floor finishing, and many other tasks are best accomplished when natural light is available. If your work is going to involve disrupting water, gas, or electric service, make sure it takes place at a time when you can arrange for emergency repairs or professional assistance. In other words, do not choose a holiday weekend or a Sunday afternoon. In addition, do not depend on supply houses (for plumbing, electrical, cabinetry items, or fixtures) being open on Saturday afternoons and Sunday mornings. Call in advance to avoid frustration.

Check your local trash pickup or dump rules and schedule. Make sure you understand the policy on what refuse materials are acceptable for the local dumping station or landfill.

The motivation for most of us tackling a home improvement project on our own is to save money by investing our time. Be careful about trying to save time as well. In the rush to get started, many homeowners have broken windows, scratched finishes, and even damaged their new materials with improper handling, storage, or installation methods.

Assess your collection of equipment and tools, and make a list of what you will need. Research the best brands and deals. Call around for the best price, keep your eyes open for a tool or material on sale, or start purchasing materials in advance to save money or to spread out the expense. (A convenient worksheet is included at the back of this book to help you list and price individual materials.) If you are looking for bargains, make sure the product is complete and of acceptable quality, and that it can still be serviced. Keep in mind that the better the quality or grade of material, the easier it is, generally, to work with. Also, prior to purchasing an item, check on any delivery charges or additional fees for special treatments that may increase the actual cost of the item.

Read and follow the instructions accompanying tools, materials, appliances, and fixtures. If the can of paint says, "let the first coat dry for eight hours before applying the second coat," wait eight hours.

Sometimes we are in such a rush to step back and enjoy the fruits of our labor that we eliminate steps such as the recommended sanding, final coat, or buffing. Overcome the urge, and you will be thankful later. The work may take longer, but it will last and look better.

An addition may add to your home's square footage. Keep in mind that only areas with a ceiling height of six feet or more in a 1-1/2 story residence are considered living area.

There is no greater satisfaction than having accomplished a major project ourselves, especially when it comes to a home improvement. But, if we all could do it ourselves, the professionals would be out of business. Recognize your limitations; plan your project and evaluate the impact of the construction on your family's comfort; read the safety section

that follows; seek expert advice; and most importantly, enjoy the experience.

Keep a journal of your project, and track the costs and time involved. Photograph the steps along the way, especially for major, time-consuming projects. This will give you a record of the "event," as well as a document that may be useful for future work on your home or for resale.

Remodeling the interior of your home presents some definite challenges, but it also promises rewards that are realized daily in the improved appearance, function, and value of your house. Being aware of what is involved in a project and knowing the costs in advance will help you to plan the best renovation.

All of the projects covered in this book start with a space enclosed by walls and located under the cover of a roof. If you are interested in building an addition to house your new room, consult *Exterior Home Improvement Costs*, which contains 64 projects ranging from room additions, garages, and greenhouses to patios, decks, and landscaping.

Note: Whether you are hiring a contractor or doing the work yourself, it is important to have a quality standard for workmanship in mind. We recommend *Residential & Light Commercial Construction Standards*, a publication that defines quality in construction, to help avoid disputes with your contractor and to answer installation questions with authority. The book is based on building code requirements and guidance from leading professional associations, product manufacturer institutes, and other recognized experts. The book is available at the contractor desk of many major home improvement centers and bookstores, or can be ordered from R.S. Means Co. at 1-800-334-3509 or www.rsmeans.com

Cost Versus Value

When you decide to undertake any home remodeling project, it's important to consider how that improvement will add to the value of your property. How much of the cost of remodeling will be recouped if you sell your house?

Traditionally, updating a kitchen and adding a bathroom have been the projects that have the highest resale value. Improvements such as these can boost a home's resale value by more than they cost as soon as one year after the job is completed. Real estate agents say that a minor kitchen remodel (average cost: $9,000) will generally recoup 98% of the cost. A major kitchen remodel (average cost: $22,000) will recoup about 85% of the cost. A bathroom addition (about $13,000) can regain 89% of its cost at sale time. The percentage varies depending on where you live. Other improvements that are almost as valuable as a kitchen remodeling or bathroom addition include adding a home office, family room, or master bedroom suite.

Remodeled homes sell better and faster in today's real estate market. Older homes can be remodeled to include the same features that are standard in new homes. However, don't "overimprove" your house – by making it considerably more elaborate than the other homes in your immediate neighborhood. When you add to your home to make it conform to the neighborhood, the cost of the improvement is easier to regain at sale time.

Quality is important. Use good quality products, and make sure the job is well done. If you are not adept at carpentry, hire a professional. A poorly executed remodeling job will not add value to your home. (See page xi for information and a recommended reference to help you define quality in construction.)

Don't base all remodeling decisions on resale value, however. You need to live in your house, and it may be several years before you even think of selling it. Your project should be primarily for your own convenience and enjoyment. Also, if you're trying to decide whether to move or improve, weigh the cost and value against the cost of selling and moving into another house.

Because we spend so much time in our homes, we tend to overlook some of the very basic cosmetics that make a home appealing: paint, wallpaper, and other finishes that are especially subject to everyday wear and tear.

Upgrading these items can make a tremendous difference in the saleability of a house, with a modest investment by the do-it-yourselfer.

Prior to undertaking any project, you should contact your local building department to clarify the need for permits or licensed contractors to do the work. Later on when you wish to sell the property, this could be a very important issue.

SAFETY TIPS

Each project description in this text features a "What To Watch Out For" section, which often includes safety tips relevant to that project. You must be prepared to do your work safely and meet local building codes. If you don't think you can attain these standards, you're better off hiring a licensed professional. Following are some additional recommendations.

- Make a point of keeping the work area neat and organized. Eliminate tripping hazards and clutter, especially in the areas used for access. Take the time to clean up and reorganize as you go along. This means not only keeping neat piles of materials, but also remembering not to leave nails sticking out; pull them or bend them over. You may be called away or run into difficulties that delay the project, and you should not leave a hazard.

- Wear the proper clothing and gear to protect yourself from possible hazards.

- Avoid loose or torn clothing if you are working with power tools. Serious injury can be caused if clothing is caught in moving parts.

- Wear heavy shoes or boots to protect your feet, and a hardhat in any situation where materials or tools could fall on your head.

- Wear safety glasses when working with power tools or in any other circumstance where there is the potential for injury to the eyes.

- Use hearing protection when operating loud machinery or when hammering in a small, enclosed space.

- Wear a dust mask to protect yourself from inhaling sawdust, insulation fibers, or other airborne particles.

- Wear suitable gloves whenever possible to minimize hand injuries.

- When lifting, always try to let your leg muscles do the work, not your back. Keep your back straight, your chin tucked in, and your stomach pulled in. Maintain the same posture when setting an item down. Seek assistance when moving heavy or awkward objects and, remember, if an object is on wheels, it is easier to push than to pull it.

- When working from a ladder, scaffold, or any temporary platform, make sure it is stable and well braced. When walking on joists, trusses, or rafters, always watch each step to see that what you are stepping on is secure.

- When working with adhesives, protective coatings, or other volatile products, be sure to follow manufacturers' recommendations on ventilation. Pay particular attention to drying times and fire hazards associated with the product. If possible, obtain from your supplier a Material Safety Data Sheet, which will clearly describe any associated hazards.

- Do not use or store flammable liquids or use gasoline-fueled tools or equipment inside of a building or enclosed area.

- When working with electricity or gas, be sure you know how to shut off the supply; then make sure it is off in a safe way by testing the outlet fixture or equipment. It may be wise to invest in a simple current-testing device to determine when electric current is present. Be sure you know how to use it properly. If you don't already have one, purchase a fire extinguisher, learn how to use it, and keep it handy. Have emergency telephone numbers and utility telephone numbers at hand.

- When using power tools, never pin back safety guards. Choose the correct cutting blade for the material you are using. Keep children or bystanders away from the work area, and never interrupt someone using a power tool or actively performing an operation. Always unplug tools when leaving them unattended or when servicing or changing blades.

A few tips on hand tools:

- Do not use any tool for a purpose other than the one for which it was designed. In other words, do not use a screwdriver as a pry bar, pliers as a hammer, etc. Not only can the tools be easily ruined, but the impact that causes the damage may also injure the user.

- Do not use any striking tool (such as a hammer or sledgehammer) that has dents, cracks, or chips; shows excessive wear; or has a damaged or loose handle. Also, do not strike a hammer with another hammer in an attempt to remove a stubborn nail, get at an awkward spot, etc. Do not strike hard objects like concrete or steel, which could chip the hammer, causing personal injury.

- Do not hold an item in one hand while using a screwdriver on it. One slip and your hand is wounded. Do not use a screwdriver for electrical testing or near a live wire.

- If you rent or borrow tools and equipment, take time to read the instructions or have an experienced person demonstrate proper usage.

Seek further advice on proper tool selection and use from your local building supply dealer, or from the Hand Tools Institute at 25 N. Broadway, Tarrytown, NY 10591 or http://www.hti.org. Always review manufacturers' instructions and warnings.

Home remodeling can be a satisfying and rewarding experience. Proper planning, common sense, and good safety practices go a long way to ensure a positive, money-saving experience. Take your time, know your limitations, get some good advice, and have fun.

SPECIAL CONSIDERATIONS FOR REMODELING AN OLDER HOME

There is something special about working on an old house. In some regions of the country it may seem like you don't really own your house until you move away from it. "The Smiths lived in the old Haywood house until they sold it to the Joneses, who now live in the old Taylor house." You can get the sense that any work you do should meet with the approval of the original owners. And what about the craftsmen who first created the home? When you work on an older home, you cannot help but feel a kinship with them. Often, you will find a signature or mark that identifies the craftsperson. In some instances, you may recognize a detail that is associated with a particular builder. True craftsmen respect the work of those that came before them, and good carpenters, masons, plumbers, and electricians are remembered for generations.

People who own old homes usually discover, at some point, that upkeep and routine maintenance generally cost more than work on newer homes. The satisfaction of preserving the home's personality – shaped by its designer, builder, and occupants – needs to outweigh the costs. Many levels of commitment are necessary to maintain the old structure exactly as it was. The same dedication is needed when you are introducing modern conveniences in a way that maintains the original character while providing an easier-to-maintain structure. You will need to decide how "pure" you want to be and to what level you wish to preserve your home's original features.

Houses constructed prior to 1940 have unique characteristics not seen in homes built today. From foundation to finishes, the differences can be significant. The casual observer might not recognize these features, but knowledgeable homeowners and building professionals can make a long list. Consider the following conditions as you contemplate a remodeling project for your old house.

Old houses are often built with true dimensional lumber. This means that a 2 x 4 is actually 2″ x 4″. Whereas the actual dimensions of a 2 x 4 purchased today are 1-1/2″ x 3-1/2″. This difference in dimensions is consistent for all framing lumber produced today. In old houses, the boards used for both rough and finished carpentry may be a full 1″ thick, compared to stock used today that is actually 3/4″ thick. If you are pursuing an exact restoration, you will need to find a mill producing the exact dimensional lumber you require. If exact reproduction is not important, then you must be prepared to introduce shims to build up narrower new members to match what is already in place. In cases where new members (both cosmetic and structural) will be exposed, true dimensional lumber will need to be used, and it can be costly.

Some older homes are balloon-framed. With this style of wood framing, the vertical structural members (the posts and studs) are continuous pieces from sill to roof plate. The intermediate floor joists are supported by ledger boards spiked to, or let into, the studs. Temporary bracing and floor support can be critical to working on both interior and exterior walls that have been balloon-framed. Adding new wires and utilities may actually be easier in a balloon-framed wall, because the studs run uninterrupted from sill to roof plate, with the exception of firestops.

Old houses usually have higher ceilings that require longer lengths of lumber for wall framing. If the walls are over 8′ high, you may incur the cost of greater waste when purchasing wall finish materials. You may also find that the rooms are smaller, and that the partitions are structural (load-bearing), requiring more material when reframing. In some cases, the original builder (or a subsequent remodeler) may have placed additional blocking or bracing within the wall structure. This may obstruct the placement of new pipes or wiring, which could lead to additional demolition and reframing.

Doors and windows are often unique shapes and sizes. Replacement may involve the services of a custom millwork company. Reworking old doors and sash can be extremely time consuming. There may also be a long lead time for ordering custom units. The different dimensions of framing members may result in varying wall thicknesses, which means stock door frames. Window frames may require extension jambs, adding to the cost and time of construction.

Door and window frames may be part of the structural components that make up a typical wall framing system and may require greater care during removal. Similarly, cabinets and bookcases are often built in place. Removing them may necessitate a fair amount of patching and matching of finishes. Lath and plaster in older homes is generally thicker than drywall used today. It can be difficult to patch this wall covering so that it matches existing surfaces, if you want to achieve a perfect finish. Often, it is necessary to reconstruct and finish an entire wall rather than patch a small area.

To match interior and exterior mouldings, you may need to have the material custom-milled. When removing existing mouldings and trim, try to determine the nailing pattern and fastening method that was used originally. Take care and try to save as much of the material as possible.

Even small pieces may come in handy for patching and replacing sections of trim, siding, mouldings, or other details. When you are trying to match existing materials, be sure to remove all paint to establish true dimensions.

Exterior trim replacement or matching of new to existing trim can sometimes be almost impossible. Many details become obscured or disappear under numerous coats of paint. Hand-planed mouldings, and those of the same shape but different size, may be costly to reproduce. Before you consider eliminating mouldings or details, take a good look at what the finished product will look like without them. Many stock mouldings can be used in combination to produce a detail that is very similar to an original design.

When rebuilding cornices and overhangs, it is a good idea to dismantle them carefully to understand how they were put together, as well as to make patterns from the original pieces. Carefully document the lengths of overhangs and flashing details so that you can reproduce

the look that is so pleasing and common to older homes. Taking things apart carefully offers another advantage, too. You will be able to fabricate parts and do some of the assembly on the ground, rather than on a scaffold. We recommend priming and applying first coats of paint prior to attaching any exterior trim or finish materials.

If your renovation involves replacing siding, it is a good idea to mark the spacing of the old siding on the cornerboards. Then, you can apply the new material to match the exact layout of the existing siding.

Many companies specialize in reproducing building components, lighting fixtures, bath and kitchen fixtures, and cabinetry. Often, they can supply almost anything else you think of that would complete your renovation in keeping with the period. Using reproductions can help you satisfy current building codes more easily, while ensuring the safety of the occupants.

There is no reason you should avoid tackling any of the projects outlined in

this guide. However, owners of older homes should consider working with professionals to address structural concerns, fireplace and chimney rebuilding, electrical, plumbing, and abatement of hazardous materials (such as lead paint or asbestos pipe covering). It may also be worth your while to visit your library or to seek the assistance of the local historical society prior to planning your projects. You may discover some interesting history regarding your house and its original design and construction. This knowledge may help you decide how many of the original details you want to preserve.

Working on an old house is special. You may do it for sentimental reasons, as a reminder of a simpler time, to preserve its aesthetic "character," or out of respect for the craftsmanship. Whatever the reason, be patient and enjoy the satisfaction that comes from knowing you have added your signature to something that will endure because you cared.

Part One
WORKING WITH A CONTRACTOR

DECIDING TO WORK WITH A CONTRACTOR

The "What to Watch Out For" section in each project description points out some of the difficulties that can emerge during the course of a home renovation. In addition, some projects indicate that the time involved for a do-it-yourselfer can be 100-150% or more than the time estimated for a professional to complete the work. This is not to say that you cannot complete the work; sometimes, however, it makes more sense to hire a professional to do all or some of the work.

You may choose to act as general contractor for the project. It is important to understand what the term "general contractor" implies: he or she has a general knowledge of the construction process and manages the tradespeople who participate in the actual construction of the project. The responsibility for scheduling, planning, quality control, and overall job performance belongs to the general contractor.

Because of the close relationship you have with your own home, you need to be completely satisfied with your contractor's commitment to the work to be done. Many people put more time and energy into choosing a contractor for their home than they do into choosing a physician for themselves! The following lists are suggestions for selecting and working with a contractor. They are by no means exhaustive, but can be helpful in beginning to think about the relationship you will be establishing if you choose to hire someone to work on your home improvement project. The following discussion assumes that you are acting as a general contractor, hiring subcontractors to work with you. Should you hire a general contractor, he or she will undoubtedly use the same principles.

SELECTING A CONTRACTOR

Hiring a contractor who is licensed guarantees that he or she has satisfied the state's requirements to perform a certain type of work–it does not, however, guarantee that the work will be high quality. You want a contractor who has a good track record and will be pleasant to work with.

- Ask friends and colleagues for recommendations when looking for a carpenter, plumber, electrician, painter, or other contractor. You may have heard about your friend's experiences (both good and bad); you may have seen the completed project; you may have even met the contractor.
- As you walk or drive around your area, pay attention to the construction that is taking place. You may be surprised both at the amount of work being done and at the number of contractors doing the work. Look for signs and names on trucks, or ask one of the workers for a business card.
- Make inquiries at your local lumber yard, building supply house, or building official's office. While they will typically not make specific recommendations, they may give you some general suggestions. Often very specialized tradespeople – stair builders, for instance – can be discovered this way.
- Look for a contractor who works with a qualified team of subcontractors and suppliers. Contractors who do repeat business with subcontractors and suppliers receive discounts on materials and labor – cost savings that will be passed on to you.
- Ask a contractor whom you trust from one trade about another. For example, ask your carpenter about plumbers; ask your plumber for the name of a good electrician. You may begin to see that, if you ask enough people, the same names keep coming up.
- Check with your local builder's association or trade associations.
- While asking for names, also ask about the contractor's reputation for follow-up and warranties.
- When choosing a contractor, age should not be a factor. Sometimes younger tradespeople have more knowledge about newer materials and methods than do professionals who have been in the business for years.
- Ask to see pictures or a list of addresses where you can see the contractor's previous work. No one should be hesitant to provide references.

ESTABLISHING A RELATIONSHIP WITH A CONTRACTOR

There is no reason to believe that hiring a contractor will be anything but a positive experience. As long as you both agree to the terms of the work and follow through on your obligations, the relationship should be productive and mutually profitable.

- Don't be shy about asking a lot of questions. Make sure your questions are answered to your satisfaction and that you and the contractor both understand the terms of the work to be done. Be sure to communicate your needs clearly, especially if you intend to request materials made by specific manufacturers or other special items.

- Your contractor should be able to gauge the approximate length of time the project will take, identify the nature of disruptions to your daily life, and estimate the impact of his or her work on existing plumbing, heating, and electrical systems and on your landscaping.
- Request a written proposal that clearly defines the scope of the work and the time in which it is to be performed.
- A professional contractor will provide you with a schedule of the job so you can track the progress of the project. The schedule should include deadlines for selections you must make, such as paint colors or flooring materials.
- Agree on a payment schedule and adhere to it.
- Be reasonable in your requests. Establish the terms of the work to be done before it begins; try not to change the terms of the agreement after the job is under way. If you do request a change, realize that it will probably involve additional time and expense.
- When establishing a time frame in which the work will be completed, realize that the forces of nature (rain, snow, earthquakes) and other extenuating circumstances (illness, availability of materials) can delay the project. Be firm but fair in your expectations.

CONTRACTUAL ARRANGEMENTS

Even a small job requires a legally binding document that spells out the terms and conditions of your agreement. A professional contractor will probably insist on a detailed contract with all aspects of the job in writing. A fair contract will address the needs of both parties. There are several types of contractual

arrangements that can be used for construction work. The following is a partial list providing basic information; you should consult a legal advisor if you need further advice.

- A *lump sum contract* stipulates a specific amount as the total payment due to the contractor for performance of the contract.
- A *cost-plus-fee contract* provides for payment of all costs associated with completion of the work, including direct and indirect costs as well as a fee for services, which may be a fixed amount or a percentage of costs.
- A *cost-plus-fee contract with not-to-exceed clause* is similar to the cost-plus-fee contract, but the profit is a set amount. The contractor guarantees that the project cost will not exceed a set amount.
- In a *labor-only contract*, the contractor is paid for labor only. The materials are furnished by the owner or others.
- In a *construction-management-fee-only contract*, the owner pays all subcontractors on the project, and pays a supervision fee to the general contractor in a lump sum or as a percentage of the work involved.

WHAT TO WATCH OUT FOR

- Don't pay any amount that does not seem justified or that was not agreed to beforehand without asking for an explanation. You should not be required to pay for a job in full before the work is complete. On the other hand, most contractors will require at least one partial payment before and/or during the course of the work.
- Ensure that the contractor you intend to work with is fully insured and licensed for the work he or she will

perform. Be sure you know the licensing requirements in your jurisdiction. (You can contact your state's department of business regulations or licensing for information.) Contractors should carry Workers' Compensation and general liability insurance. Ask him or her for a current certificate of insurance.
- Be sure to ask for all warranties, manuals, and other literature related to new equipment or materials being installed. Request documentation of brands being specified.
- If you intend to do some of the work yourself, you and your contractor should both understand where (and when) your work ends and the contractor's work begins.
- Maintain a businesslike relationship with the contractor. If you become too friendly or casual, the terms of the agreement may become blurred, time may be wasted, and disagreements may ensue.
- Make sure you understand *who* will be doing the actual work. The person you make arrangements with may not be the same person who will do the physical work. Again, ask plenty of questions.

ETHICAL CONSIDERATIONS

- When soliciting bids for the job, provide equal information to all parties. Provide the same opportunity for site visits to all bidders as well.
- Once the decision has been made to hire a certain contractor, inform unsuccessful bidders (keeping in mind that you may wish to solicit them in the future).

- As much as possible, stay out of the workers' way while they are doing their jobs. Be available to answer questions or make last-minute decisions, but try not to create distractions or engage in needless conversation. You do not want to be responsible for accidents or delays in the completion of the project.
- If you have agreed to do certain prep work yourself, make sure it is complete before the workers arrive to do their jobs.
- Respect safety issues and job site conditions. If a contractor asks you to move all breakable items and keep small children out of the area, there is good reason to do so.
- Do not withhold information that could be valuable to the contractor. If you know of any condition or situation that could pose a problem during the course of the work, let the contractor know.
- Realize that the contractor is there to perform the agreed-upon work; don't ask for "freebies." For example, if the contractor is there to install a new window, don't expect him or her to also repair other windows in the room for no additional cost.

CLOSING OUT THE PROJECT

When you decide to complete a home improvement project on your own, you really have no obligation to finish the project. Many of us occupy a new space before it is complete. If you hire an outside contractor, however, you assume that the work will be completed. The following items should not be overlooked.

- Ensure that any necessary inspections are completed.
- Obtain occupancy permits if necessary.
- Have in your possession all warranties, manuals, and other literature associated with new equipment or materials that have been installed.
- Ask the contractor about materials that may be left over from the job; e.g., if you had a ceramic tile floor installed, you will want to have some extra tiles on hand for repairs. If you have no need for additional materials and you have paid for them in advance, ask about credit for materials not used.
- Ensure that items remaining on the punch list have been completed. A punch list is comprised of items that remain to be replaced or completed at the time of substantial completion, in accordance with the requirements of the contract.

PROJECT FORMS

The following pages contain forms that can be helpful as you begin making decisions about your home improvement project. The first form, "Questions for the Building Department," might be completed over the telephone for a small project. For larger projects, you will probably have to make an appointment with your local building department. Many homeowners wrongly assume that permits and inspections are not required for small projects, or that the permitting process will cause more aggravation than it is worth. This is a false assumption. The purpose of establishing and enforcing codes is to promote public safety and adherence to minimum building standards. When in doubt, do not hesitate to inquire with your local building officials. Obtaining answers to the questions on the form will enhance your understanding of the building process.

The second set of forms, the "Contractor Evaluation Sheet," reflects important information that you should have for every contractor you consider hiring. You may need to make several copies of this document.

NOTES

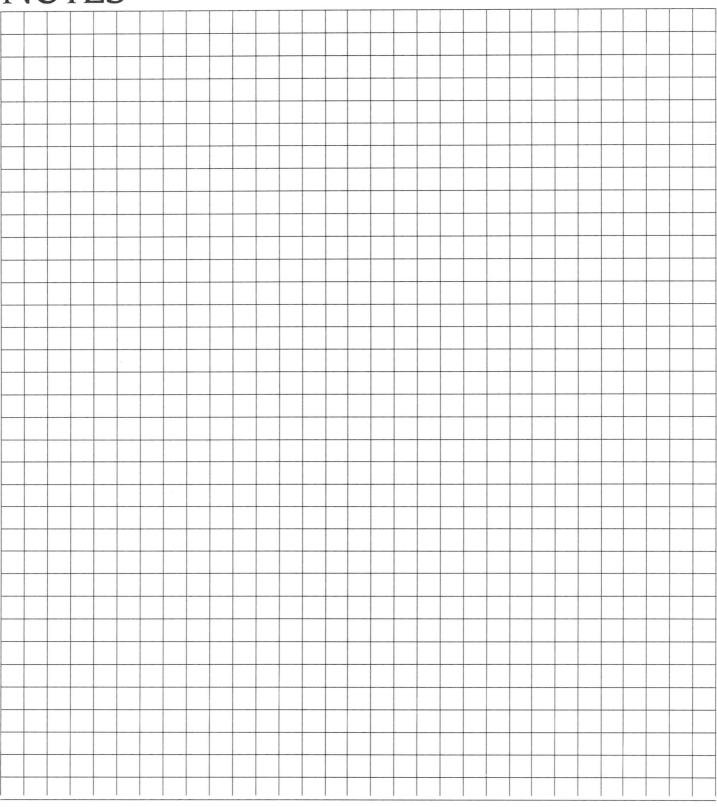

QUESTIONS FOR THE
BUILDING DEPARTMENT

1. What general information do you require on permit forms?

5. Is a septic plan and evaluation/ inspection required? ☐ yes ☐ no

    Is one or the other required for the building permit? ☐ yes ☐ no

2. Do I need to submit plans? ☐ yes ☐ no

    If yes, how many copies do I need to file for permit?

    Do the plans need to be prepared and/or stamped by an architect or engineer? ☐ yes ☐ no

6. Are there fire alarm requirements? ☐ yes ☐ no

    If yes, do I obtain that information from the fire department? ☐ yes ☐ no

3. Do you require the following on the blueprints?

7. How much advance notice is required to schedule an inspection?

    1/4" scale floor plan? ☐ yes ☐ no
    Exterior elevations? ☐ yes ☐ no
    Floor framing? ☐ yes ☐ no
    Roof framing? ☐ yes ☐ no
    Interior door schedule? ☐ yes ☐ no
    Exterior door schedule? ☐ yes ☐ no
    Window schedule? ☐ yes ☐ no
    Interior elevation? ☐ yes ☐ no
    Wiring schematic? ☐ yes ☐ no
    Lighting plan? ☐ yes ☐ no
    Heating or air conditioning plan? ☐ yes ☐ no
    Labor specifications? ☐ yes ☐ no
    Material specifications? ☐ yes ☐ no

8. Does the town require soil compression tests? ☐ yes ☐ no

    If yes, by whom is the testing done?

9. Does the town require an "as built" drawing upon completion of the foundation? ☐ yes ☐ no

4. Do I need a certified plot plan by a registered engineer? ☐ yes ☐ no

    If yes, how many copies?

    Is the plot plan required to include septic design? ☐ yes ☐ no

    Site utility locations? ☐ yes ☐ no

    What elements need to be shown?

10. Does the town require that the foundation be sited by a registered engineer before pouring of the concrete? ☐ yes ☐ no

Courtesy of Cornerstone Consulting, Inc., Weymouth, MA

11. Does the town require the foundation to be a specified height above the crown of the street? ☐ yes ☐ no

    If yes, what is the height?

12. Does the town require a footing for the foundation? ☐ yes ☐ no

    If yes, does the town require a footing inspection? ☐ yes ☐ no

13. Does the town require a foundation inspection before backfilling? ☐ yes ☐ no

14. Does the town require a post-footing inspection? ☐ yes ☐ no

15. Does the town require that the electrical and plumbing work be completed before framing inspection? ☐ yes ☐ no

16. Does the town require an insulation inspection? ☐ yes ☐ no

17. Does the town require fireproofing? ☐ yes ☐ no

    If yes, solid blocking? ☐ yes ☐ no

    If yes, mineral wool? ☐ yes ☐ no

If yes, fireproof caulking for electrical and plumbing holes? ☐ yes ☐ no

If an attached garage is being installed are there fire-related material requirements?

    For sheetrock on wall between garage and house? ☐ yes ☐ no

    For sheetrock on ceiling of garage? ☐ yes ☐ no

    Fire-rated garage door? ☐ yes ☐ no

If a new heating system is being installed are there fire-rated material requirements?

    For sheetrock on walls in system area? ☐ yes ☐ no

    For sheetrock on ceiling in system area? ☐ yes ☐ no

18. What are the requirements for a Certificate of Occupancy?

19. What is the fee for the building permit?

20. How long does it take to get the building permit?

21. How long does the building permit remain in effect?

Courtesy of Cornerstone Consulting, Inc., Weymouth, MA

CONTRACTOR EVALUATION SHEET

Part 1: Telephone Interview

Company: _____

Address: _____

Town: _____

State: _____ Zip: _____

Telephone: _____

Beeper: _____

Main contact: _____

If a post office box is given for the address above, ask for a residential address.

Address: _____

Town: _____

State: _____ Zip: _____

POINTS TO COVER WITH A SUBCONTRACTOR:

1. Are you required to have a license? ☐ yes ☐ no
   If yes, what is your license number?

   If yes, what type of license?

2. Are you required by the state to be registered? ☐ yes ☐ no

3. Do you carry Workers' Compensation? ☐ yes ☐ no

4. Do you carry Liability Insurance? ☐ yes ☐ no

5. Can your insurance company send me a Certificate of Insurance? ☐ yes ☐ no

6. Is this business your sole means of support? ☐ yes ☐ no
   If no, what percentage of your time is dedicated to your other interests?

   If no, of the remaining time, how much will be dedicated to my project?

7. How long have you operated your business?

8. How many years of experience in the industry do you have?

9. Do you belong to any professional organizations? ☐ yes ☐ no
   If yes, what organizations?

10. Do you subscribe to and read any professional publications? ☐ yes ☐ no
    If yes, what publications?

Courtesy of Cornerstone Consulting, Inc., Weymouth, MA

11. Are you involved personally with the work on a physical basis? ☐ yes ☐ no
    If yes, what portions?

19. Do you or your employees work on holidays? ☐ yes ☐ no

12. Will you be involved in other projects while this one is underway? ☐ yes ☐ no
    If yes, what percentage of the time?

20. In the event you cannot be at the work site as scheduled, how will I know?

13. Will you be able to start and complete the work before you undertake another project? ☐ yes ☐ no

21. What is your policy on changes I request that may add or subtract from the work?

14. Does your work require you to respond to emergencies? ☐ yes ☐ no
    If yes, what percentage of the time?

22. What is your policy on changes initiated by you that may add or subtract from the work?

15. What are your daily starting and ending times?

23. What is your policy on my giving instructions to your employees?

16. What are your operating and office hours?

24. What is your policy on cleaning up after yourself for the work that you or your workers do?

17. Will you or your employees take vacations during the project? ☐ yes ☐ no

18. Do you or your employees work on weekends? ☐ yes ☐ no

Courtesy of Cornerstone Consulting, Inc., Weymouth, MA

25. What is your policy on being a resource to me for work outside your expertise?

26. Do you provide warranties for your workmanship?   ☐ yes ☐ no
    If so, please explain your warranty coverage.

27. Do you provide warranties from manufacturers?   ☐ yes ☐ no
    If yes, please explain how these warranties are implemented.

28. Are you or your employees required by any manufacturer to have special training in order to install its product?   ☐ yes ☐ no
    If yes, what products?

    If yes, is the warranty affected without this training?

29. Do you provide me with a "Release from Payment" from your suppliers?   ☐ yes ☐ no

30. Do you allow the homeowner to provide his or her own materials?   ☐ yes ☐ no

31. Do you subcontract any of the work?   ☐ yes ☐ no
    If yes, what parts?

    If yes, do you provide me with a "Release from Payment" from your subcontractor(s)?   ☐ yes ☐ no

32. What is your policy regarding decisions I make that have a negative effect on you, such as a delay in making a decision?

33. Are your workers made aware of your policies and procedures?   ☐ yes ☐ no

34. When are payments made?

35. What form of payment do you accept?

36. Are all checks for payment made out to your company?   ☐ yes ☐ no
    If no, what is your Social Security number?

37. Are you willing to be paid according to a bank's schedule?   ☐ yes ☐ no

38. Have you previously done work similar to this project?
☐ yes ☐ no

    If no, how are you qualified to work on this project?

39. Do we agree on who is supplying which materials under what circumstances?
☐ yes ☐ no

40. Do you have any other policies or procedures I need to be aware of?
☐ yes ☐ no

    If yes, what policies?

    If yes, what procedures?

41. As references, please provide me with the names and telephone numbers of three of your former customers.

    Name _____

    Telephone _____

    Time to Call _____

    Name _____

    Telephone _____

    Time to Call _____

    Name _____

    Telephone _____

    Time to Call _____

42. As references, please provide me with the names and telephone numbers of three of your professional contacts.

    Name _____

    Telephone _____

    Time to Call _____

    Name _____

    Telephone _____

    Time to Call _____

    Name _____

    Telephone _____

    Time to Call _____

43. As references, please provide me with the names and telephone numbers of three of your suppliers.

    Name _____

    Telephone _____

    Time to Call _____

    Name _____

    Telephone _____

    Time to Call _____

    Name _____

    Telephone _____

    Time to Call _____

Courtesy of Cornerstone Consulting, Inc., Weymouth, MA

CONTRACTOR EVALUATION SHEET
Part 2: Points to Cover Before Signing a Contract

44. Do we agree on the basic costs of the work? ☐ yes ☐ no
    If no, what costs need to be resolved?

45. Do we agree on the optional costs
    of the work? ☐ yes ☐ no
    If no, what optional costs need to be
    resolved?

46. Do we agree on those items of work
    that are not included? ☐ yes ☐ no
    If yes, what are those items of work that
    are not included?

47. Do you have any miscellaneous charges ☐ yes ☐ no
    of which I need to be aware?
    If yes, what charges?

48. Please indicate your payment schedule:
    Payment 1:
    Payment 2:
    Payment 3:
    Payment 4:
    Payment 5:
    Payment 6:

49. Please indicate what sections of the work
    each payment covers:
    Payment 1:
    Payment 2:
    Payment 3:
    Payment 4:
    Payment 5:
    Payment 6:

50. Are there any unresolved issues? ☐ yes ☐ no
    If yes, what issues are unresolved?

    If yes, whose responsibility is it
    to resolve each issue?

51. Do you have two copies of ☐ yes ☐ no
    the specifications?

52. Have you attached to your contract ☐ yes ☐ no
    a copy of the specifications?

53. Does your contract read: "Refer to Attached Plans and Specifications"? ☐ yes ☐ no

54. Have you signed both copies of the contract? ☐ yes ☐ no

55. Do you have my signature on both copies of the contract? ☐ yes ☐ no

56. Do I have a copy of the signed contract? ☐ yes ☐ no

FOR THE HOMEOWNER:

The above points were discussed with the contractor as a basis of understanding and do not represent contractual commitments by either party.

Signed _____

Signed _____

Date _____

FOR THE SUBCONTRACTOR:

The above points were discussed with the client as a basis of understanding and do not represent contractual commitments by either party.

Signed _____

Title _____

Date _____

Courtesy of Cornerstone Consulting, Inc., Weymouth, MA

WORKING WITH AN INTERIOR DESIGNER

An interior designer has the expertise to create a customized environment to meet your needs. The designer's job is to plan spaces and solve problems, drawing on a vast array of product resources and services. The interior design process considers both the human environment and your own life experience within the framework of a given space.

The benefits of working with an interior designer include the following:

- Professional interior design results in a setting that combines creativity and good taste with comfort and function.
- A designer can work within your time frame and your budget.
- Design project management addresses your specific needs. The designer's goal is to improve the quality of your environment.
- A designer knows the right questions to ask to prevent you from making costly mistakes.
- Working with a designer gives you the opportunity to define and enhance your sense of style through the design of your environment.
- A designer creates a home environment that reflects your personal tastes and preferences, while allowing for changing individual and family needs.
- Professional interior designers keep abreast of environmental safety laws, building and fire codes, access design regulations, and historic preservation. Thus a designer can select suitable, attractive, and durable furnishings that meet fire codes and flammability ratings.

CHOOSING A DESIGNER

Before contacting an interior designer, ask yourself the following questions:

- What feeling do I wish the completed project to communicate?
- For whom is the space to be designed?
- What activities will take place here?
- What is my time frame?
- What is my budget?

Answering these questions allows you to prepare for an interview with a designer. The interview is your opportunity to acquaint the designer with the nature of the project. The more information you can provide, the more successfully your designer will be able to meet your expectations. It is also a good time to hear the designer's observations and insights, to assess your mutual compatibility, and to review the designer's portfolio.

Topics you may wish to discuss with a prospective interior designer include:

- The designer's references, educational background, professional affiliation (for example, American Society of Interior Designers), experience, and portfolio.
- The scope of the designer's services.
- Your deadlines and budget, including the method of compensation.
- Details regarding a contract or letter of agreement that outlines the project specifics and obligations to be performed.

Be sure to interview a number of interior designers. You will be working closely with this person, and you will want to hire someone with whom you can feel comfortable.

This guidance is based on recommendations of the American Society of Interior Designers (ASID), the country's oldest and largest interior design association with a rich resource of 20,000 educated design professionals throughout the world. The mission of ASID is to encourage excellence in the practice of interior design, assist its members to professionally serve the public, and work for a favorable environment for the practice of interior design. ASID designers receive the most current information on appropriate materials, technology, building codes, government regulations, flammability standards, design psychology, and product performance. With this knowledge, ASID designers protect the safety and welfare of people in interior environments. For further information and complete recommendations, visit the Web site at **http: // www.asid.org**

NOTES

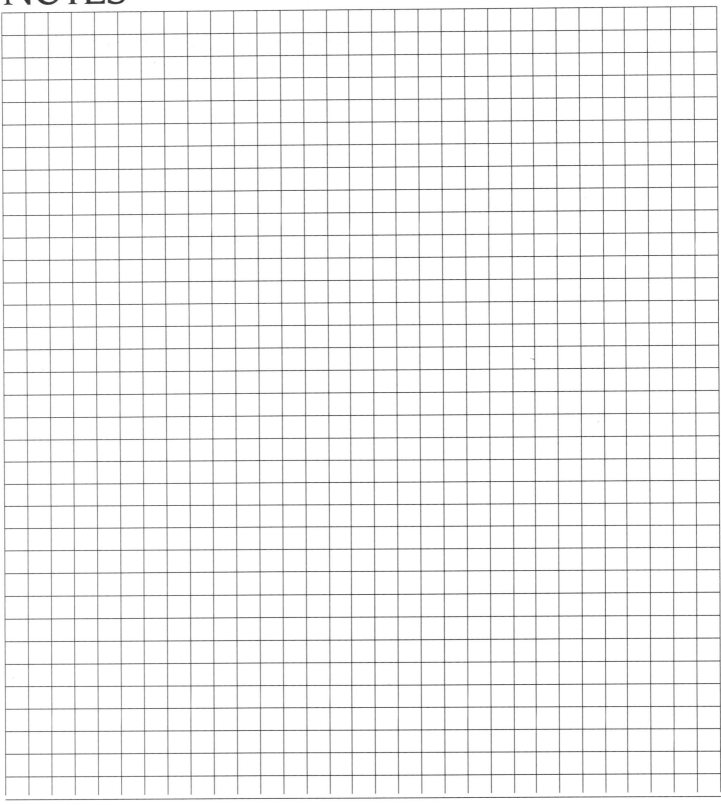

Part Two
INTERIOR PROJECTS

HOW TO USE PART TWO

Part Two, "Interior Projects," provides descriptions and cost estimates for complete home improvement projects. Part Three, "Details," provides cost information for individual construction items. Tips for using "Details" can be found at the beginning of that section.

Each project plan in Part Two contains three types of information to help you organize the job. First, a detailed illustration shows the finished project and the relationship of its components. Second, a general description of the plan includes a review of the materials and the installation process. It also evaluates the level of difficulty of the project. The

"What To Watch Out For" section highlights ways to enhance the project or to cope with particularly difficult installation procedures. Third, a detailed project estimate chart lists the required materials, the professional installation times, and their corresponding costs. The components of this chart are described in detail below.

Key to Abbreviations

C.Y. – cubic yard
Ea. – each
L.F. – linear foot
Pr. – pair
Sq. – square (100 square feet of area)
S.F. – square foot
S.Y. – square yard
V.L.F. – vertical linear foot

Description of material items used in the project.

Quantity/Unit of each material and its standard unit of measure are given in this figure. (See the Key to Abbreviations, also on this page).

Labor-Hours represents the estimated time it takes a professional tradesperson to perform each of the installations within the project. The total number of contractor's hours required for a complete project is given at the bottom of this column.

Material is an estimated cost for each item in the "Description" column. This figure represents the price of a single unit, multiplied by the required number of units shown in the "Quantity/Unit" column. The prices reflect the national average of what you can expect to pay at a building supply retailer for the components used in the project. The total material cost for the project can be found at the bottom of this column. (Sales tax has not been included in these figures and will have to be added for a more accurate total.) To tailor these prices to your specific area of the country, refer to "Adjusting Project Costs to Your Location" section at the back of this book.

Deluxe Master Bath

Description	Quantity/Unit	Labor-Hours	Material
Partition wall for shower, 2 x 4 plates & studs, 16" O.C., 8' high	152 L.F.	2.2	65.66
Drywall, 1/2" on walls, water-resistant, taped & finished, 4' x 8'	160 S.F.	2.7	49.92
Paint, ceiling, walls and door, primer	330 S.F.	1.3	15.84
Paint, ceiling, walls and door, 1 coat	330 S.F.	2.0	19.80
Deluxe vanity base cabinet, 2 door, 72" wide	1 Ea.	1.7	414.96
Lavatory, solid surface material with integral bowl, 22" x 73"	1 Ea.	1.0	750.00
Fittings for lavatory	2 Sets	13.9	177.60
Shower stall, terrazzo receptor, 36" x 36"	1 Ea.	8.0	828.00
Shower door, tempered glass, deluxe	1 Ea.	1.3	366.00
Bathtub, 5', recessed, porc. enamel on CI, w/trim, mat bottom	1 Ea.	3.6	444.00
Fittings for tub	1 Set	7.7	147.60
Fittings for shower, thermostatic	1 Ea.	1.0	271.20
Inlet strainer for shower	1 Ea.		46.20
Walls, 4-1/4" x 4-1/4" ceramic tile, stall & wainscoting, thin set	170 S.F.	14.3	426.36
Flooring, porcelain tile, 1 color, color group 2, 2" x 2"	40 S.F.	3.4	190.56
Toilet, tank-type, vitreous china, floor mounted, 1 piece, color	1 Ea.	3.9	772.20
Fittings for toilet	1 Ea.	5.3	151.20
Towel bar, ceramic	4 Ea.	0.8	45.60
Toilet tissue dispenser, ceramic	1 Ea.	0.2	11.40
Electrical, light fixtures with wiring	4 Ea.	1.1	168.00
Electrical, 2 light switches	2 Ea.	2.8	42.72
Mirror, plate glass, 34" high x 60" wide	14 S.F.	1.4	100.80
Electrical, one GFI outlet	1 Ea.	1.7	52.80
Ceiling light/fan/heat unit	1 Ea.	0.8	92.40
Totals		82.1	$5,650.82

Project Size	8' x 10'	Contractor's Fee Including Materials	**$12,048**

Key to Abbreviations
C.Y.–cubic yard   Ea.–each   L.F.–linear foot   Pr.–pair   Sq.–square (100 square feet of area)
S.F.–square foot   S.Y.–square yard   V.L.F.–vertical linear foot   M.S.F.–thousand square feet

Project Size gives the basic dimensions of each interior project. The cost estimates are based on these measurements.

Section One
CONVERSIONS TO LIVING SPACE

The primary considerations in converting a space such as an attic to living space are a large enough area, adequate head room, and accessibility. Following are some additional considerations.

- Many home improvement centers and specialty suppliers offer workshops that can be extremely helpful in the planning stages of a project. Attending these sessions should give you a better "feel" for whether to tackle the project yourself or hire a professional. These workshops should also help you figure out what tools you need to rent or purchase.

- Before you purchase materials from a home center or lumberyard, be sure to clarify whether delivery is available and if there are extra charges.

- Check with local officials regarding restrictions on the disposal of rubbish and debris that will be generated by your project.

- In attic renovations, try to integrate the stairs into the living space in order to maximize the feeling of open space. Also, do as much as you can to insulate the living space from the temperature extremes of the roof, without sacrificing too much head room.

- Check the size of floor joists and framing members to make sure you have proper support for the proposed project.

- If you are opening up a roof or wall to accomplish the conversion, you may want to move in equipment and other large items (e.g., premolded shower stalls, cabinetry) at that time.

- Always consider measures to save heating and cooling costs as part of the job. For example, consider adding new or additional weatherstripping or insulation. Minimum amounts required may be specified in your local building code. Since most insulating and weatherstripping products are relatively inexpensive, their cost can be recovered through energy savings in a short period of time.

- Before starting your project, make sure the conversion will not eliminate or diminish access to utilities, piping, circuit breakers, etc.

- Check local building codes for requirements, including egress, window height, ventilation, framing, wiring, plumbing, etc.

- Don't forget to install smoke detectors in the new living space. Your home should have at least one detector per floor.

- Check other interior projects for specific tips for various elements of this project. You may want to paint adjoining rooms to solve decoration or space use problems in conjunction with this project.

FULL ATTIC

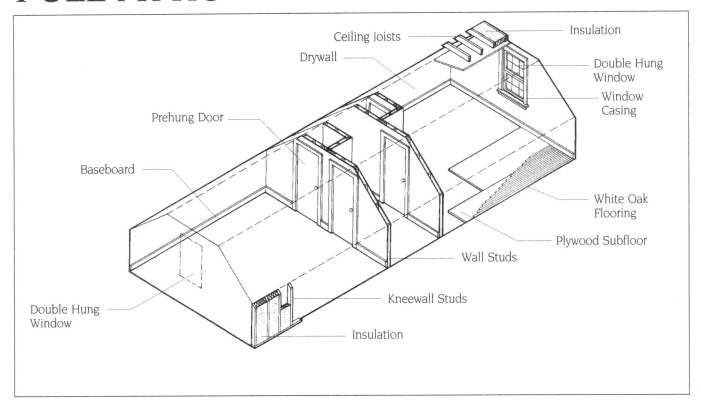

Ceiling Joists — Insulation — Double Hung Window — Window Casing — Drywall — Prehung Door — Baseboard — White Oak Flooring — Plywood Subfloor — Wall Studs — Double Hung Window — Kneewall Studs — Insulation

Expanding your home living space by using your attic requires careful consideration. Basic physical restrictions can cause problems or even prohibit attic renovation without major structural changes to the existing house. For example, inadequate head room results in the additional cost of adding dormers or raising the ridge and changing the roof line. If you have enough head room and can give up the floor area required for a staircase, then your attic can be converted into a comfortable, pleasant living area.

MATERIALS

The materials used for attic renovation are basically the same as those used for interior work anywhere in your home. One advantage of attic renovation, assuming that dormers or changes in the roof line are not required, is that the area is weathertight and protected at the start. Standard construction-grade framing materials, including 2 x 4s for wall studs and plates, and 2 x 6s for ceiling joists, can be used to support 1/2″ sheetrock. If you plan on installing wood floors, economical 1/2″ sheathing plywood can be used for the subflooring.

Insulation costs will vary according to recommended R-values for your area and preference for paperfaced, foilfaced, or unfaced fiberglass insulation. Windows for the gable ends of the attic room can be simple double-hung units or more expensive thermopaned casement models. Extra money invested at the time of construction for quality windows will be returned to you in energy savings and lower maintenance costs over the years.

The electrical materials used in the project include basic outlets, switches, light fixtures, an electric baseboard heater, and the wiring required to connect the various components. Make sure that you check local codes for proper installation if you are doing the electrical work on your own. Also, have the work inspected by a qualified person after you have installed it.

The finish work for your attic will vary according to your personal taste and the anticipated use of the living space. Taping and finishing the sheetrock and applying two coats of paint will finish the ceiling and walls. Hardwood flooring, carpeting, or other materials can be used for floor coverings. Basic prehung, hollow-core lauan doors or more

expensive paneled or louvered units can be used for the doorway and closets. For trim, ranch or colonial casing may be installed around doors, windows, and floors.

LEVEL OF DIFFICULTY

Major renovation of an attic is a challenging project for the most experienced intermediate, even if dormers or other structural changes are not required. Accessibility alone can cause problems in getting materials to a second- or third-floor attic location. If you are to do some of the work alone, plan ahead as the project progresses and arrange to have someone assist you for an hour or two in carrying sheetrock, plywood, and other bulky items upstairs. Much of the work can be done by the homeowner, from easier jobs like insulating and painting to more skilled operations like framing, flooring, and window installation. Specialized work, especially electrical, should be left to the experts or contractors. If dormers are required, consider bringing in a professional contractor to build and close in the attic area. As a general rule, beginners with little or no building experience should double the estimated labor-hours for the project; intermediates with some experience and skill in the use of tools should add an additional 50% to the labor-hours estimate, and experts, about 20%. Beginners should seek professional help and advice throughout the various stages of the construction process.

WHAT TO WATCH OUT FOR

The primary considerations in any remodeling project are space and attic accessibility. Before you plan the renovation, check to see that the head clearance in your attic will allow for a 7'-6" or higher ceiling at the center of the room and that you have enough floor space to make the project worthwhile. Then check on the accessibility of the attic. Consider the staircase in terms of additional cost as well as the reduction of living or storage area that it causes in the floor below. As a final preliminary check, make sure that the prospective floor joists for the attic room are sufficient to bear the added weight of the new living area. If they are not adequate, they will have to be strengthened at additional cost and time by adding floor joists before the project is begun.

If the new rooms are to be used as bedrooms, be sure that the window(s) meet the minimum egress requirements of the building code – usually 20" × 24" of clear opening.

If insulation is removed from the attic floor and placed above the ceiling between roof rafters, you will need to ensure that your roof is properly vented. This can be accomplished with soffit vents and a vented ridge strip, or gable louvers above your new ceiling. If you are unsure of this aspect of the work, seek professional advice, as lack of roof ventilation can cause serious structural damage to your home.

Full Attic

Description	Quantity/ Unit	Labor-Hours	Material
Flooring, underlayment-grade plywood, 4' x 8' sheets, 1/2" thick	544 S.F.	6.0	417.79
Ceiling joists, 2 x 6, 16" O.C., 7' long	29 Ea.	2.6	138.85
Plates, top and bottom, 2 x 4, 12' lengths	21 Ea.	5.0	108.86
Kneewall studs, 2 x 4, 16" O.C., 4' long	58 Ea.	3.4	100.22
Wall studs, 2 x 4, 16" O.C., 8' long	44 Ea.	5.1	152.06
Blocking, misc., to wood construction, 2 x 4 x 8'	5 Ea.	1.3	17.28
Insul., R-11, 3-1/2" thick, 15" wide foil-faced blanket, 88 S.F./roll	12 Rolls	5.3	430.85
Insulation, R-19, 6" thick, 15" wide foil-faced blanket, 88 S.F./roll	8 Rolls	1.3	129.89
Windows, double-hung, plastic-clad, insulating, 3' x 4'	2 Ea.	1.8	592.80
Drywall, 1/2" thick, taped and finished, 4' x 8' sheets	1,920 S.F.	31.8	576.00
Doors, flush, HC, lauan finish, 2'-6" x 6'-8", prehung with casing	4 Ea.	3.2	542.40
Trim, casing for window, pine, 11/16" x 2-1/2"	30 L.F.	1.0	28.08
Trim, for baseboard, pine, 9/16" x 3-1/2"	210 L.F.	7.0	287.28
Handrail, pine, 1-1/2" x 1-3/4" x 10' long	2 Ea.	0.7	52.56
Flooring, prefinished, white oak, prime grade, 2-1/4" wide	550 S.F.	25.9	3,861.00
Painting, primer, wall and trim	2,100 S.F.	8.2	100.80
Painting, 1 coat, wall and trim	2,100 S.F.	12.9	126.00
Lockset, residential, interior door, ordinary	4 Ea.	2.0	60.48
Electrical, hall light	1 Ea.	1.5	19.62
Electrical, two switched closet lights	2 Ea.	0.5	84.00
Electrical, 5 plugs per room	10 Ea.	14.0	213.60
Electrical, baseboard heaters, 10' long	2 Ea.	4.9	328.80
Electrical, baseboard heater, 3' long	1 Ea.	1.0	50.40
Totals		146.4	$8,419.62

Project Size	16' x 36'	Contractor's Fee Including Materials	**$19,439**

Key to Abbreviations
C.Y.–cubic yard   Ea.–each   L.F.–linear foot   Pr.–pair   Sq.–square (100 square feet of area)
S.F.–square foot   S.Y.–square yard   V.L.F.–vertical linear foot   M.S.F.–thousand square feet

SUMMARY

The use of attic space can be an efficient way to expand the living area of your home, especially when a steep-pitched roof provides enough room without the addition of dormers. If your house meets the requirements of area and accessibility, the attic can become an attractive source of new living space.

For other options or further details regarding options shown, see

   *Attic ventilation**

   *Dormers**

   Interior doors

   Painting and wallpapering

   *Skylights**

   Stairways

   *Standard window installation**

   Whole-House fan

   * In Exterior Home Improvement
     Costs

GARAGE CONVERSION

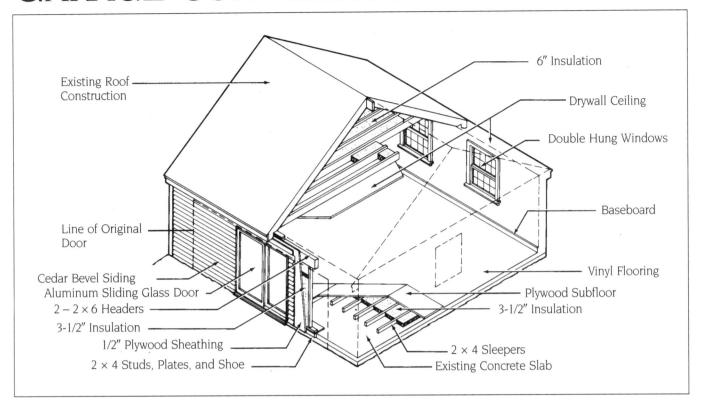

Existing Roof Construction

6" Insulation

Drywall Ceiling

Double Hung Windows

Line of Original Door

Baseboard

Cedar Bevel Siding
Aluminum Sliding Glass Door
2 – 2 x 6 Headers
3-1/2" Insulation
1/2" Plywood Sheathing
2 x 4 Studs, Plates, and Shoe

Vinyl Flooring

Plywood Subfloor

3-1/2" Insulation

2 x 4 Sleepers

Existing Concrete Slab

Garages can be economically converted to living, work, or studio space. Because the expenses for excavation, foundation work, framing, and closing in the new living area have already been taken care of, your only concerns will be the costs of preparing the interior, adding the windows and a door, and closing in, when necessary, portions of the exterior walls.

MATERIALS

The primary materials used for the garage conversion consist of framing lumber, various sheet goods for the floor, walls, and ceiling, several windows, sliding glass doors, and the finish materials for the inside and the outside of the exterior walls. Insulation, electrical materials, and various incidentals are also required and will vary in type, amount, and size, depending on such factors as the geographical area of your home and the particular location of the garage in relation to the main structure.

The minimum framing and structural lumber required in the garage conversion includes 2 x 4s and a few 2 x 8s. The 2 x 4s used as sleepers for the support system of the new floor should be

pressure-treated lumber. Conventional 2 x 4s are used as studs in the exterior walls to frame in the new windows and sliding glass door unit and to frame the new wall that fills the old garage door opening. The new 2 x 4 bottom plate should be pressure treated. Sheathing plywood that is 5/8" thick serves as the rough covering for the sleepers and the exterior of the new wall framing. Be sure to insulate the floor and place a vapor barrier on top of the 2 x 4 floor sleepers before laying the plywood. Additional framing may be required at extra cost if the floor of the new room has to be raised or the ceiling lowered for aesthetic or practical reasons.

The wall studs and ceiling joists of the new room can be covered with various types of materials, such as drywall or wood paneling. An economical choice is 1/2" drywall. If the garage is already drywalled, but not insulated, hire an insulating contractor to blow insulation into the walls and, if necessary, the ceiling. If your garage is situated under the living area of your house, it may already be insulated. While the walls are open, the rough electrical work should also be installed. As for the exterior, these walls should be finished with matching siding around

the new window and door installations and in the wall space that fills the void left by the garage door. Make sure that you lace the new siding into the old, prime all of the new materials, and then put a uniform finish coat on all three sides of the new room's exterior.

The windows and door selected for the new room should match or complement the style and design of other units in your house. The type of windows suggested in the plan are of standard size and design, but many other styles and sizes are available at varying costs. The sliding glass door unit, too, is a standard item, but many other options in both sliding and conventional door styles can be purchased over a wide price range. One general rule to follow when purchasing window and door units is that quality products will usually return the homeowner's investment in energy savings, durability, and low maintenance costs over the life of the unit.

The finish work in the new room should reflect its anticipated use. The trim for the windows and door, as well as the baseboard, can be painted, stained, or left natural and finished with a clear sealer. A wide range of flooring materials can be installed, from hardwood to carpeting or, as in this plan, durable vinyl sheet goods. Make sure that you vent the subflooring support system so that it can breathe. Also, if you plan to use vinyl or carpeting for the finished floor, it might be worth the additional expense and time to install a layer of 3/8" particle board underlayment over the sheathing plywood. Not only will the surface material look and wear better, but the floor will be stronger, more even, quieter, and warmer.

LEVEL OF DIFFICULTY

The conversion of a garage into living space is a major undertaking; but because of the convenient location and accessibility of the work site, it is the type of project that can be done in stages, over a longer period of time than other remodeling jobs. Many of the tasks involved in the final stages of the project can be accomplished by beginners who seek advice and assistance from knowledgeable people. Novices, however, should leave the window, door, and filler framing jobs to a professional. The installation of the floor sleepers, too, can be a tricky job, especially in garages

with sloped or uneven concrete floors. Shimming is necessary in constructing the support frame, and often the 2 x 4 sleepers have to be ripped to size so that they can be set on the sloped concrete. These operations demand skill in the use of tools and the basic know-how of someone who has done this work before. Beginners and most intermediates should leave the electrical work to a competent tradesperson. As a general rule, if you are a beginner, add 90% to the labor-hour estimates for those tasks that you feel you can handle. If you are a reasonably skilled intermediate, add 40% to the estimate and more if you plan to lay the sleeper support system, rough in the windows and sliding doors, and fill in the garage door opening. Experts should add about 10% to all tasks involved in the conversion.

WHAT TO WATCH OUT FOR

Problems with the floor and ceiling heights can be corrected with careful design and planning. Garage floors, for example, are often located below grade level or are pitched dramatically for drainage. Take stock of the existing floor's location and slope before you determine the level of the new floor and the design for its support. You should also make provisions to divert surface water via drains or other means so that run-off or icing won't penetrate under the floor of the new room and cause damage. If there is an existing floor drain, be sure to close it off to eliminate the possibility of fumes or odors. Garage ceilings vary in height, so plan on additional expense if new joists have to be installed to decrease the height, or other modifications made to increase it.

SUMMARY

A garage conversion is another way in which you can use an existing sheltered area to create new living space in your home. This often underutilized area can be conveniently remodeled to become a comfortable and functional room.

Garage Conversion

Description	Quantity/ Unit	Labor- Hours	Material
Flooring, joists 2 x 4, 16" O.C., on existing concrete slab	525 L.F.	6.7	226.80
Plywood, CDX, 5/8" thick, 4' x 8' sheets for floors & 1 exterior wall	704 S.F.	10.7	481.54
Plates, top, 2 x 4, 12' lengths	27 L.F.	0.5	11.66
Plate, bottom, treated 2 x 4, 12' lengths	13 L.F.	0.4	7.80
Wall studs at door and new windows, 2 x 4 x 8' long, 16" O.C.	192 L.F.	2.8	82.94
Headers over openings, double 2 x 6	28 L.F.	1.2	19.15
Siding, cedar bevel, A-grade, 1/2" x 6"	155 S.F.	5.0	306.90
Insul., fbgls batt, paper-backed, 3-1/2", R-11, floor & walls	1,000 S.F.	11.4	336.00
Insul., fbgls batt, paper-backed, 6", R-19, for ceiling	500 S.F.	6.7	228.00
Windows, dbl.hung, vinyl-clad, insul., 3'x4', frame, screen & trim	3 Ea.	2.7	889.20
Doors, sliding glass, aluminum, 5/8" insulated glass, 6' wide	1 Ea.	4.0	1,530.00
Drywall, 1/2", walls & ceiling, standard, taped & finished, 4' x 8'	1,216 S.F.	20.2	379.39
Trim, casing for windows & door, 11/16" x 2-1/2" wide	50 L.F.	1.7	46.80
Trim, baseboard, 9/16" x 3-1/2"	90 L.F.	3.0	123.12
Flooring, vinyl sheet goods, 0.125" thick	500 S.F.	17.4	1,350.00
Paint, wall, ceiling & trim, primer	1,200 S.F.	4.7	57.60
Paint, wall, ceiling & trim, 1 coat	1,200 S.F.	7.4	72.00
Paint, exterior siding, oil-base primer	185 S.F.	2.3	17.76
Paint, exterior siding, oil-base, 2 coats, brushwork	185 S.F.	3.7	28.86
Electrical, duplex receptacles	8 Ea.	12.0	156.96
Electrical, switch	1 Ea.	1.4	21.36
Electrical, baseboard heaters, 10' long	4 Ea.	9.7	657.60
Totals		135.6	$7,031.44

Project Size	22' x 22'	Contractor's Fee Including Materials	$16,821

Key to Abbreviations
C.Y.–cubic yard Ea.–each L.F.–linear foot Pr.–pair Sq.–square (100 square feet of area)
S.F.–square foot S.Y.–square yard V.L.F.–vertical linear foot M.S.F.–thousand square feet

For other options or further details regarding options shown, see

Flooring

Main entry door*

Painting and wallpapering

Patio & sliding glass doors*

Standard window installation*

*In Exterior Home Improvement Costs

HALF ATTIC

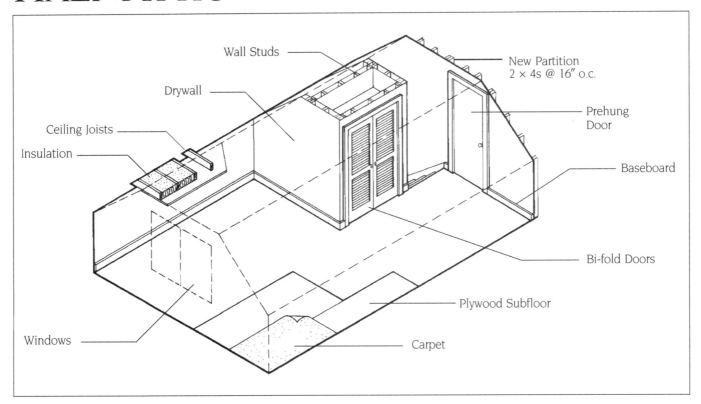

Wall Studs

Drywall

Ceiling Joists

Insulation

New Partition
2 × 4s @ 16″ o.c.

Prehung
Door

Baseboard

Bi-fold Doors

Plywood Subfloor

Windows

Carpet

If space is limited in your attic, or if you want to continue using a portion of it for storage, then a partial renovation may accommodate your needs. Follow the same guidelines as in a full attic plan when you consider the feasibility of expanding a half attic space.

MATERIALS

The materials used for a half attic renovation include the common interior construction components employed in most house systems. The framing members consist of construction grade 2 x 4s for the walls and closet, and 2 x 6s for ceiling joists. Conventional 1/2″ sheetrock, taped and finished, or paneling may be installed at varying costs for the walls and ceilings. Care should be taken in selecting the R-value of the insulation material so that the attic room and the rest of the house are efficiently protected from heat loss and gain in the winter and summer. An alternative to the 5/8″ subfloor for the carpet, in cases where more strength is needed, consists of a layer of 1/2″ plywood sheathing covered by a layer of 3/8″ particleboard. Try to run the particleboard at right angles to the plywood and avoid coincidental seams of

the two layers of material. This two-layer underlayment costs more and takes more time to install, but the extra money and effort will give your carpeting and the new floor a more even and durable surface.

The window unit used in this half attic is a vinyl-clad casement type, which is more expensive than a conventional all-wood uninsulated window; but the extra initial cost will be returned over time in savings on fuel and maintenance. The door on the interior wall of the new room is a standard hollow-core lauan prehung unit, but you can purchase paneled and other solid or decorative doors at additional cost. The closet door used in this plan is a bifolding, louvered unit that provides total accessibility to the 4′ closet opening.

The electrical materials include five duplex receptacles, two light fixtures, an electric baseboard heater, and the wiring required to supply and connect the components. If you want to add to the electrical work, make sure that you plan ahead and do it before the sheetrock is installed. Also, check to see that your electric service has the capacity to allow for the new room's circuit. There is a good chance that the new supply wiring

will have to be fished up from the service panel, so unless you are experienced in electrical work, call an electrician to perform this task.

The cutting, staining or painting, and installation of trim boards may involve a few mistakes at first, but with a simple miter box, careful measuring, and some patience, even beginners will catch on quickly. You can stain or paint the trim boards in your back yard or cellar before you install them, which allows you to paint the walls through to the finish coat. Any marks made during installation of the trim can be easily touched up in the final stage. Taping and finishing the sheetrock may take some time if you have never done it before; but the right tools, sound advice, and a quick lesson from a friend who has done it can give you professional results while saving you money.

LEVEL OF DIFFICULTY

Like a full attic conversion, a half attic renovation project is a major challenge for homeowners who are doing much of the work on their own. Although the project appears to be relatively routine, many different building skills are required to complete the job from framing through to the finish work. Beginners should not jump into a project of this size without ample professional assistance and advice. If you are a beginner and want to tackle a half or full attic conversion, plan on adding 100% to the professional labor-hours estimate. If you are acquainted with tools and have limited experience in building, add 50% to the professional time. If you have extensive experience in building, anticipate an additional 15% above the professional time.

WHAT TO WATCH OUT FOR

The attic space of most old or new dwellings can be used for renovation if there is a minimum of 8' from the floor to the ridge board. If your attic does not have sufficient height and area to accommodate a livable room, then dormers may have to be constructed or other costly structural alterations may have to be made. Some state building codes require 9" or more of ceiling insulation in finished rooms. Make sure you are aware of such requirements. Accessibility is another major consideration. Remember that a conventional staircase will take up precious living or storage space not only in the new attic room, but also in the floor below. Remember, too, that in most cases, building materials will have to be carried to the second- or third-floor attic site, so carefully measure the stairway or other means of access to the work area to head off any materials access problems. It is also a good idea to check the floor joists in the attic for general condition and spacing. The ceiling joists in many older homes are spaced at odd intervals, rather than at 16", and, therefore, need additional strengthening to support the proposed living space above.

If the new rooms are to be used as bedrooms be sure that the window(s) meet the minimum egress requirements of the building code – usually 20" × 24" of clear opening.

If insulation is removed from the attic floor and placed above the ceiling between roof rafters, you will need to ensure that your roof is properly vented. This can be accomplished with soffit vents and a vented ridge strip, or gable louvers above your new ceiling. If you are unsure of this aspect of the work, seek professional advice, as lack of roof ventilation can cause serious structural damage to your home.

Half Attic

Description	Quantity/Unit	Labor-Hours	Material
Flooring, underlayment grade plywood, 4' x 8' sheets, 5/8" thick	288 S.F.	3.3	241.92
Ceiling joists, 2 x 6, 7' long, 16" O.C.	17 Ea.	1.5	81.40
Plates, top and bottom, 2 x 4, 12' lengths	10 Ea.	2.4	51.84
Kneewall studs, 2 x 4, 4' long, 16" O.C.	28 Ea.	1.6	48.38
Wall studs, 2 x 4, 8' long, 16" O.C.	34 Ea.	4.0	117.50
Blocking, misc. to wood construction, 2 x 4 x 8'	2 Ea.	0.5	6.91
Insul., R-11, 3-1/2" thick, 15" wide, foil-faced blanket, 88 S.F./roll	8 Rolls	3.5	287.23
Insulation, R-19, 6" thick, 15" wide, foil-faced blanket, 88 S.F./roll	2 Rolls	0.9	86.59
Window, casement-type, vinyl-clad, double window, 4' x 4'	1 Ea.	0.9	313.20
Drywall, 1/2" thick, taped and finished, 4' x 8' sheets	896 S.F.	14.9	268.80
Door, HC prehung flush lauan, 2'-6" x 6'-8", w/passage set	1 Ea.	0.8	135.60
Closet door, bi-folding, louvered, 4' x 6'-8"	1 Ea.	1.3	214.80
Trim for closet jams, 1 x 6	20 L.F.	0.8	20.88
Trim, casing, for window and closet door, pine, 11/16" x 2-1/2"	40 L.F.	1.3	37.44
Trim for baseboard, pine, 9/16" x 3-1/2"	76 L.F.	2.5	103.97
Handrail, pine, 1-1/2" x 1-3/4", 2 pcs, 10' long	20 L.F.	0.7	52.56
Carpet, foam-backed, nylon, medium traffic	32 S.Y.	5.2	559.87
Paint, primer, including wall and trim	1,050 S.F.	4.1	50.40
Paint, 1 coat, including wall and trim	1,050 S.F.	6.5	63.00
Electrical, wiring duplex outlet, 15 amp., EMT and wire	5 Ea.	7.5	98.10
Electrical, recessed downlight, round, prewired, 50 or 70 watt	1 Ea.	0.3	42.00
Electrical, switch, single pole, 15 amp., EMT and wire	1 Ea.	1.4	21.36
Electrical, baseboard heaters, 10' long	2 Ea.	4.9	328.80
Totals		70.8	$3,232.55

Project Size	16' x 20'

Contractor's Fee Including Materials	**$8,180**

WALL FINISH OPTIONS
Cost per Square Foot, Installed

Description	
Sheetrock, taped and finished	$0.99
Blueboard and skim coat	$0.99
Plaster - 3 Coat	$2.66
1/4" Paneling, birch faced	$2.59
Board paneling, knotty pine	$3.73

SUMMARY

If your house meets the basic height, space, and accessibility conditions, a half attic renovation may be an ideal way to convert under-used space to living area. Because this space is already under a roof and protected from the elements, the construction of a new room can be accomplished easily and economically.

For other options or further details regarding options shown, see

> Attic ventilation*
>
> Dormers*
>
> Flooring
>
> Interior doors
>
> Painting and wallpapering
>
> Skylights*
>
> Stairways
>
> * In Exterior Home Improvement Costs.

Key to Abbreviations
C.Y.–cubic yard   Ea.–each   L.F.–linear foot   Pr.–pair   Sq.–square (100 square feet of area)
S.F.–square foot   S.Y.–square yard   V.L.F.–vertical linear foot   M.S.F.–thousand square feet

HOME OFFICE/COMPUTER ROOM

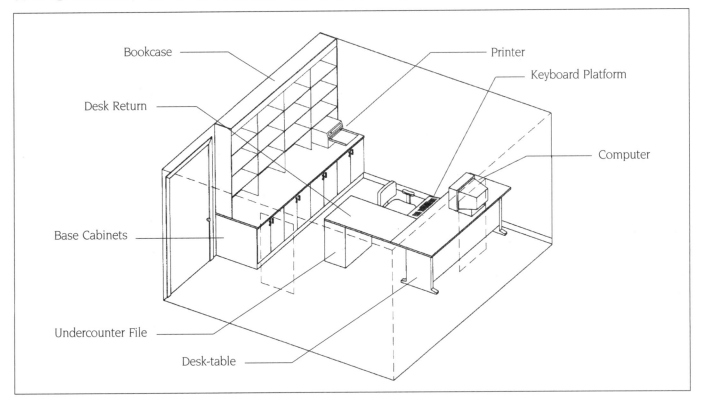

Bookcase

Desk Return

Base Cabinets

Undercounter File

Desk-table

Printer

Keyboard Platform

Computer

Advances in the computer and communications industries have allowed more and more people to work at home. Home offices tend to be thrown together haphazardly, which can make working at home an unpleasant experience and reduce your productivity dramatically. Creating an attractive, efficient, and comfortable home office does not have to be expensive or difficult. Your basic requirements will be a work surface, a chair, some kind of storage for paperwork and/or supplies, and whatever equipment and materials are required for your work.

MATERIALS

Your first step is to decide on a location. There are many advantages to having your work located in a separate room: it prevents family members from regularly borrowing office supplies or disturbing work in progress; it confines the mess to one area of the house; and you can leave the room and close the door, leaving your work behind. You may have an extra room or unused space in your attic, basement, garage, or even a roomy walk-in closet. The space you use depends on the amount and type of work you'll be doing. If you work full time from home,

you'll have more complex work-space requirements than someone whose main need is for a place to deal with household paperwork.

The cabinet system in this scheme is comprised of stock hardwood base units and bookcases, complete with all trim. Because many cabinet options are available, it pays to do some research and to shop for a system that meets your decorative and durability standards while remaining within your budget limitations.

The desk included in this project is a modular, free-standing system, 30″ x 72″, with a 24″ x 48″ return. Also included are an undercounter, 2-drawer file and a keyboard platform. The choice of desk – and of all the materials in this home office – will depend on your budget, how the office will be used, and the image you wish to present. There is a variety of styles large enough to fit any budget.

The electrical materials used in this project include basic outlets, switches, light fixtures, and the wiring required to connect the various components. If you want to add to the electrical work, be

sure to plan ahead and do it before any new sheetrock is installed.

Don't depend too heavily on natural light. Even if your home office has an abundance of windows, you'll need efficient task lighting. Task lighting options include the traditional desk lamp, track lighting, clamp lights, and pendant lamps.

Be sure to choose a well-designed chair that provides good support and ease of movement. A good chair should have a padded adjustable back and seat, a stable base, a swivel mechanism, and padded arm rests if you desire them. Depending on the kind of work you will be doing, you may also need one or more chairs for visitors.

Get one or more separate phone lines for your home office. That way, when the phone rings after hours, you'll know whether it's a business call or a personal call. You can turn the phone off or have a machine pick up when you don't want to be interrupted by business calls. You can also purchase a voice mail system from the phone company. If your computer has a modem you'll want a phone line for it as well.

LEVEL OF DIFFICULTY

If the cabinet retailer does not provide installation, a qualified carpenter can do the work. If you plan to install them on your own, consider the expense of the materials and the skillful placement required for the job, and work slowly and patiently.

Check to see that your electrical service has the capacity to allow for the new wiring. There is a good chance that the new supply wiring will have to be fished up from the service panel, so unless you are experienced in electrical installations, call an electrician to perform this task.

Generally, a beginner who tackles this home office project should add 150% to the time estimates provided here. Intermediates should add 100%, and experts, 50%.

WHAT TO WATCH OUT FOR

The home office must function as an office but must be flexible enough to be used as a different type of room. You don't want to turn off future homebuyers who may want the room for something else.

When choosing a space for your home office, pay attention to privacy, sound, and views. Computers and office equipment needs are changing and increasing rapidly, so be flexible in your design. Don't limit yourself to only the requirements of your present equipment. Keep in mind that for most purposes, you'll need plenty of horizontal surfaces – a computer, fax, phone, and answering machine use up lots of desk space. You may also need space for a desktop photocopier or a coffee machine.

The number of cords and plugs that are usually found in a modern work area are a potential hazard. Trailing wires from typical office equipment – phone, answering machine, fax, computer components, etc. – can be treacherous and can overload a single power outlet. Be sure to plug your computer and other sensitive equipment into a surge suppressor as protection in the event of a power outage.

Home Office/Computer Room

Description	Quantity/Unit	Labor-Hours	Material
Base cabinets, hardwood, 24" - 27" wide	4 Ea.	2.9	1,219.20
Countertop, plastic laminate	1 Ea.	2.4	221.40
Bookcase, 4 bay, 96" wide	1 Ea.	8.9	438.00
Desk-table, wood, 30" x 72"	1 Ea.		504.00
Desk return, wood, 24" x 48"	1 Ea.		357.60
Undercounter file	1 Ea.		420.00
Keyboard platform	1 Ea.		264.00
Electrical, light fixtures	4 Ea.	1.1	168.00
Electrical, light switches	2 Ea.	3.2	48.00
Telephone receptacle	2 Ea.	0.6	16.32
Electrical, duplex outlets	4 Ea.	6.0	78.48
Totals		25.1	$3,735.00

Project Size	Average Room	Contractor's Fee Including Materials	$6,636

Key to Abbreviations
C.Y.–cubic yard   Ea.–each   L.F.–linear foot   Pr.–pair   Sq.–square (100 square feet of area)
S.F.–square foot   S.Y.–square yard   V.L.F.–vertical linear foot   M.S.F.–thousand square feet

SUMMARY

A tidy, organized work environment produces a feeling of calm and control. Even if you work just a few hours a week at home, it's a good idea to have a separate, well-equipped space in which to work. Also, remember that your space should be an enjoyable place to work. Experiment with colors and creative storage space.

For other options or further details regarding options shown, see

Electrical system: Fixtures and receptacles

Full attic

Half attic

Painting and wallpapering

Walk-out basement

PROFESSIONAL HOME OFFICE

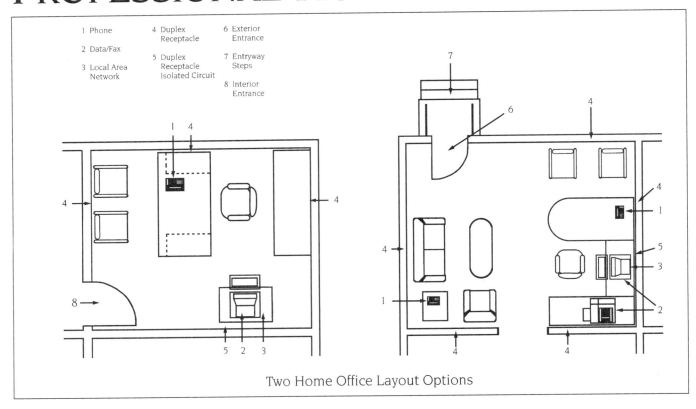

1 Phone

2 Data/Fax

3 Local Area Network

4 Duplex Receptacle

5 Duplex Receptacle Isolated Circuit

6 Exterior Entrance

7 Entryway Steps

8 Interior Entrance

Two Home Office Layout Options

The desk in the corner of the basement no longer fits today's home office needs. What was an area Mom and Dad used to pay bills and do "home work" is now becoming a professional workspace, as more and more people carry on full-time occupations from home. A truly professional home office includes lighting for all work areas, storage facilities, an optimum size work surface, seating for visitors, possibly a private entrance, and the all-important equipment, including telephone and fax machine, printer and copier, conveniently available power sources, and computer cabling. Consider the need for quiet, privacy, and security in selecting your furnishings, location, and layout.

Proper planning up-front is crucial to this project since it involves so many components. We recommend consulting one of the available publications on home office planning, or getting advice from office furniture manufacturers, your local phone service or computer system provider, and your data transmission vendor.

This project involves several separate projects, such as installation of lighting, wiring and cabling, and a drop ceiling, as well as wall and floor finishes. Other

sections in this book (such as Conversion to Living Space, Closets and Storage, Electrical Systems and Fixtures and Interior Finishes) may apply to your home office project. Note: Dormers, Exterior Doors and Windows, and Room Additions are other home office options found in the *Exterior Home Improvement Costs Guide*.

MATERIALS

This project includes a suspended ceiling system and new wainscoting, both of which provide opportunities to run new wires without cutting into existing ceilings and walls. (A space can be provided behind the new wainscot to run wires, and the suspended ceiling system gives you accessible space for installing current and future lighting and wiring.) Plan the location of all equipment (including task lighting), and provide for power requirements. Don't forget isolated electrical circuits or surge protection for computers and fax machines.

If you are purchasing office furniture, consider your work process, storage and equipment needs. Plan a layout that gives you easy access to frequently used files or equipment. Build in flexibility for future upgrades. Storage within the

work area should be task-oriented, while supplies and materials used less often can be elsewhere. Allow enough surface area for writing and meetings. (Don't overlook corner and specialized shapes that can give you more depth for computers or other equipment.) Keep in mind that flecked or speckled patterns disguise marks, and a light color surface minimizes glare. Don't forget to check on the ergonomic qualities of furnishings.

Once you know the size of your office space, you can determine the foot cost for items such as the suspended ceiling and painting from the projects shown in this book. (Remember to factor the cost to your location, using the Location Factors at the back of the book.) For example, the suspended ceiling cost may equate to approximately $4.00 per square foot. Multiplying your room's actual area by this amount should give you a good ballpark figure. You can also refer to the "Details" section at the back of the book for square foot or "per each" costs for most items.

LEVEL OF DIFFICULTY

This project differs from others in this book because it has many parts, some of which (like computer cabling and electrical) will probably have to be done by specialists. (Remember, if you hired a professional builder to create this home office, he or she would probably subcontract the electrical and computer wiring.) Other projects in this book that may be incorporated into this multi-task office project have their own guidance on level of difficulty.

WHAT TO WATCH OUT FOR

Time spent properly planning this project is well invested. Carefully measure your space and draw a plan (with a consistent scale for accuracy) that includes all features of the room itself, like windows and doors. Furniture and equipment

should also be sized and positioned. You may want to make to-scale cut-outs of proposed furniture and equipment to "try out" the space. They are easier to move than the real thing, and using this technique will ensure that you have taken everything into consideration.

Coordinating deliveries and scheduling construction work can be a full-time job. Make sure you have allowed plenty of lead time for procuring materials, so you don't get held up waiting for one component. Plan to have space ready to store equipment and materials procured in advance. Consider the need for dust protection and protection of existing surfaces, as well as cutting and patching to match the surrounding materials already in place. These issues can add significant time to a job. Beginners should add 100% to the time for construction, and intermediates and experts 50% and 20% respectively, for the planning, layout and construction process.

SUMMARY

No one can predict exactly how a professional home office will alter your lifestyle, but it surely will. Some of the suggested elements of this project require structural alterations to existing spaces and the introduction of materials that are often thought of as commercial. Before you start, keep in mind that you are going to be spending a lot of time in this space. Make it comfortable and efficient. Remember your goal should be to create an office that has the look and feel of a space designed by professionals — especially for you.

For other options or further details regarding options shown, see

> *Elevated deck Elevated L-shaped deck Entryway steps Redwood hot tub*

Professional Home Office

Description	Quantity/ Unit	Labor- Hours	Material
Main entry door	4 Ea.	2.9	1,219.20
Entry steps	1 Ea.	2.4	221.40
Wainscote & chair rail	1 Ea.	8.9	438.00
Oak strip floor	1 Ea.		504.00
Suspended ceiling	1 Ea.		357.60
Painting	1 Ea.		420.00
Fluorescent recessed light fixture	1 Ea.		264.00
Duplex recepticals	4 Ea.	1.1	168.00
Lighting wiring	2 Ea.	3.2	48.00
Lighting wiring conduit	2 Ea.	0.6	16.32
Switches, three way	4 Ea.	6.0	78.48
Totals		25.1	$3,735.00

Project Size	Average Room	Contractor's Fee Including Materials	$6,636

Key to Abbreviations
C.Y.–cubic yard   Ea.–each   L.F.–linear foot   Pr.–pair   Sq.–square (100 square feet of area)
S.F.–square foot   S.Y.–square yard   V.L.F.–vertical linear foot   M.S.F.–thousand square feet

IN-LAW APARTMENT

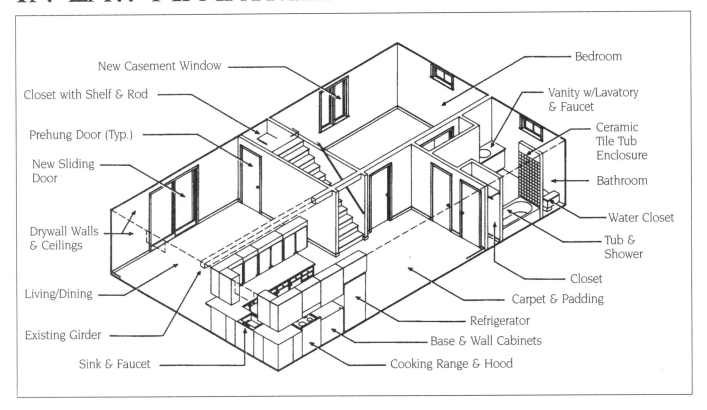

New Casement Window

Closet with Shelf & Rod

Prehung Door (Typ.)

New Sliding Door

Drywall Walls & Ceilings

Living/Dining

Existing Girder

Sink & Faucet

Bedroom

Vanity w/Lavatory & Faucet

Ceramic Tile Tub Enclosure

Bathroom

Water Closet

Tub & Shower

Closet

Carpet & Padding

Refrigerator

Base & Wall Cabinets

Cooking Range & Hood

Basements can often be economically converted to comfortable living areas. This project presents an in-law apartment arrangement that can also be used as an au pair suite or studio apartment.

MATERIALS

Drywalling and framing a basement room is a challenging but reasonable undertaking for even inexperienced homeowners. You will need some help bringing the sheetrock into the cellar and installing it on the ceiling, but with a little patience, you can hang the sheets on the walls alone. The taping and painting of the drywall can also be done by most beginners.

The plumbing fixtures in the full bath consist of the sink, toilet, and shower and tub unit. The sink is a standard-sized porcelain-on-cast-iron unit, with an 18" round bowl and appropriate trim. If you use a colored sink instead of white, add about 25% to the estimated cost of the fixture. The two-piece floor-mounted toilet also costs about 25%-35% more for color than for the white unit shown here. The fiberglass bathtub and modular shower surround come as a ready-to-

assemble unit, which can provide years of trouble-free service when properly installed and maintained. The fixture costs for the sink and tub units include the faucets, control, and shower head. The various fittings required to tie the plumbing fixtures into the water supply and drainage systems are listed separately and include such items as the shut-off valves and rough plumbing. In homes with septic systems, there may be a need for a pump to bring wastewater up to the level of the septic system from the basement apartment. This can add considerably to the plumbing costs for this project.

The bathroom accessories, vanity, and medicine cabinet provide options of varying costs for the homeowner. A standard two-door, 30" vanity with a durable laminate top serves as a support for the sink while providing necessary storage. The medicine cabinet should be large enough to accommodate the needs of the apartment's occupant(s).

The kitchen cabinets are made of quality hardwood, include durable hardware, and come completely finished from the factory or shop. Before you purchase the cabinets, do some research, shop extensively, and know precisely what you are buying.

The surface coverings used in the kitchen are economical materials that are attractive, durable, and easily maintained. You can purchase the laminate countertop as a prefabricated unit from the cabinet retailer or from specialized countertop suppliers. If you decide to install the countertop on your own, don't cut the opening for the sink until you have consulted your plumber. He or she may prefer to do the entire sink installation, including the cutting of the opening, at one time. The vinyl flooring material is also a proven product for kitchen use. Many decorative options are available in both sheet and tile format. Sheet goods cost slightly more; and they require more skill to install than tiles. The materials are available in various ranges of durability (as defined by the length of their warranties) and are priced accordingly. Because sheet flooring is seamless, it is less prone to water penetration than individually laid vinyl tiles.

The appliances for the kitchen have been selected for economy. Undersized appliances are available for situations where space is a critical concern, but they are usually special-order units and cost more. If you can fit standard-sized appliances into the new space, you should do so whenever possible.

LEVEL OF DIFFICULTY

If you have not worked with drywall before, seek assistance, and purchase or borrow the correct tools. Good advice or instruction can head off time-consuming correcting of errors and enhance the appearance of the finished product.

Beginners should hire a plumber to install the bathroom fixtures, and even intermediates should consider professional assistance in installing the fixtures.

If you plan to install the kitchen cabinets, get some advice beforehand and work slowly. Beginners should leave the cabinet installation to a professional, but experts and skilled intermediates should be able to handle the installation if they are given some professional advice. Experts can lay the floor in this kitchen, as can skilled intermediates with professional guidance.

WHAT TO WATCH OUT FOR

Even if the occupant of this apartment is healthy and relatively young now, keep future conditions in mind when you plan your remodeling project. Before converting space into living quarters for a loved one, you should consider the following questions:

- Is the person physically challenged or likely to develop some functional limitations with age while living in the apartment? Are there special features, such as lever door handles or grab bars, that should be built into the project?
- Is the person able to move freely and independently, or will special services and equipment be needed?
- Will doorways need to be widened to accommodate a wheelchair?
- Will landscaping need to be done to provide a convenient, safe path from the apartment to its parking area?

You may also want to use slip-resistant flooring in the bath and kitchen. In addition, both interior and exterior lighting should be appropriate and adequate.

Before you begin, be sure that you have sufficient headroom to meet the requirements of the local building code.

Building codes have stringent requirements for egress from bedrooms. Be sure that any room to be used as a bedroom has a window that meets these minimums.

The layout of the rooms will depend on local building department requirements and existing plumbing locations. A professional plumber should be brought in early in the process to assist in the placement of fixtures.

If you plan to rent the apartment now or in the future, you might want to consider installing separate electrical, hot water, and heating systems so that these utilities can be metered.

In-law Apartment

Description	Quantity/ Unit	Labor-Hours	Material
Partition walls, 2 x 4 plates & studs, 16" O.C., 8' high	822 L.F.	12.0	355.10
Blocking, miscellaneous, nailers for walls	63 L.F.	2.0	27.22
Cut masonry openings, 6' x 6'-8" and 4' x 4'	2 Ea.	1.5	
Furring strips for ceilings & walls, 1 x 3, 16" O.C.	1,400 L.F.	22.6	319.20
Insulation, rigid molded bead board, 1/2" thick, R-3.85	1,020 S.F.	10.2	171.36
Drywall, 1/2" thick, taped & finished, 4' x 8' sheets	2,760 S.F.	45.8	828.00
Window, casement, metal clad, double window, 4' x 4'	1 Ea.	0.9	246.00
Door, sliding glass, wood, premium, 6' wide	1 Ea.	4.0	1,146.00
Doors, flush HC, lauan finish, 2'-6" wide x 6'-8" high, pre-hung	4 Ea.	3.2	542.40
Doors, bi-passing closet, including hardware, 5'-0" x 6'-8"	1 Ea.	1.5	205.20
Locksets, non-keyed, privacy	4 Ea.	2.7	216.00
Trim, jambs for doors & windows, 1 x 8 pine	100 L.F.	3.6	163.20
Trim, baseboard, pine, 9/16" x 3-1/2"	170 L.F.	5.7	232.56
Trim, window casing, pine, 11/16" x 2-1/2"	48 L.F.	1.6	44.93
Railing, oak, built-up	24 L.F.	3.2	187.20
Paint, ceilings & walls, primer	2,130 S.F.	8.4	102.24
Paint, ceilings & walls, 1 coat	2,130 S.F.	13.1	127.80
Paint, trim, wood, incl. puttying, primer	320 L.F.	3.9	7.68
Paint, trim, wood, 1 coat	320 L.F.	3.9	11.52
Padding for carpet, sponge rubber cushion	754 S.F.	4.5	280.49
Carpet, nylon, 20 oz., medium traffic	754 S.F.	13.6	1,402.44
Standard full bathroom (see bathroom section for full detail)	1 Ea.	47.1	2,588.79
Standard U-shaped kitchen (see kitchen section for full detail)	1 Ea.	52.3	6,183.02
Electrical, light fixtures with wiring	4 Ea.	1.1	168.00
Electrical, light switches	4 Ea.	5.6	85.44
Electrical, duplex receptacles	15 Ea.	22.5	294.30
Totals		296.5	$15,936.09

Project Size	24' x 40'	Contractor's Fee Including Materials	$37,535

Key to Abbreviations
C.Y.–cubic yard   Ea.–each   L.F.–linear foot   Pr.–pair   Sq.–square (100 square feet of area)
S.F.–square foot   S.Y.–square yard   V.L.F.–vertical linear foot   M.S.F.–thousand square feet

SUMMARY

Many families find themselves in need of living space for older family members or child caregivers. A dry basement with a walk-out doorway can be a good solution. This is a major project that requires detailed planning up-front to meet both its current and long-term requirements.

For other options or further details regarding options shown, see

> Electrical system: Light fixtures/Receptacles
> Frame and finish partition wall
> Full attic conversion
> Garage conversion
> Half attic conversion
> Interior doors
> Painting and wallpapering
> Standard full bath

STANDARD BASEMENT

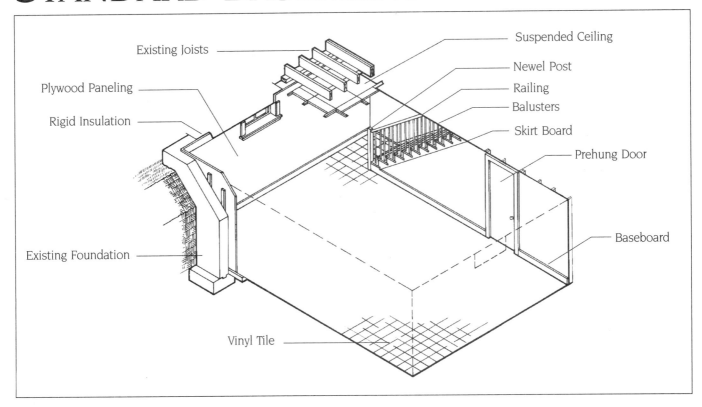

Existing Joists

Plywood Paneling

Rigid Insulation

Existing Foundation

Vinyl Tile

Suspended Ceiling

Newel Post

Railing

Balusters

Skirt Board

Prehung Door

Baseboard

Even if your basement is not constructed with a walk-out design, it can be converted into efficient living space with a minimum of time, effort, and cost. As long as the area is dry and the existing walls, ceiling, and floor are in good shape, a basement room is an efficient and cost-effective place to expand your home's living space.

MATERIALS

With the house's foundation serving as a solid exterior wall system, a limited amount of partitioning is required to prepare the room for the finished walls. The area under the stairwell requires 2 x 4 framing members to back the drywall and to frame the understairs closet. In most cases, you will be able to use the existing cellar stair stringers, but if you have to construct the staircase from scratch, you can figure on extra cost for the stringer material and other staircase support modifications. The concrete walls have to be prepared with 1 x 3 furring strips, which are placed vertically at 16" on center and horizontally across the top and bottom of the wall. The furring strips can be fastened to the concrete with masonry nails or by firing special nails

into the concrete with a stud gun. This procedure can be tricky and sometimes dangerous, so seek professional advice if you have not done it before. After you have completed all of the partitioning, the rough electrical work should be performed. Chances are, especially in new homes, that electrical circuits are readily available from the service panel in the basement. If you are unfamiliar with electrical installations, hire a professional to do the work or seek assistance from a qualified person. The styrofoam insulation material used on the exterior walls should match the thickness of the wall furring strips.

Drywalling a basement room is a challenging but reasonable undertaking for even inexperienced homeowners. You will need some help in bringing the sheetrock into the cellar, but with a little ingenuity, you can hang the sheets on the walls alone.

The taping and painting of the drywall can also be done by most beginners. If you have not worked with drywall before, seek some assistance and purchase or borrow the correct tools. Some good advice or instruction can head off much time-consuming sanding and finish work

and enhance the appearance of the finished product.

The finished ceiling and floor can be constructed from prefinished materials, all of them readily available from building supply outlets.

The ceiling consists of 2' x 2' prefinished fiber tiles, which are suspended on a prepackaged system of hangers, carriers, and runners. These systems are usually accompanied by instructions for installation by homeowners. The resilient tile flooring can be adhered directly to the concrete floor. Take time to prepare the surface so that it is dry, clean, and smooth. If the concrete floor is sloped or in rough shape, a subfloor may have to be constructed from 2 x 4 sleepers and sheathing. This installation requires some know-how and involves some extra cost, so get some advice and assistance if it has to be done. Individual tiles or a variety of sheet goods are available for resilient flooring. Check with a flooring contractor or consult someone who has done this kind of work before you select the type of floor for your basement. Follow the manufacturers' instructions on the flooring material and adhesives.

Finishing the basement will require extensive work on the staircase. Even if you are a fairly accomplished intermediate, you might need assistance in this part of the project. It might be a good idea to hire a carpenter to place the risers, treads, bannisters, and handrail. The materials are expensive, and precision is required in calculations and cutting. For the door under the stairs, you can choose a modest prehung lauan unit, or a more expensive solid-core or paneled design. Pine, colonial, or ranch design trim can be used for the casings around the windows and doors and at the base of the walls.

LEVEL OF DIFFICULTY

Except for the staircase and the electrical work, the project can be completed by most homeowners. If you are a beginner, seek the advice and assistance of experts as the building progresses. If you follow directions, work patiently, and learn from your mistakes, you should get very good results at a modest cost. Add 80% to the labor-hours if you are a beginner or have limited building experience. If you are competent with tools and house projects, then add 40% to the estimated professional time. If you are an expert, an additional 10% should do it.

WHAT TO WATCH OUT FOR

One of the major problems with renovations of standard 75%-below-grade basements is moisture. Even though the basement may be dry and free from flooding or seepage during the wet times of year, ground moisture or condensation can be trapped in closed basements and cause dampness in your new room. If you sense that present or future moisture problems may affect the basement room, apply a sealer to the walls and floor as the first step in construction. If your basement is prone to occasional flooding, don't despair. Systems are available to correct the problem from both inside and outside the foundation, but they can be costly. If your dwelling has these problems, get professional advice and assistance before proceeding with the conversion.

Unfinished basements usually provide easy access to plumbing and electrical wiring. Finishing the basement may require that you move shut-offs or pre-wire for future projects before closing up the walls and ceilings.

Another option for finishing the walls is paneling. Today's creative designs offer a variety of finishes. These include designs such as Victorian bead board. When installing paneling, be sure to locate all electrical outlets, switches, and thermostats in advance so that you do not cover them with paneling. Also ensure that all of the paneling comes from the same run or lot. Check panels for match or repeat pattern before installing. Finally, when using adhesive be sure to have clean-up solvent handy and plenty of bracing to temporarily hold the panel while the adhesive sets.

SUMMARY

A dry basement that meets height and accessibility requirements may be a source of new living area, and its renovation may help you to put this normally under-utilized space to better use. Because the rough walls, ceiling, and floor are already in place, the finishing of a basement room is a good way to expand your environment with a limited amount of cost and effort.

For other options or further details regarding options shown, see

> Bulkhead door*
>
> Closets and storage
>
> Flooring
>
> Interior doors
>
> Painting and wallpapering
>
> Stairways
>
> * In Exterior Home Improvement Costs

Standard Basement

Description	Quantity/ Unit	Labor- Hours	Material
Partition wall for stairway, 2 x 4 plates & studs, 16" O.C., 8' high	180 L.F.	3.6	77.76
Blocking, misc., nailer for walls	12 L.F.	0.3	5.62
Furring strips for walls, 1 x 3, 16" O.C.	665 L.F.	10.8	151.62
Insulation, rigid molded bead board, 1/2" thick	615 S.F.	6.2	103.32
Drywall, 1/2" thick, taped and finished, 4' x 8' sheets	832 S.F.	13.8	249.60
Suspended ceiling system, 2' x 2' or 2' x 4' panels	440 S.F.	9.3	934.56
Door, flush, HC, lauan finish, 2'-6" wide x 6'-8" high, prehung	1 Ea.	0.8	135.60
Lockset, keyed, for flush door	1 Ea.	0.7	54.00
Stairway additions: skirt board, 1 x 10	12 L.F.	1.8	45.36
Stair treads, oak	13 Ea.	5.8	358.80
Newel posts, 3'-4" wide, starting	1 Ea.	1.6	88.20
Balusters, turned, pine, 30" high	21 Ea.	6.0	103.57
Railing, oak, built-up	24 L.F.	3.2	187.20
Trim, jambs for doors and windows, 1 x 8 pine	16 L.F.	0.6	26.11
Trim, for baseboard, pine, 9/16" x 3-1/2"	96 L.F.	3.2	131.33
Trim, casing, pine, 11/16" x 2-1/2" wide	10 L.F.	0.4	10.30
Paint, trim, incl. puttying, primer	125 L.F.	1.6	3.12
Paint, trim, incl. puttying, 1 coat	125 L.F.	1.6 *48*	4.68
Flooring, vinyl composition tile, 12" x 12", 1/16" thick	440 S.F.	7.0	385.44
Electrical, light fixtures with wiring	2 Ea.	3.0	39.24
Electrical, switches	2 Ea.	0.5	84.00
Electrical, duplex receptacles	6 Ea.	8.4	128.16
Electrical, baseboard heaters, 10' long	2 Ea.	4.9	328.80
Electrical, baseboard heater, 5' long	1 Ea.	1.4	69.60
Totals		96.5	$3,705.99

Project Size	24' x 20'	Contractor's Fee Including Materials	**$10,197**

Key to Abbreviations
C.Y.–cubic yard   Ea.–each   L.F.–linear foot   Pr.–pair   Sq.–square (100 square feet of area)
S.F.–square foot   S.Y.–square yard   V.L.F.–vertical linear foot   M.S.F.–thousand square feet

WALK-OUT BASEMENT

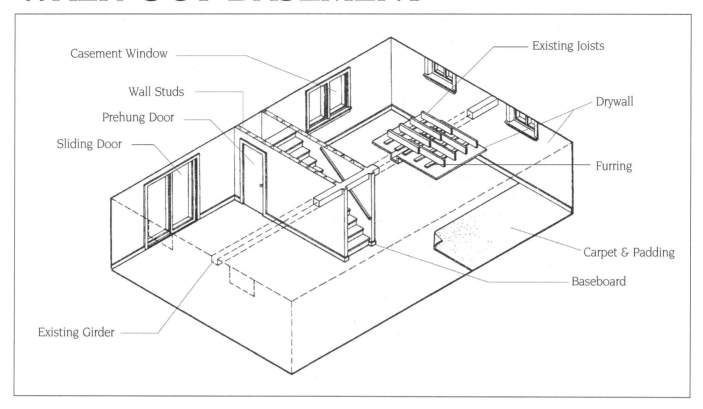

Casement Window — Wall Studs — Prehung Door — Sliding Door — Existing Girder — Existing Joists — Drywall — Furring — Carpet & Padding — Baseboard

Basements can be economically converted to comfortable living areas, turning under-utilized space into recreation, home office, or utility rooms. If your basement area is dry and already equipped with a walk-out doorway, a conversion can be easily undertaken.

MATERIALS

Because your foundation can serve as the rough wall system for the new room, 2 x 4 framing members are required only to enclose the entry staircase. Additional framing materials will be necessary at extra cost if you want to add a closet or storage area or erect a partition for a second room. The support system for 1/2" drywall begins with 1 x 3 furring strips secured vertically at 16" on center to the existing foundation wall. Consult someone who has experience in hammering concrete nails or firing specialized nails with a stud gun. If you plan to use 1/2" sheetrock on the ceiling, it might be a good idea to put the furring on the joists first to offset any settling or bowing that has occurred. The additional cost of this procedure is minimal, and the improvement in the appearance of your ceiling can make it worthwhile. Once the

walls and ceiling have been furred, the rough electrical work can be done and the insulation and sheetrock put into place. If you are inexperienced in electrical installations, call a contractor or arrange to have a qualified person assist you in running the wires and installing the receptacles, switches, and fixtures. Drywalling a basement room is a challenging but reasonable undertaking for even inexperienced homeowners. You will need some help in bringing the sheetrock into the cellar and installing it on the ceiling; but with a little ingenuity, you can hang the sheets on the walls alone.

The taping and painting of the drywall can also be done by most beginners. If you have not worked with drywall before, seek some assistance and purchase or borrow the correct tools. Some good advice or instruction can head off much time-consuming sanding and finish work and enhance the appearance of the finished product.

If your foundation has been constructed to accommodate sliding doors and large window units, vinyl-clad, aluminum, or paintable wood door and window units can be installed after the old units have been removed. If, as in this example,

concrete or block has to be removed to enlarge existing openings or to create new ones, you should consult a masonry contractor. When existing openings do not match the desired door and window design, you may choose to modify your plan somewhat for the sake of economy.

The cost of your project will vary according to the amount of professional assistance required. The door for the understairs closet can be an economical prehung hollow-core lauan model or a more expensive paneled or decorative unit. Pine boards in ranch or colonial trim are available for the jambs, casings, and interior facings of windows and doors. The baseboard can be obtained in the same styles and materials.

As for the carpet, a foam-backed type is used in this model, and can be installed directly on a clean, smooth concrete floor in a dry basement. If you have no experience in carpet laying, it might be a good idea to contact a professional.

LEVEL OF DIFFICULTY

Except for the electrical work and the foundation preparation for windows and doors, most aspects of this project are relatively straightforward and uncomplicated. There is a limited amount of framing involved in the conversion and, except for ceiling obstacles like existing plumbing and electrical fixtures, there are few problems to keep you from direct and rapid progress. If you are a beginner, double the estimated time for the professional labor-hours. Add about 40% to the labor-hour estimates if you have had some experience with tools and renovation, and 10% if you are thoroughly skilled in these areas.

WHAT TO WATCH OUT FOR

Walk-out basements are usually accessible, but consider the size of your entranceways, as you will need to bring in some large materials. Also, double-check the ceiling height to see that it meets minimum requirements for he Remember that you will lose s when you add furring to the j more with a layer of sheetrock fixtures installed in the ceiling require safe clearance. If you h around low-hanging plumbing, lose some headroom in those a Lally columns that support the h primary structural beams are anoth consideration. They can be boxed in, ᴜ if they have to be removed or relocated, proper support systems must be installed to prevent expensive structural problems. There are a number of new composite wood products that make greater spans possible. Consult your local building supplier for suggestions. These additional operations can affect remodeling costs significantly. It is best to check with a professional contractor before you make any changes that involve the structural support system.

SUMMARY

A dry and accessible walk-out basement is a good source of new living area. Because the space is already enclosed, much of the project can be accomplished by the do-it-yourselfer for a modest investment.

For other options or further details regarding options shown, see

> Closets and storage
>
> Exterior doors, windows*
>
> Flooring
>
> In-law apartment
>
> Interior doors
>
> Interior partitions
>
> Main entry door*
>
> Painting and wallpapering
>
> Patio & sliding glass doors*
>
> * In Exterior Home Improvement Costs

Walk-Out Basement

Description	Quantity/ Unit	Labor-Hours	Material
Partition wall for stairway, 2 x 4 plates & studs, 16" O.C., 8' high	312 L.F.	6.2	134.78
Blocking, miscellaneous, nailers for walls	24 L.F.	0.6	11.23
Cut masonry openings, 6' x 6'-8" and 4' x 4'	2 Ea.	1.5	
Furring strips for ceiling & walls, 1 x 3, 16" O.C.	1,400 L.F.	22.6	319.20
Insulation, rigid molded bead board, 1/2" thick, R-3.85	1,020 S.F.	10.2	171.36
Drywall, 1/2" thick, plasterboard, taped and finished, 4' x 8' sheets	2,368 S.F.	39.3	710.40
Windows, casement, metal-clad, double window, 4' x 4'	1 Ea.	0.9	246.00
Door, sliding glass, wood, premium, 6' wide	1 Ea.	4.0	1,146.00
Door, flush HC, lauan finish, 2'-6" wide x 6'-8" high, pre-hung	1 Ea.	0.8	135.60
Lockset, keyed, for flush door	1 Ea.	0.7	54.00
Trim, jambs for doors & windows, 1 x 8 pine	50 L.F.	1.8	81.60
Trim, for baseboard, pine, 9/16" x 3-1/2"	160 L.F.	5.3	218.88
Trim, window casing, pine, 11/16" x 2-1/2"	48 L.F.	1.6	44.93
Railing, oak, built-up	24 L.F.	3.2	187.20
Paint, ceiling & walls, primer	2,370 S.F.	9.3	113.76
Paint, ceiling & walls, 1 coat	2,370 S.F.	14.6	142.20
Paint, trim, wood, incl. puttying, primer	260 L.F.	3.2	6.24
Paint, trim, wood, incl. puttying, 1 coat	260 L.F.	3.2	9.36
Padding for carpet, sponge rubber cushion	910 S.F.	5.4	338.52
Carpet nylon, 20 oz. medium traffic	910 S.F.	16.4	1,692.60
Electrical, light fixtures with wiring	2 Ea.	3.0	39.24
Electrical, light switches	2 Ea.	0.5	84.00
Electrical, duplex receptacles	10 Ea.	14.0	213.60
Electrical, baseboard heaters, 10' long	4 Ea.	9.7	657.60
Totals		178.0	$6,758.30

Project Size	24' x 40'	Contractor's Fee Including Materials	$18,576

Key to Abbreviations
C.Y.–cubic yard   Ea.–each   L.F.–linear foot   Pr.–pair   Sq.–square (100 square feet of area)
S.F.–square foot   S.Y.–square yard   V.L.F.–vertical linear foot   M.S.F.–thousand square feet

Section Two

BATHROOMS AND SPAS

Before embarking on any bathroom project, read through all the steps carefully. Always use good judgment and when necessary, appropriate safety equipment when attempting any of these projects. Check with your local building department for information on permits, codes, and other laws that might apply to your project. Additional tips for completing your bathroom project follow.

- Enameled cast iron is the most common material in bathtubs; cast acrylic and fiberglass are also popular. Enameled cast iron can withstand much more abuse than plastic alternatives, but plastic is lighter and manufacturers offer a wider variety of forms and surface features in plastic.

- The most common size for tubs that back up to a wall is 32″ x 60″. Widths from 24″ to 42″ are available. Corner tub units are about 48″ on the wall side.

- Tubs can be placed at floor level, raised on a platform, or sunk into the floor.

- Whirlpools are made of the same materials and in the same colors as tubs. Rectangular whirlpools range from the size of an average bathtub up to 48″ x 84″. A whirlpool contains a pump to circulate the water through jets positioned around the sides of the unit. Whirlpools that don't come with a heater will draw hot water from your home's water heater.

- Check the total weight of the whirlpool when in use. If it exceeds 30 pounds per square foot, you may need to strengthen the framing under it to support the additional load.

- The floor covering for a bathroom should be able to withstand regular exposure to water, be safe to walk on when wet, and be easy to clean.

- In small bathrooms, a large mirror and chrome fixtures can help to provide a feeling of spaciousness.

- All new bathroom fixtures require venting. Rooftop vents let air into the drain-waste-vent system. This helps waste water flow freely into the sewer line.

- In a new bathroom, it is a good idea to provide access panels to piping in the wall.

- If new piping is to be buried, care should be taken to protect, wrap, and provide easy access to the pipe for service in the event that pipes fail.

- Be sure to have your new sink and toilet standing by before tearing out the old units.

- When removing a sink, be sure to have a bucket ready because the trap will have water in it.

- If your project involves soldering pipes, take the time to organize your work area, remove all inflammable materials and have a fire extinguisher handy. Observe the area around the piping for any signs of smoke or fire, especially in a confined area or where hot material could drop into a wall or floor cavity.

- Pre-assemble piping components whenever practical to make final installations easier.

- Depending on the age of your home, it is a good idea to check the existing supply and waste piping before you begin any plumbing installations. You might uncover some situations that should be rectified, such as different-sized water pipes or improperly sized waste and vents. Remember, a previous owner may not have been so conscientious.

STANDARD HALF BATH (UPGRADE EXISTING)

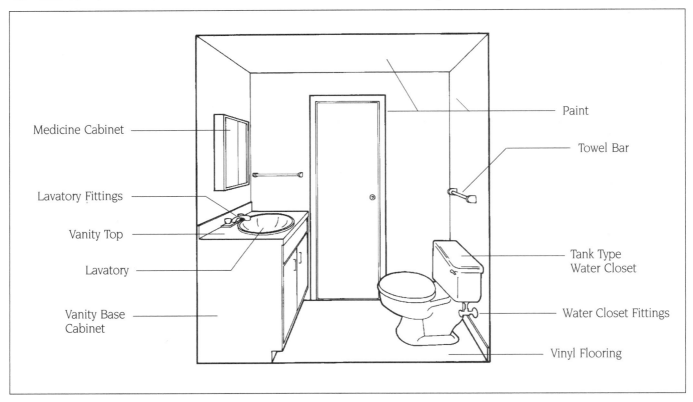

Medicine Cabinet

Lavatory Fittings

Vanity Top

Lavatory

Vanity Base Cabinet

Paint

Towel Bar

Tank Type Water Closet

Water Closet Fittings

Vinyl Flooring

One of the most challenging yet gratifying ways to improve your home is to remodel an existing bathroom. In some cases, the room has to be redone because of wear and age, but a renovation might also be undertaken to head off problems or simply to spruce it up. With today's fixtures and materials, remodeling a half bath may be easier and less expensive than you think, especially if you opt to do some or all of the work on your own.

MATERIALS

The plan for the standard half bath renovation includes the complete replacement of the basic plumbing fixtures, flooring, vanity cabinet, and accessories. The two primary plumbing fixtures in any half bath are the sink and the toilet. The size of the sink may vary according to personal taste, but 18″ round is a good, functional size for a porcelain-coated cast iron unit. A standard, two-piece, floor-mounted tank-type toilet, manufactured from vitreous china, is a basic stock item at any plumbing supply outlet. In many locations a water-saver toilet is required for all replacements. This will increase cost slightly. If you want colored fixtures rather than white, they

will cost about 25% more. New chrome-plated fittings, including faucets and shut-off valves, complete the plumbing installation.

The flooring is a critical item in any room, but especially in a bathroom where water and moisture are ever present. When you take up the old floor, make sure that you give the existing subflooring and floor joists a careful inspection for signs of rot and general deterioration. If the subfloor is solid and dry, leave it in place to be used as a base for the new floor. If the subfloor looks questionable, even if only in one area, you should take the time to tear it up and replace it with new underlayment, comprised of 3/16″ hardboard. The new finished floor of vinyl covering can be laid after the underlayment has been properly leveled and prepared to receive it. Before the new floor is put into place, you might want to clean and apply two coats of paint to the walls and ceiling. This makes the whole process easier since you won't have to cover the floor when you paint. Wallpapering is another alternative, and while it may be slightly more expensive, it can go a long way to brighten a small room. Water-resistant, washable paper is available in a wide variety of patterns and colors.

A vanity cabinet to support the sink, a new medicine cabinet, and basic accessories add the finishing touches to the half bath. A 30″ vanity with a laminated plastic (Formica) top provides some additional storage and shelf space. A basic metal medicine cabinet involves a modest investment that will be returned many times over in the convenience it provides. The stainless steel towel bars and tissue dispenser used in this example could be replaced by models made of other materials.

LEVEL OF DIFFICULTY

Even if you have only limited experience with plumbing or electrical installation, a small half-bath renovation is a reasonable undertaking. Beginners might be better off leaving the actual installation and tying in of the fixtures and fittings to a professional, but the rest of the project, including the floor, involves manageable tasks that can be accomplished by novices with some professional advice and assistance.

Beginners should plan on twice the estimated professional labor-hours to complete the project. Even intermediates and experts should tread carefully in the plumbing tasks, but the flooring, wall, and accessories installations should pose no real problems. Intermediates should add 50% to the professional labor-hours estimate; experts, about 20%.

WHAT TO WATCH OUT FOR

One of the biggest problems in any bathroom renovation involves replacement of rotted and/or deteriorated subflooring and the structural members underneath or adjacent to it. In half baths, the problem occurs most frequently around the base of the toilet, where moisture from seepage or condensation collects and does not evaporate because of poor ventilation. If the rot or deterioration is extensive and joists or other structural elements have been affected, then a professional should be called in, at least to inspect the damage. Beginners, especially, should seek assistance if the situation requires correction. Intermediates and experts should be able to handle the problem by removing the affected sections and adding replacement support. Ensuring that the floor is solid and installing the new fixtures with care should keep future problems to a minimum.

SUMMARY

Renovating an existing half bath is a good way to improve the value and appearance of your home. It is an approachable project in that it does not tie up the only bath in the house and only two fixtures are involved. Beginners and intermediates can add to their plumbing and remodeling experience by tackling this project, but they should seek professional advice along the way.

For other options or further details regarding options shown, see

> *Ceramic tile floor*
> *Deluxe half bath*
> *Electrical system: Light fixtures*
> *Electrical system: Receptacles*
> *Painting and wallpapering*
> *Tile walls*
> *Ventilation system*
> *Vinyl sheet flooring*

Standard Half Bath

Description	Quantity/ Unit	Labor- Hours	Material
Rough in frame for medicine cabinet, 2 x 4 stock	8 L.F.	0.3	3.46
Flooring, underlayment-grade hardboard, 3/16", 4' x 4' sheets	32 S.F.	0.3	15.74
Paint, ceiling, walls & door, primer, 1 coat	162 S.F.	0.6	7.78
Medicine cabinet w/mirror, 16" x 22", unlighted	1 Ea.	0.6	82.20
Vanity base cabinet, 2 door, 30" wide	1 Ea.	1.0	222.00
Vanity top, plastic laminated, basic	1 Ea.	0.3	9.18
Lavatory, with trim, porcelain enamel cast iron, 18" round	1 Ea.	2.5	180.00
Fittings for lavatory	1 Set	7.0	88.80
Towel bars, stainless steel, 18" long	1 Ea.	0.4	38.40
Toilet tissue dispenser, surface mounted, stainless steel	1 Ea.	0.3	14.10
Flooring, vinyl sheet goods, backed, 0.080" thick	20 S.F.	0.7	48.00
Toilet, tank-type, vitreous china, floor-mounted, 2 piece, white	1 Ea.	3.0	168.00
Fittings for toilet	1 Set	5.3	151.20
Electrical, light fixture with wiring	1 Ea.	0.3	42.00
Electrical, light switch	1 Ea.	1.4	21.36
Electrical, GFI outlet	1 Ea.	1.7	52.80
Totals		25.7	$1,145.02

Project Size	4' x 6'	Contractor's Fee Including Materials	$2,921

Key to Abbreviations
C.Y.–cubic yard   Ea.–each   L.F.–linear foot   Pr.–pair   Sq.–square (100 square feet of area)
S.F.–square foot   S.Y.–square yard   V.L.F.–vertical linear foot   M.S.F.–thousand square feet

DELUXE HALF BATH

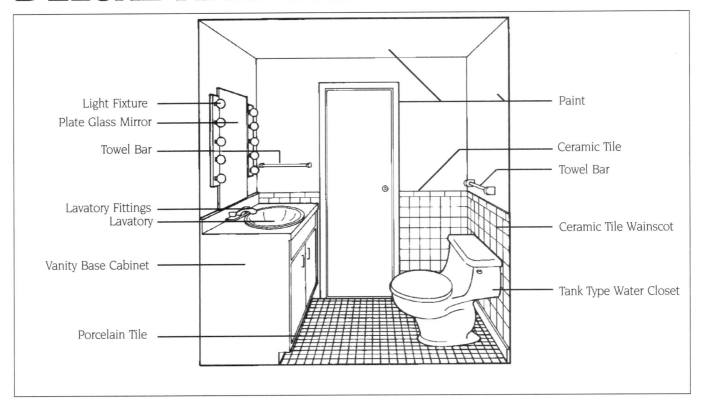

Light Fixture
Plate Glass Mirror
Towel Bar
Lavatory Fittings
Lavatory
Vanity Base Cabinet
Porcelain Tile

Paint
Ceramic Tile
Towel Bar
Ceramic Tile Wainscot
Tank Type Water Closet

Often a half bath is located in an area of your home where more decorative and deluxe features are desired – perhaps off a room where you do much of your entertaining. A deluxe half bath may also be in order simply because you want to upgrade the facility in keeping with other improvements throughout your home.

MATERIALS

The toilet in this bath is a floor-mounted, one piece unit of colored vitreous china. The lavatory is a trimmed 18"-round porcelain-on-cast-iron model, also in color. If you want white fixtures, you can deduct about 30% of the estimated cost of the toilet and 25% from the price of the sink. Chrome-plated fittings have been selected for the sink and toilet in this model.

The deluxe half bath also features ceramic tile on both the floor and the lower part of the wall. Working with ceramic floor and wall materials requires some skill and the right tools, but with some advice and assistance at the start, beginners can produce pleasing results. Ceramic floor tile is more durable and

easily maintained than vinyl and most other floor coverings. Ceramic wainscoting has the same benefits and also gives the room a neat look. The ceiling, upper part of walls, and door can be finished with a primer and finish coat of paint.

Washable wallpaper can also enhance the room's appearance. The choice of the vanity and other accessories for the deluxe half bath should suit your personal taste and coordinate with the style of the room. A deluxe 30" two-door vanity with a marble top provides extra storage space, and high-quality hardwood construction offers an extra attractive support for the sink. A lighted medicine cabinet with a built-in mirror is another feature. Stainless steel towel bars and tissue dispenser are one option, but other accessories are available in a variety of materials at varying costs.

LEVEL OF DIFFICULTY

The deluxe half bath is a fairly demanding project for even experienced do-it-yourselfers. The three most difficult tasks involved in the renovation are the plumbing procedures, electrical work, and the floor and wall ceramic tile installations. All of these jobs require specialized skills and a considerable degree of expertise. Beginners can handle the ceramic work, but only if they are given ample instruction beforehand and guidance during the installation. If you have not done any plumbing work in the past, you should leave the fixture installations to professionals, even if you are an accomplished do-it-yourselfer. Intermediates and experts should be able to complete a great deal of this project, including tile installation, but they will need to obtain specialized tools for ceramic work and seek some instruction at the start. In using the professional labor-hour estimates, add 100% to the time if you are a beginner, 50% if you are an experienced intermediate, and 20% if you are an expert.

WHAT TO WATCH OUT FOR

Ceramic tile, like any other wall or floor covering, requires a dry and clean surface for mounting. Be sure to check the condition of the floor underlayment and the sheetrock before laying the tile. The rough floor surface must be cleaned, leveled, and smoothed out before the tile flooring sections are laid. The sheetrock on the walls must be clean, dry, and solid to provide adequate backing for the wall tiles. Bathrooms are notorious collectors of moisture, and the adhesives used in ceramic tile installations will not bond effectively to damp or moist surfaces. The extra cost involved in preparing or replacing the existing floor or walls is money well spent to ensure a quality installation.

BATHROOM FLOOR OPTIONS
Cost per Square Foot, Installed

Description	
Sheet Vinyl	$ 4.46
Vinyl Tile	$ 1.51
Ceramic Tile	$ 7.20
Marble Tile	$17.55

SUMMARY

Upgrading a basic half bath to deluxe standards can enhance the value and overall appearance of your home. Because a half bath is not the primary facility in a house, its renovation does not involve as much inconvenience as that of a full bath. As a result, you may be able to perform much of this project yourself to make it a more economical undertaking.

For other options or further details regarding options shown, see

> *Ceramic tile floor*
> *Closets and storage*
> *Electrical system: Light fixtures*
> *Electrical system: Receptacles*
> *Painting and wallpapering*
> *Standard half bath*
> *Tile walls*
> *Vinyl sheet flooring*

Deluxe Half Bath

Description	Quantity/ Unit	Labor- Hours	Material
Paint, ceiling, walls & door, primer	142 S.F.	0.6	6.82
Paint, ceiling, walls & door, 1 coat	142 S.F.	0.9	8.52
Mirror, plate glass, 30" wide x 34" high	7 S.F.	0.7	50.40
Deluxe vanity base cabinet, 2 door, 30" wide cabinet	1 Ea.	1.4	310.80
Vanity top, marble, 3/4" thick, maximum	1 Ea.	1.5	228.00
Lavatory, with trim, porc. enamel on cast iron, 18" round, in color	1 Ea.	2.5	180.00
Fittings for lavatory	1 Set	7.0	88.80
Towel bars, ceramic, 18" long	2 Ea.	0.4	22.80
Toilet tissue dispenser, surface mounted, ceramic	1 Ea.	0.2	11.40
Walls, ceramic tile wainscoting, thin set, 4-1/4" x 4-1/4" tile	20 S.F.	1.7	50.16
Flooring, porcelain-type, 1 color, color group 2, 2" x 2" tile	20 S.F.	1.7	95.28
Toilet, tank-type, vitreous china, floor mounted, in color, 1 piece	1 Ea.	3.9	772.20
Fittings for toilet	1 Set	5.3	151.20
Electrical, light fixture with wiring	3 Ea.	1.0	185.40
Electrical, light switch	2 Ea.	2.8	42.72
Electrical, GFI outlet	1 Ea.	1.7	52.80
Totals		33.3	$2,257.30

Project Size	4' x 6'	Contractor's Fee Including Materials	$4,865

Key to Abbreviations
C.Y.–cubic yard   Ea.–each   L.F.–linear foot   Pr.–pair   Sq.–square (100 square feet of area)
S.F.–square foot   S.Y.–square yard   V.L.F.–vertical linear foot   M.S.F.–thousand square feet

ECONOMY FULL BATH

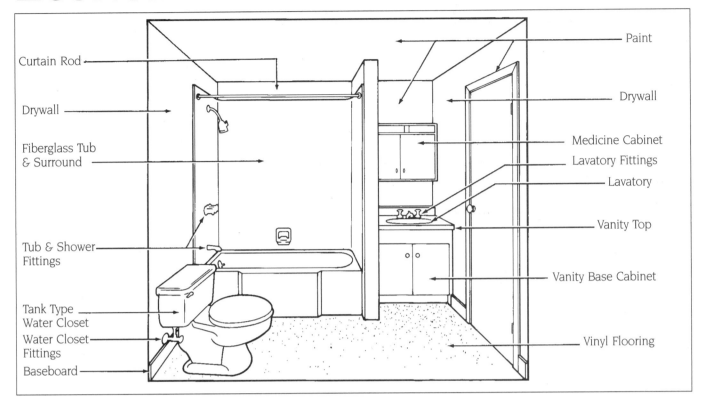

Curtain Rod

Drywall

Fiberglass Tub
& Surround

Tub & Shower
Fittings

Tank Type
Water Closet

Water Closet
Fittings

Baseboard

Paint

Drywall

Medicine Cabinet

Lavatory Fittings

Lavatory

Vanity Top

Vanity Base Cabinet

Vinyl Flooring

Even though the renovation of a full bath involves adding only one more fixture than that of a half bath, the full bath project is a significantly more challenging undertaking. If your house has just one bathroom, expediency in the plumbing installation becomes especially critical. Despite the inconvenience caused while the work is carried out, a modernizing or remodeling project like this can result in a more attractive, low-maintenance bathroom for years to come.

MATERIALS

The materials employed in the complete renovation of a full bath consist of the three basic plumbing fixtures; the flooring, wall, and ceiling coverings; and the vanity and accessories. The plumbing fixtures consist of replacement sink, toilet, and shower and tub unit. For the sink, you probably should stay with a standard-sized porcelain-on-cast-iron unit, with an 18" round bowl and appropriate trim. If you want the sink to be colored instead of white, add about 25to the estimated cost of the fixture. The two-piece floor-mounted toilet also costs about 25%-30% more for color than the estimated unit price. A water-saver toilet, required in

some locations, will also increase costs slightly. The fiberglass bathtub and modular shower surround come as a ready-to-assemble unit, which can provide years of trouble-free service when properly installed and maintained. The fixture costs for the sink and tub units listed in the chart include the faucets, control, and shower head. The various fittings required to tie in the plumbing fixtures to the water supply and drainage systems are listed separately and include such items as the shut-off valves and rough plumbing. Beginners should hire a plumber to install the fixtures, and even intermediates should consider professional assistance in taking out the old fixtures and installing the new ones.

The installation of the floor, wall, and ceiling materials should be manageable tasks for most homeowners. In any bathroom remodeling project, the floor is a critical concern because it is exposed to splashed water and other forms of moisture. The condition of the subfloor and its supports should be carefully determined, particularly in the area around the tub and toilet. Take a few minutes after you have removed the old tub and toilet to poke the underlayment, subfloor, and floor joists with a screwdriver or awl to locate any soft spots, wet areas,

or decay. If you find places where the flooring or joists are in questionable condition, they will have to be replaced at additional cost. If the problem is extensive, seek professional assistance before going ahead, even if you are reasonably experienced. After the new underlayment has been placed, vinyl sheet flooring can be laid by a specialty contractor or on your own, if you have an experienced helper.

A small framing job is required if no partition exists between the foot of the tub and the vanity. Beginners should seek some assistance here. The cavity surrounding the tub and shower module should be walled with 1/2" water-resistant drywall, which must be taped and finished after the shower module is placed. The other walls, ceiling, and door can be painted with a primer and finish coat of durable, washable paint.

The conventional bathroom accessories, vanity, and medicine cabinet provide options of varying costs for the homeowner. A standard two-door, 30" vanity with a durable Formica-type top serves as a support for the sink while providing necessary storage. The medicine cabinet should also be large enough to accommodate the needs of your family. If you want a cabinet with built-in lighting, the cost is higher than the estimate given.

LEVEL OF DIFFICULTY

Tackling a full bath renovation is a big job for even experienced and competent intermediates. A full range of building skills is required, from simple tasks like painting to specialized plumbing and electrical work. Even the most experienced do-it-yourselfers should think twice about carrying out the plumbing installations, especially if old fixtures have to be removed before the new ones are installed. Beginners should not attempt the extensive plumbing required, especially the installation of tub and shower fixtures. They should double the estimated man-hours for other tasks in the project. Intermediates and experts should add about 50% and 20%, respectively, to the professional time estimates throughout.

WHAT TO WATCH OUT FOR

If you plan to install the fiberglass tub and shower module on your own, take some time to read up on the intricacies of the task and seek some assistance from a qualified person. Even small discrepancies in alignment, caused by a settled floor or an out-of-plumb wall, can create major obstacles during installation, and they should be allowed for or corrected beforehand. Routine jobs like cutting the openings for the rough plumbing for the shower valve and tub fill can become difficult if the overall alignment of the unit is not true. Make sure that the manufacturer's guidelines and instructions are followed closely during the installation. The size of the tub is also a factor to consider, especially in terms of doorway clearances within the house.

SUMMARY

An economy full bath renovation or installation is a major project, but with the right materials and plumbing fixtures, it is an undertaking that can be at least partially completed by most experienced homeowners. When the job is done properly, it will add value to your home and provide years of dependable service.

For other options or further details regarding options shown, see

  Closets and storage

  Electrical system: Light fixtures

  Electrical system: Receptacles

  Flooring

  Painting and wallpapering

  Tile walls

  Ventilation system

Economy Full Bath

Description	Quantity/ Unit	Labor- Hours	Material
Partition wall, 2 x 4, plates & studs, 16" O.C., 8' high	64 L.F.	0.9	27.65
Rough in frame for medicine cabinet, 2 x 4 stock	10 L.F.	0.3	4.32
Drywall, 1/2" thick, water-resistant, taped & finished, 4' x 8' sheets	64 S.F.	1.1	19.97
Flooring, underlayment-grade hardboard, 3/16", 4' x 4' sheets	128 S.F.	1.4	62.98
Painting, ceiling, walls & door, primer	310 S.F.	1.2	14.88
Painting, ceiling, walls & door, 1 coat	310 S.F.	1.9	18.60
Vanity base cabinet, 2 door, 30" wide	1 Ea.	1.0	222.00
Vanity top, plastic laminated, basic	1 Ea.	0.7	22.95
Lavatory, with trim, porcelain enamel on cast iron, 18" round	1 Ea.	2.5	180.00
Fittings for lavatory	1 Set	7.0	88.80
Bathtub, module & shower wall surround, molded fbgls., 5' long	1 Ea.	4.0	840.00
Fittings for tub & shower	1 Set	7.7	147.60
Curtain rod, stainless steel, 5' long, 1" diameter	1 Ea.	0.6	31.80
Medicine cabinet, sliding mirror doors, 36" x 22", unlighted	1 Ea.	1.1	102.00
Flooring, vinyl sheet goods, backed 0.080" thick	36 S.F.	1.3	86.40
Toilet, tank-type, vitreous china, floor mounted, 2 piece, white	1 Ea.	3.0	168.00
Fittings for toilet	1 Set	5.3	151.20
Toilet tissue dispenser, surface mounted, stainless steel	1 Ea.	0.3	14.10
Towel bar, stainless steel, 30" long	1 Ea.	0.4	38.40
Trim, baseboard, 9/16" x 3-1/2" wide, pine	12 L.F.	0.4	16.42
Electrical, 2 light fixtures with wiring	2 Ea.	0.5	84.00
Electrical, 2 light switches	2 Ea.	2.8	42.72
Electrical, one GFI outlet	1 Ea.	1.7	52.80
Totals		47.1	$2,437.59

Project Size	7' x 8'	Contractor's Fee Including Materials	$5,830

Key to Abbreviations
C.Y.–cubic yard   Ea.–each   L.F.–linear foot   Pr.–pair   Sq.–square (100 square feet of area)
S.F.–square foot   S.Y.–square yard   V.L.F.–vertical linear foot   M.S.F.–thousand square feet

STANDARD FULL BATH

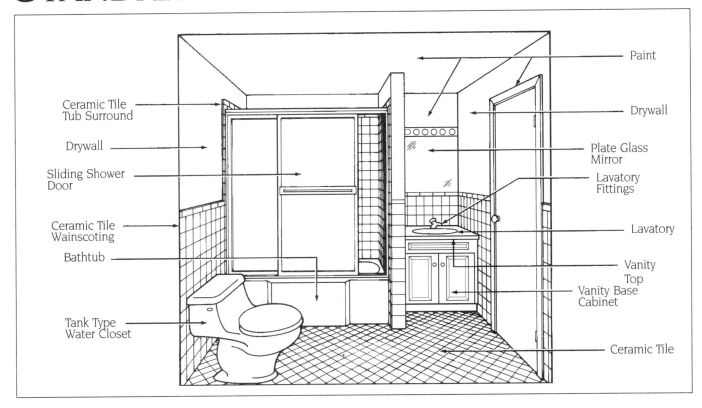

Ceramic Tile Tub Surround
Drywall
Sliding Shower Door
Ceramic Tile Wainscoting
Bathtub
Tank Type Water Closet

Paint
Drywall
Plate Glass Mirror
Lavatory Fittings
Lavatory
Vanity Top
Vanity Base Cabinet
Ceramic Tile

The standard full bath includes the amenities and upgraded fixtures to provide an attractive and durable addition to any home. For the additional expense of features like a one-piece toilet, a sliding shower door, and ceramic tile, this bath offers first quality materials for a relatively modest investment, especially if you opt to do most of the work on your own. This project provides a quality family bath in limited space.

MATERIALS

The full bath features quality fixtures, durable ceramic wall and floor tiling, and attractive accessories. The fixtures used in the bath include matching colored 18" lavatory, one-piece, floor-mounted toilet, and porcelain-on-cast-iron recessed tub with a nonskid mat bottom. Complementing the fixtures and included in the estimated cost are quality faucets, valves and diverters, and shower head. The fittings for the fixtures, including the minimum rough plumbing and supply valves, are listed as a separate cost, which may vary according to the amount of rearranging and plumbing modifications made for the particular facility. If you are considering doing the standard bath

remodeling project on your own, you may want to hire a plumber for the removal and installation of the primary fixtures, especially the tub and shower, which require advanced plumbing skills.

The wall and flooring systems of the standard bath feature ceramic tile. Although the wall area looks relatively small in comparison to say, a bedroom, don't let it fool you. And even though the ceramic tile covers only the lower portion of the walls outside of the tub area, there is quite a bit of effort and expense involved in tiling this room. If you haven't done tile work before, check with an experienced person for advice; be sure to get the right tools for the job. The grouting process alone requires some know-how and patience. Give particular attention to the ceramic work around the tub and shower. As for the floor, the ceramic installation is a little easier and faster because the material comes in sections, usually 1' x 1', which can be trimmed and cut where necessary as the floor is laid. The partition between the tub and vanity is constructed of 2 x 4s and should be carefully plumbed and set precisely at a right angle to the line of the tub. If it is out of plumb even slightly, the ensuing drywall and, particularly,

ceramic tile installations become more difficult. The upper part of the walls and the ceiling require two coats of paint (one primer and one finish coat) to complete the job correctly. The door and any trim work should also be refinished with paint, or stained and then sealed with two coats of polyurethane.

The vanity and other accessory items for the standard bath can vary with individual taste, but a 30" two-door standard vanity cabinet with a solid surface top adds an attractive and functional support for the lavatory. The medicine cabinet can include a built-in lighting fixture and twin sliding mirror doors at reasonable cost. The towel bar and tissue dispenser can be of conventional design, and selected from various materials such as stainless steel, brass, chrome, and ceramic. Instead of a basic shower curtain and support rod, this bath includes a sliding shower door made from tempered glass.

LEVEL OF DIFFICULTY

The standard full bath project poses certain challenges, and expediency is a prime consideration, especially if your home has just one bathroom. For all but expert do-it-yourselfers, a plumber should be hired to do the fixture installations, and an electrician should be hired for the wiring. Beginners should also hire someone to do the tile work for this project. In time, a poor tiling job will cost you many times over the installer's fee in leakage and water problems around the tub and shower. Beginners can figure on 150% added to the labor-hour estimates on other parts of the project. They are advised to seek professional assistance along the way. Intermediates should contract the plumbing and add

70% to the labor-hour estimates for all other operations. Experts who can handle the plumbing should be able to complete the entire project with about 30% added to the professional time estimate.

WHAT TO WATCH OUT FOR

There are many critical parts to any bathroom renovation, but perhaps the most important task outside of the actual plumbing work is the drywalling and

ceramic work. Be sure to install 1/2" water-resistant sheetrock for the tile backing and then tape and finish it thoroughly to eliminate moisture penetration at the seams. Thoroughly taping and finishing the drywall seams and nail or screw head indentations also provides a flush, continuously flat surface for the ceramic tile. Install the tile carefully, following the adhesive manufacturer's instructions and recommendations. After the tile has been installed, grout it thoroughly, paying careful attention to the vertical corner nearest the shower head to make sure that no voids or open spaces exist in the line of grout. If you have included a ceramic soap dish or handle fixture in the tub and shower enclosure, then grout around it carefully as well. A bead of flexible caulking or tub sealer along the seam at the top of the tub will add further protection against water penetration and subsequent damage to the floor and wall in the tub area.

Ceramic tile and custom color fixtures are long-lasting items. Thoughtful consideration should be used when selecting colors and styles. It may also be wise to purchase 10 percent extra material in case you need to replace any tiles in the future.

SUMMARY

Whether you are modernizing an old bath or constructing a new one, the value and comfort of your home will be enhanced. You can save on labor costs while learning and doing some of the work on your own, and the investment of your time and money will be returned in future years.

For other options or further details regarding options shown, see

Closets and storage

Electrical system: Light fixtures

Electrical system: Receptacles

Flooring

Painting and wallpapering

Tile walls

Ventilation system

Standard Full Bath

Description	Quantity/ Unit	Labor- Hours	Material
Partition wall, 2 x 4 plates & studs, 16" O.C., 8' high	64 L.F.	0.9	27.65
Drywall, 1/2" thick, water-resistant, taped & finished, 4' x 8' sheets	64 S.F.	1.1	19.97
Paint, ceiling, walls, and door, primer	230 S.F.	0.9	11.04
Paint, ceiling, walls, and door, 1 coat	230 S.F.	1.4	13.80
Vanity base cabinet, deluxe, 2 door, 30" wide	1 Ea.	1.4	310.80
Vanity top, solid surface material, with center bowl 17" x 19"	1 Ea.	2.0	801.00
Fittings for lavatory	1 Set	7.0	88.80
Bathtub, recessed, P.E. on C.I. w/trim, mat bottom, 5' long, color	1 Ea.	3.6	444.00
Fittings for tub and shower	1 Set	7.7	147.60
Sliding shower door, deluxe, tempered glass	1 Ea.	1.0	283.20
Mirror, plate glass, 30" x 34"	7 S.F.	0.7	50.40
Walls, 4-1/4" x 4-1/4" cer. tile, shower enc. & wainscoting, thinset	140 S.F.	11.8	351.12
Flooring, porcelain tile, 1 color, 4-1/4" x 4-1/4" tiles	36 S.F.	3.2	386.64
Toilet, tank-type, vitreous china, floor mounted, one piece color	1 Ea.	3.9	772.20
Fittings for toilet	1 Set	5.3	151.20
Toilet tissue dispenser, surface mounted, stainless steel	1 Ea.	0.3	14.10
Towel bar, stainless steel, 30" long	1 Ea.	0.4	38.40
Electrical, light fixtures with wiring	4 Ea.	1.1	168.00
Electrical, light switches	2 Ea.	2.8	42.72
Electrical, GFI outlet	1 Ea.	1.7	52.80
Totals		58.2	$4,175.44

Project Size	7' x 8'	Contractor's Fee Including Materials	**$8,770**

Key to Abbreviations
C.Y.–cubic yard   Ea.–each   L.F.–linear foot   Pr.–pair   Sq.–square (100 square feet of area)
S.F.–square foot   S.Y.–square yard   V.L.F.–vertical linear foot   M.S.F.–thousand square feet

STANDARD MASTER BATH

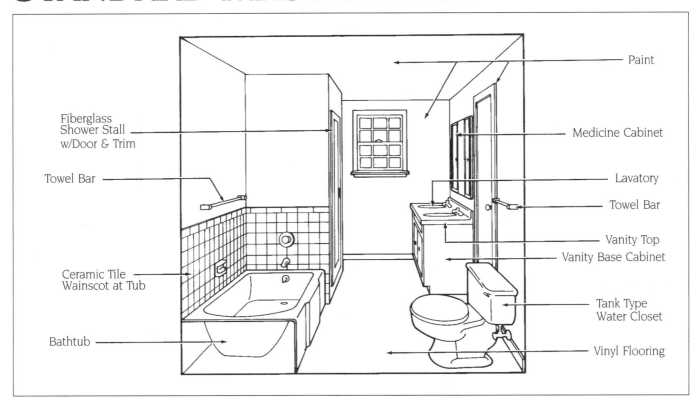

Fiberglass Shower Stall w/Door & Trim

Towel Bar

Ceramic Tile Wainscot at Tub

Bathtub

Paint

Medicine Cabinet

Lavatory

Towel Bar

Vanity Top

Vanity Base Cabinet

Tank Type Water Closet

Vinyl Flooring

A standard master bath includes the three basic plumbing fixtures of most standard full baths and adds a separate shower unit and a second sink. The 8' x 10' master bath is designed for large homes that provide the space for such a facility off the master suite. The addition of the two plumbing fixtures and the larger size of the master bath make this remodeling project that much more time-consuming and costly than the renovation of a standard half or full bath.

MATERIALS

The materials used in the master bath include the five primary plumbing fixtures, typical bathroom floor and wall coverings, and the amenities and accessories found in most baths. The lavatories are standard porcelain-on-cast-iron units in white with 18" bowls, and the toilet is a common two-piece, floor-mounted vitreous china model, also in white. A water-saver toilet, required in some locations, will increase costs slightly. Add about 30% to the estimated cost of these fixtures, as well as to the price of the tub, if you want them in color. The tub is a standard 5' recessed enamel-on-cast-iron unit. The head of the tub backs

to a partition, which separates it from a one-piece fiberglass shower stall. It is important to note that backing the tub to the shower creates savings in the rough plumbing cost and installation time for these fixtures. The rough plumbing for both fixtures fits easily into the 2 x 4 partition that separates the two units. The flooring, wall, and ceiling coverings for the master bath are available in many options and varying price ranges. If you choose to economize, you can install a quality grade of low-maintenance vinyl flooring and place a limited amount of ceramic tile around the tub area. Both of these materials require some knowledge and skill in their respective installations; but because they are relatively small flooring and tiling jobs, intermediates and advanced beginners can accomplish them with some advice, guidance, and the right tools. The ceramic work involves just two half-wall sections at the side and head of the tub; the flooring installation requires only two cutouts, one for the vanity and one for the toilet flange. As with other bathroom flooring jobs, make sure that the subflooring and its supports are in good shape before you begin laying the new underlayment and vinyl finish material. If you neglect moisture problems at the time of

installation, you will wind up paying for them in aggravation and premature replacement costs.

The vanity, medicine cabinet, and other accessories should be in keeping with the decor of a master bath. A full 72" vanity with matching Formica top provides ample storage and counter space around the sink. A smaller, more economical medicine cabinet can be installed, but the recommended two-cabinet model is commensurate with master bath standards. You can also select, at varying costs, tissue dispensers and towel bars in several designs and of several different materials.

LEVEL OF DIFFICULTY

Like the other three- or four-fixture baths, the standard master bath requires significant remodeling skills and competence with tools. Beginners and intermediates should not tackle the extensive plumbing required in installing the fixtures or the electrical wiring. If the rough plumbing is already in place and functioning properly, the fixture installation will be easier, but still challenging. In situations where relocation of the fixtures or new rough plumbing is called for, a plumber should do the work. Even skilled or expert do-it-yourselfers should consider professional help for fixture relocations. For all tasks involved in this bathroom project, exclusive of the plumbing, beginners should add 100% to the labor-hour estimates, intermediates should add 50%, and experts about 20%.

WHAT TO WATCH OUT FOR

Installing sheet flooring requires knowledge of the vinyl material, several important tools, and an awareness of several "tricks of the trade." However, with a few suggestions and advice, most intermediates and advanced beginners can complete the finished flooring work for the standard master bath project for the cost of the material, adhesive, and incidental expenses. First, make sure that the underlayment is thoroughly nailed, dry, and clean. All seams and indentations should be filled and sanded to surface level with a floor leveler. The next step involves making a template for the floor area from building paper (e.g., felt roofing paper), with the cutouts and precise edge cuts included. Strips of the felt paper attached to the large pieces of the template with masking tape provide a precise replica of the entire floor area, including all irregularities along the edges. After the design of the template has been transferred by tracing onto the vinyl, it can be carefully cut with a good pair of scissors or a sharp utility knife. Before spreading the adhesive with a grooved trowel, double-check the fit of the new cut piece by laying it on the bathroom floor and pressing it into corners and along the edges. Follow the directions of the adhesive manufacturer for spreading the material and laying the floor. Use a borrowed or rented floor roller to complete the installation. This method of placing the vinyl applies to a one-piece installation; a two-piece job requires more expertise and several additional steps.

WALL FINISH OPTIONS
Cost per Square Foot, Installed

Description	
Ceramic Tile, plain	$ 5.71
Ceramic Tile, decorative	$12.45
Paint	$ 0.54
Hardboard Panels	$ 2.45
Marble Tile	$18.50

SUMMARY

If the value and size of your home justify installing or renovating a master bath, you can approach the project knowing that your investment will be returned. Carrying out this enterprise can be a major undertaking for even skilled intermediates, but you can complete much of the work on your own if you have the time and seek advice as you proceed.

For other options or further details regarding options shown, see

> *Electrical system: Light fixtures*
>
> *Electrical system: Receptacles*
>
> *Flooring*
>
> *Painting and wallpapering*
>
> *Standard window installation**
>
> *Tile walls*
>
> *Ventilation system*
>
> * In Exterior Home Improvement Costs

Standard Master Bath

Description	Quantity/ Unit	Labor-Hours	Material
Partition for shower stall, 2 x 4 plates & studs, 16" O.C., 8' high	152 L.F.	2.2	65.66
Rough-in frame for medicine cabinet, 2 x 4 stock	16 L.F.	0.5	6.91
Drywall, 1/2", water-resistant, taped & finished, 4' x 8' sheets	96 S.F.	1.6	29.95
Flooring, underlayment grade, hdbd., 3/16" thick, 4' x 4' sheets	128 S.F.	1.4	62.98
Paint, ceiling, walls & door, primer	330 S.F.	1.3	15.84
Paint, ceiling, walls & door, 1 coat	330 S.F.	2.0	19.80
Vanity base cabinet, 2 door, 72" wide	1 Ea.	1.2	296.40
Vanity top, plastic laminate	1 Ea.	1.6	55.08
Lavatory with trim & faucet, porc. enamel on CI, 18" round, white	2 Ea.	5.0	360.00
Fittings for lavatory	2 Sets	13.9	177.60
Shower stall w/door & trim, fbgls., 1 piece, 3 walls, 32" x 32"	1 Ea.	6.7	402.00
Bathtub, recessed, P.E. on cast iron, w/ trim, mat bottom, 5' long	1 Ea.	3.6	444.00
Fittings for tub and shower	1 Set	7.7	147.60
Medicine cabinet, center mirror, 2 end cabinets, 72" long	1 Ea.	1.1	333.60
Walls, ceramic tile at tub, thin set, 4-1/2" x 4-1/2"	12 S.F.	1.0	30.10
Flooring, vinyl sheet goods, backed, 0.080" thick	50 S.F.	1.7	120.00
Toilet, tank-type, vitreous china, floor mounted, 2 piece	1 Ea.	3.0	168.00
Fittings for toilet	1 Set	5.3	151.20
Towel bar, stainless steel, 30" long	2 Ea.	0.7	76.80
Toilet tissue dispenser, surface mounted, stainless	1 Ea.	0.3	14.10
Trim baseboard, 9/16" x 3-1/2" wide, pine	16 L.F.	0.5	21.89
Electrical, 2 light fixtures with wiring	2 Ea.	0.5	84.00
Electrical, 2 light switches	2 Ea.	2.8	42.72
Electrical, one GFI outlet	1 Ea.	1.7	52.80
Totals		67.3	$3,179.03

Project Size	8' x 10'	Contractor's Fee Including Materials	**$7,906**

Key to Abbreviations
C.Y.–cubic yard   Ea.–each   L.F.–linear foot   Pr.–pair   Sq.–square (100 square feet of area)
S.F.–square foot   S.Y.–square yard   V.L.F.–vertical linear foot   M.S.F.–thousand square feet

DELUXE MASTER BATH

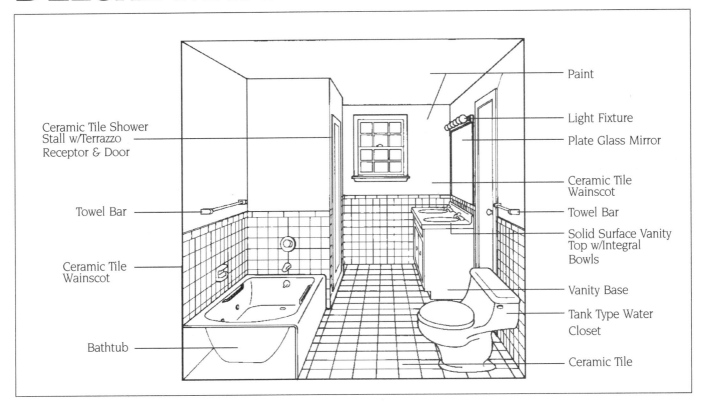

Ceramic Tile Shower Stall w/Terrazzo Receptor & Door

Towel Bar

Ceramic Tile Wainscot

Bathtub

Paint

Light Fixture

Plate Glass Mirror

Ceramic Tile Wainscot

Towel Bar

Solid Surface Vanity Top w/Integral Bowls

Vanity Base

Tank Type Water Closet

Ceramic Tile

The extra accessories and amenities offered in this deluxe master bath are designed to reflect the quality of larger, higher-priced homes. The five-fixture project offered in this plan involves a considerable commitment of both money and time. If you plan to accomplish some or all of this renovation on your own, don't hesitate to seek the assistance of knowledgeable people. Good advice can save you money and prevent damage to expensive fixtures and materials.

MATERIALS

The fixtures and accessories selected for this room are first quality and appropriate for a deluxe bathroom facility. The five plumbing fixtures include a porcelain-on-cast-iron recessed tub with a nonskid bottom and a separate shower stall with a terrazzo receptor. The toilet is a quality one-piece, floor-mounted unit made from vitreous china. Both the tub and toilet are available in white or colored styles, with the colored fixtures costing about 25% to 30% more than the white units. The basic rough-in for these fixtures may cost more than the estimated price if relocation or replacement of the existing plumbing is required. Only expert

do-it-yourselfers and plumbers should tackle the plumbing tasks encountered in this project.

Ceramic tile is the primary wall and floor covering throughout the deluxe master bath. If you are considering doing the tiling in this bath, be prepared to make a major commitment to the task. All of the walls are wainscoted to a height of about 4', and the shower stall above the terrazzo receptor is completely tiled. If you haven't done any tile work before, you should hire a tilesetter to do the job. A poor tiling job will not only harm the appearance of the bathroom, but also invite major problems from water penetration later on. The shower stall and tub area, in particular, require precise tile installation if they are to remain watertight for a long period of time.

Placing the ceramic flooring is a bit easier than installing the ceramic wall tile, but it also requires patience and a reasonable level of skill. Like any other flooring job, one of your first and most important concerns should be the thorough inspection and proper reconditioning and preparation of the subfloor.

The accessories and amenities included in the plan for the deluxe master bath can be altered or changed within a wide

range of options. Basic items like the toilet tissue dispenser and towel bar, for example, are available in ceramic, chrome, wood, brass, and stainless steel. Different styles of deluxe and standard vanity cabinets can also be purchased at varying costs to suit the decor of your bath.

LEVEL OF DIFFICULTY

Installing this bathroom is a major job for even the accomplished remodeler. The extensive tile work and complexity of the plumbing fixtures and electrical work dictate that even the most experienced worker will need some advice from a professional. In this case, we suggest that the expert add 25% to the labor-hours given. The intermediate should have the plumbing contracted and get assistance on the tile work. Jobs like constructing the partition at the head of the tub and drywalling and finishing the shower stall are well within the realm of most experienced do-it-yourselfers. Intermediates should add 60% to the labor-hours for these and the other remaining tasks. Beginners could attempt the medicine cabinet, towel bars, and toilet accessories installation, and can also save on labor costs by painting the walls, ceiling, door, and trim. They should add 100% to the labor-hours for these items and leave the rest of the work to professional contractors.

WHAT TO WATCH OUT FOR

The deluxe master bath includes some specialized items that require careful treatment during installation. Two of these are the cultured marble vanity top and the terrazzo receptor for the shower stall. Both units will provide years of service when properly installed. They are, however, prone to cracking and chipping if they are not handled carefully during placement or if they are improperly supported or fastened. The marble top should be carefully placed and then leveled on the vanity cabinet. Follow the adhesive manufacturer's instructions when securing the top. The terrazzo receptor, or pan for the base of the shower, should be carefully aligned and leveled at the base of the preframed shower stall before the drain is permanently set and secured.

SUMMARY

Renovating a master bath or expanding a full bath to a five-fixture deluxe facility improves the value of your home as well as its comfort. If this plan is in keeping with the size and style of your home, it can be a rewarding remodeling project in terms of everyday enjoyment and a return in resale value.

For other options or further details regarding options shown, see

> Closets and storage
>
> Electrical system: Light fixtures
>
> Electrical system: Receptacles
>
> Flooring
>
> Painting and wallpapering
>
> Skylights*
>
> Standard window installation*
>
> Tile walls
>
> Ventilation system
>
> * In Exterior Home Improvement Costs

Deluxe Master Bath

Description	Quantity/ Unit	Labor- Hours	Material
Partition wall for shower, 2 x 4 plates & studs, 16" O.C., 8' high	152 L.F.	2.2	65.66
Drywall, 1/2" on walls, water-resistant, taped & finished, 4' x 8'	160 S.F.	2.7	49.92
Paint, ceiling, walls and door, primer	330 S.F.	1.3	15.84
Paint, ceiling, walls and door, 1 coat	330 S.F.	2.0	19.80
Deluxe vanity base cabinet, 2 door, 72" wide	1 Ea.	1.7	414.96
Lavatory, solid surface material with integral bowl, 22" x 73"	1 Ea.	1.0	750.00
Fittings for lavatory	2 Sets	13.9	177.60
Shower stall, terrazzo receptor, 36" x 36"	1 Ea.	8.0	828.00
Shower door, tempered glass, deluxe	1 Ea.	1.3	366.00
Bathtub, 5', recessed, porc. enamel on CI, w/trim, mat bottom	1 Ea.	3.6	444.00
Fittings for tub	1 Set	7.7	147.60
Fittings for shower, thermostatic	1 Ea.	1.0	271.20
Inlet strainer for shower	1 Ea.		46.20
Walls, 4-1/4" x 4-1/4" ceramic tile, stall & wainscoting, thin set	170 S.F.	14.3	426.36
Flooring, porcelain tile, 1 color, color group 2, 2" x 2"	40 S.F.	3.4	190.56
Toilet, tank-type, vitreous china, floor mounted, 1 piece, color	1 Ea.	3.9	772.20
Fittings for toilet	1 Ea.	5.3	151.20
Towel bar, ceramic	4 Ea.	0.8	45.60
Toilet tissue dispenser, ceramic	1 Ea.	0.2	11.40
Electrical, light fixtures with wiring	4 Ea.	1.1	168.00
Electrical, 2 light switches	2 Ea.	2.8	42.72
Mirror, plate glass, 34" high x 60" wide	14 S.F.	1.4	100.80
Electrical, one GFI outlet	1 Ea.	1.7	52.80
Ceiling light/fan/heat unit	1 Ea.	0.8	92.40
Totals		82.1	$5,650.82

Project Size	8' x 10'	Contractor's Fee Including Materials	$12,048

Key to Abbreviations
C.Y.–cubic yard   Ea.–each   L.F.–linear foot   Pr.–pair   Sq.–square (100 square feet of area)
S.F.–square foot   S.Y.–square yard   V.L.F.–vertical linear foot   M.S.F.–thousand square feet

DELUXE BATH/SPA

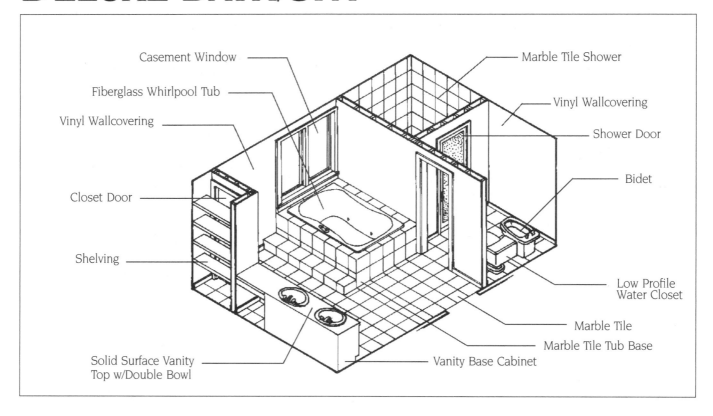

Casement Window

Fiberglass Whirlpool Tub

Vinyl Wallcovering

Closet Door

Shelving

Solid Surface Vanity
Top w/Double Bowl

Marble Tile Shower

Vinyl Wallcovering

Shower Door

Bidet

Low Profile
Water Closet

Marble Tile

Marble Tile Tub Base

Vanity Base Cabinet

If your home has the space to accommodate a deluxe bath, or if you are building an addition large enough for such a room, this remodeling plan will help you in selecting materials and designing the facility. The project is a costly one, as only top-of-the-line fixtures and materials are recommended to complement the size and intended comfort and luxury of this room. Many different layouts are possible in the design of a six-fixture deluxe bathroom, as long as the available floor area allows for them and for the various amenities and accessories. If your house or addition is not large enough to accommodate the facility or if you do not have the minimum 12' x 18' area recommended for a deluxe bath, then you might reconsider doing the project or arrange to alter the floor plan in the area where the facility is to be located. Because of the magnitude of the project and the high cost of the fixtures and materials, do-it-yourselfers are advised to take a realistic look at the renovation and seriously consider hiring a contractor to install this facility.

MATERIALS

The materials included in the deluxe bath are of the highest available quality short of custom-made items. The six plumbing fixtures include a one-piece, low-profile, colored toilet and matching bidet, a double-bowl lavatory cast into the cultured marble vanity top, a 5' x 5' marble shower stall, and a 46" x 56" molded fiberglass tub with whirlpool. The estimated cost of these fixtures includes the appropriate trim like faucets, valves, diverters, and shower heads. Additional expense may be added to the estimated cost of the fittings if the rough plumbing is not in place or requires relocation. Added expense will also be incurred if the deluxe bath project is a renovation of an existing facility and old fixtures have to be removed.

The floor is comprised of 1/2" cement board over the subfloor, covered with high-quality marble tile. If the deluxe bath is a brand new installation or part of a larger renovation project, the ceiling should be leveled and covered with 1/2" drywall. After it is taped and finished, it should be painted with a primer and a finish coat of high-quality paint. The walls

should also be covered with 1/2" sheetrock. Be sure to use 1/2" water-resistant drywall on the wall areas near the tub, on the tub support, and around the shower stall. After the walls have been taped and finished, they can be painted or covered with a washable vinyl wallpaper. Marble tile should be installed on all walls in the shower stall and on the tub support. The partition between the shower and tub compartments and the framing for the storage closet, shower stall, and the tub support should be built of construction-grade 2 x 4s. If you are doing some or all of the floor, ceiling, and wall installations yourself, it's best to take the time to do them thoroughly, and seek advice along the way. The deluxe quality of the fixtures and materials should be backed up by solid workmanship in the installation process. Beginners and less experienced intermediates should hire a contractor to perform most of these tasks.

The exterior wall can be enhanced with a vinyl-clad thermopane casement or sliding window unit. Six-panel or other decorative solid doors between compartments and on the closet maintain the deluxe character of the facility. Louvered units are also available if more ventilation is desired within the bath and the closet. The prefinished shelving for the closet and the deluxe vanity cabinet offer functional and attractive amenities to the facility. A large mirror on the wall area above the lavatory and vanity unit provides a practical substitute for a mirrored medicine cabinet while contributing to the feeling of spaciousness of the bath.

LEVEL OF DIFFICULTY

The deluxe bath is a luxurious facility comprised of expensive plumbing fixtures and other costly building materials. If these fixtures and amenities are within your budget and you have a home large enough to accommodate them, you might find it most expedient to hire a contractor to do the installations. Tasks like the painting and wallpapering can be completed by the homeowner to cut some of the cost, and beginners can accomplish these jobs without much difficulty. Intermediates and expert do-it-yourselfers who want to have a hand in constructing the facility should leave the plumbing, electrical, and ceramic installations to the professionals, but they should be able to complete most of the other jobs involved in the project. Experts should add about 30% to the professional time estimates for those tasks they intend to do; intermediates, about 70%.

WHAT TO WATCH OUT FOR

The finish materials applied to the floor, ceiling, and walls of any bath should be durable, water- and moisture-resistant, and easy to maintain. For these reasons, marble tile has proven to be one of the most effective coverings for luxurious bathroom installations. The portions of walls, ceilings, and trim that require painting should be carefully prepared; primed and then finish-coated with a high quality paint. Gloss or semi-gloss finishes generally hold up better than flat finishes and are easier to clean. If you prefer wallpaper, select a product that is labeled as washable. Even though vinyl wallcoverings, for example, may cost more than conventional wallpaper, they are worth the extra initial expense because of their durability and low maintenance.

SUMMARY

Adding or remodeling a bath to deluxe standards is a way to increase your comfort while boosting the overall value and atmosphere of your house. If a deluxe bathroom fits in with your lifestyle and your home is spacious enough to accommodate it, this project can pay dividends in the coming years.

Deluxe Bath/Spa

Description	Quantity/Unit	Labor-Hours	Material
Partition walls, 2 x 4 plates and studs, 16" O.C., 8' high	400 L.F.	5.8	172.80
Framing around tub, 2 x 4 stock	80 L.F.	2.6	34.56
Drywall, 1/2", walls & ceilings, water-res., taped & finished, 4' x 8'	1,024 S.F.	17.0	319.49
Sub-flooring, cementitious backerboard, 1/2" thick, 3' x 6' sheets	192 S.F.	5.9	188.93
Vanity, deluxe base cabinet, 4 doors, 21" x 76"	1 Ea.	2.0	495.60
Vanity top, cultured marble	1 Ea.	0.5	36.00
Lavatory, cultured marble, double bowl	1 Ea.	2.5	123.60
Fittings for lavatory	2 Sets	13.9	177.60
Tub, whirlpool, molded fiberglass, 66" x 48" x 24"	1 Ea.	16.0	2,520.00
Fittings for bathtub	1 Set	7.7	147.60
Shower receptor, custom, 12" x 12" marble tile, 6' x 6'	36 S.F.	9.6	289.44
Shower, rough-in, supply, waste and vent	1 Ea.	7.7	147.60
Shower fittings, thermostatic valve	1 Ea.	1.0	271.20
Shower door, tempered glass, deluxe	1 Ea.	1.0	283.20
Window, 4' x 4'-6", vinyl clad, casement type	1 Ea.	0.9	333.60
Doors, 6 panel, pine, 2'-6" x 6'-8"	2 Ea.	1.6	523.20
Trim for doors, window & baseboard, 9/16" x 3-1/2"	180 L.F.	6.0	246.24
Paint, ceiling & trim, primer	800 S.F.	3.1	38.40
Paint, ceiling & trim, 1 coat	800 S.F.	4.9	48.00
Tile, marble, shower & tub, 12" x 12" x 3/8"	210 S.F.	64.6	2,242.80
Wallpaper, vinyl, fabric-backed, medium-weight	560 S.F.	9.3	456.96
Shelving, 10" wide, prefinished, 4' long	4 Ea.	1.8	78.14
Toilet, one piece, low profile	1 Ea.	3.0	480.00
Fittings for toilet	1 Set	5.3	151.20
Bidet, vitreous china, white trim	1 Ea.	3.2	654.00
Fittings for bidet	1 Set	5.3	151.20
Flooring, marble, 12" x 12" x 3/8" tile, thin set	216 S.F.	66.5	2,306.88
Mirror over vanity, plate glass, 6' x 4'	1 Ea.	2.4	172.80
Electrical, light fixtures	8 Ea.	2.1	336.00
Electrical, light switches	6 Ea.	8.4	128.16
Electrical, 2 GFI outlets	3 Ea.	5.1	158.40
Total		286.7	$13,713.60

Project Size	12' x 18'	Contractor's Fee Including Materials	**$32,813**

For other options or further details regarding options shown, see

Closets and storage

Electrical system: Light fixtures

Electrical system: Receptacles

Flooring

Painting and wallpapering

*Standard window installation**

Tile walls

Ventilation system

* *In Exterior Home Improvement Costs*

Key to Abbreviations
C.Y.–cubic yard   Ea.–each   L.F.–linear foot   Pr.–pair   Sq.–square (100 square feet of area)
S.F.–square foot   S.Y.–square yard   V.L.F.–vertical linear foot   M.S.F.–thousand square feet

SAUNA ROOM

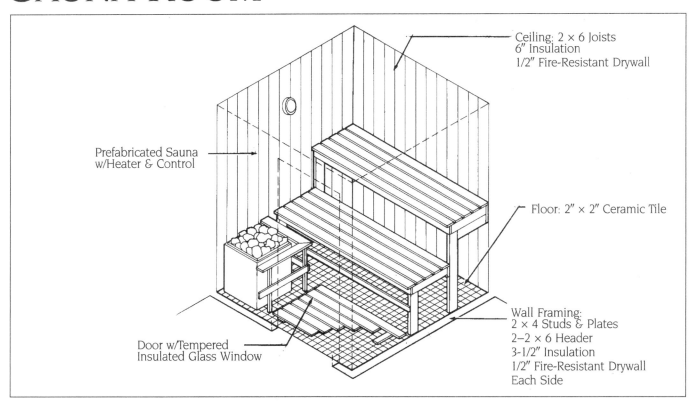

Ceiling: 2 × 6 Joists
6″ Insulation
1/2″ Fire-Resistant Drywall

Prefabricated Sauna
w/Heater & Control

Floor: 2″ × 2″ Ceramic Tile

Wall Framing:
2 × 4 Studs & Plates
2–2 × 6 Header
3-1/2″ Insulation
1/2″ Fire-Resistant Drywall
Each Side

Door w/Tempered
Insulated Glass Window

Saunas are a popular home improvement, not only as a health and luxury item, but also because they require little space and are only moderately difficult to install. Other than some basic electrical hook-ups, there is no specialized trade work involved. Unlike a steam bath, no plumbing or floor drain installation is required, and ceramic tile or concrete are the desirable floor covering materials. This plan outlines a sauna project that can be completed by most experienced intermediates. The installation features a 6′ x 6′ factory-prepared sauna room that is delivered complete with ready-to-assemble walls, ceiling, benches, heater, and basic accessories.

MATERIALS

In addition to the prefabricated sauna room, standard framing, surface covering, and insulating materials are needed for the exterior shell of the facility. These basic construction products can be purchased from most building supply retailers at competitive prices. The sauna room itself is a specialty item. Many brand names are presently on the market and, as with other prefabricated building products, the costs vary widely. The design features,

types of materials, and levels of quality also have a wide range.

The first phase of this project involves determining the location and layout of the sauna room and then constructing its exterior shell. Because the prefabricated units are not very heavy and can be ordered in a wide range of sizes and configurations, they can be located almost anywhere in the house. If you have the space off a bedroom and near a shower area, a second-floor location is ideal. However, first-floor, basement, and even exterior areas can also be utilized. This plan allows for enough material to construct the shell of a sauna room in a corner of a basement where two walls are already in place. The two new walls must be framed with 2 x 4s and then covered with 1/2″ fire-resistant drywall on both sides. A ceiling must also be framed with 2 x 6s and similarly covered on the interior side.

Insulation is required on the new walls and ceiling to ensure energy savings. Some additional expense may be incurred if the existing walls must also be insulated or covered with fireproof material. During the framing of the shell, follow the manufacturer's recommended rough measurement guidelines and

maintain precise level, plumb, and square. Inaccurate framing will cause misalignment of the prefabricated components and, consequently, heat-losing cracks and spaces at the corners and between panels.

The second phase of the project involves assembling the sauna room and its accessories. Various designs and materials are employed in sauna units, but most of them feature softwood walls and ceilings. Economical units may employ pine, spruce, and hemlock, for example; and the more expensive ones, cedar and redwood. The benches, safety heater railing, and optional duckboards and back rests are usually manufactured from wood that matches the ceiling and walls.

At the heart of the installation is the heater, which has been carefully engineered to provide the correct output for its particular size of sauna room. A light fixture, switch, and controls for the heater are also normally included in the basic sauna package. Most manufacturers feature prewiring for these electrical devices within the wall panels of their products. No special electrical service is required for the heater; and under normal conditions, a certified electrician can routinely tie the lead wires from the sauna into your house's electrical supply. Throughout the assembly of the sauna unit, follow the manufacturer's instructions and guidelines. Work patiently and follow accepted carpentry practices, as the naturally finished softwood is particularly susceptible to marks and dents caused by careless hammering or handling.

Although a floor drain has not been included in this project, you may wish to add one to your sauna. This may require cutting and patching the existing floor to gain access to a drain line.

Once the sauna unit has been assembled and fastened, the finished floor can be laid and the special prehung sauna door installed. This plan includes attractive low-maintenance ceramic floor tile. If you intend to install ceramic flooring, be sure to use appropriate grout and to seal it thoroughly, as frequent floor cleaning is recommended by most manufacturers. Duckboards are available as an accessory

flooring material at extra cost from most sauna manufacturers. Depending on the brand of sauna, several styles of prehung doors are available. All of them are insulated and have wood interior handles, and some include small or full-size insulated glass windows and exterior decoration. The door is usually included in the sauna unit price, but is sometimes an extra. Many optional accessories are available, usually at extra cost. These include recessed light fixtures, thermostat and time controls, hygrometers, and convenience items such as back rests, towel bars, and robe racks.

LEVEL OF DIFFICULTY

This project calls for moderate to advanced carpentry skills and know-how. Although the sauna facility measures only 6′ x 6′, it requires an accurate layout and a precise fit of both the exterior shell and interior wall and ceiling

assembly. Careless workmanship can be problematic if the exterior shell is misaligned or inaccurately plumbed or squared. Beginners, therefore, should seek assistance in the framing phase of the installation. They may also consider hiring a professional for the sauna assembly after weighing the cost of the unit and the chance of damaging it against the potential savings in labor. Beginners should add 100% to the professional time for insulating, finishing the drywall, and painting. If they tackle the ceramic floor installation, they should add slightly more time to allow for careful work. Intermediates and experts should add 50% and 20%, respectively, to the estimated time for all basic carpentry procedures, and more for the sauna assembly, door installation, and ceramic floor.

WHAT TO WATCH OUT FOR

Some manufacturers of prefabricated sauna units provide completely finished rooms that can be erected without a shell. These units come with insulated walls, interior and exterior finish, and the full range of electrical and convenience accessories. In some areas of the country, they also provide the services of expert installers at extra cost. As long as the enclosure for the facility is well insulated and weathertight, interior sauna kits can also be safely and effectively adapted to an exterior setting. The complete room kit, and especially the outdoor kit, will increase the cost of the project.

SUMMARY

For a reasonable investment of time and money, the unique pleasure of a sauna room can become part of your home life. Skilled do-it-yourselfers can substantially reduce the cost of this amenity by building it on their own.

For other options or further details regarding options shown, see

> *Flooring*
>
> *Insulation*

Sauna Room

Description	Quantity/ Unit	Labor- Hours	Material
Wall framing, plates, 2 x 4, top and bottom	36 L.F.	0.7	15.55
Studs, 2 x 4 x 8′, 16″ O.C.	112 L.F.	1.6	48.38
Blocking, misc., 2 x 4	20 L.F.	0.6	8.64
Header over door opening, 2 x 6, doubled	8 L.F.	0.4	5.47
Ceiling joists, 2 x 6, 16″ O.C.	60 L.F.	0.8	41.04
Insulation, walls, 3-1/2″ thick, R-11, 15″ wide	170 S.F.	0.9	69.36
Insulation, ceiling, 6″ thick, R-19, 15″ wide	36 S.F.	0.2	17.71
Drywall, 1/2″ on ceiling & walls, fire-res., taped & finished	352 S.F.	5.8	109.82
Sauna, 6′ x 6′ prefabricated kit, including heater and controls	1 Ea.	14.4	5,370.00
Door, prehung w/tempered insul. glass, incl. jambs & hardware	1 Ea.	1.3	384.00
Flooring, ceramic tile, 1 color, 2″ x 2″	36 S.F.	3.0	171.50
Paint, exterior walls, primer	85 S.F.	0.3	4.08
Paint, exterior walls, 1 coat	85 S.F.	0.5	5.10
Totals		30.5	$6,250.65

Project Size   6′ x 6′

Contractor's Fee Including Materials **$10,320**

Key to Abbreviations
C.Y.–cubic yard   Ea.–each   L.F.–linear foot   Pr.–pair   Sq.–square (100 square feet of area)
S.F.–square foot   S.Y.–square yard   V.L.F.–vertical linear foot   M.S.F.–thousand square feet

TILE WALLS & FLOORS

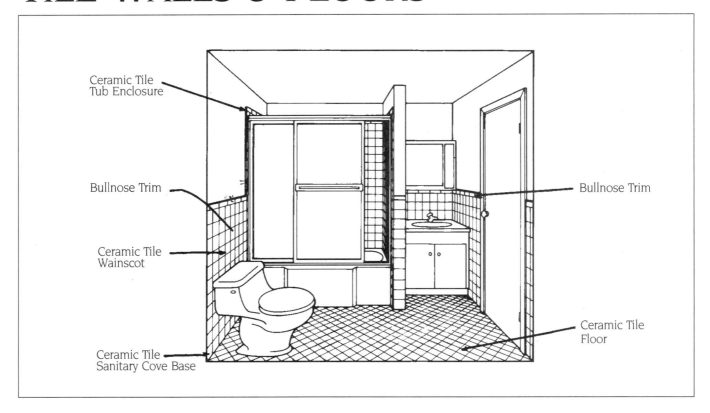

Ceramic Tile Tub Enclosure

Bullnose Trim

Ceramic Tile Wainscot

Ceramic Tile Sanitary Cove Base

Bullnose Trim

Ceramic Tile Floor

A complete bathroom renovation is never cheap. The cost of an entire set of standard fixtures for a full bath, in addition to the new floor and wall coverings, runs into hundreds of dollars before a dime is spent on labor. If your fixtures are still good-looking and functional but you want to upgrade, you may want to consider the less expensive option of retiling the walls and/or floor. Indeed, many bathrooms require tile repair or replacement because of leaks, particularly in the tub and shower or toilet area. Such leaks often go unnoticed for years and can cause considerable damage, extending even to the framing members. Conditions like these make a project of this sort obligatory, not optional.

Retiling and related repair work will interfere with the use of the bathroom and, as in the case of tiling the tub and shower area, certain fixtures may be out of action for several days. If you have no backup bath facilities, schedule the work to be done while you are away or, if you are doing it yourself, make arrangements to use a neighbor's bathroom during the project.

MATERIALS

A tiling project, no matter how small, calls for a good deal of careful planning and preparation. Tiles vary in size from 2" x 2" up to 12" square. You can select smaller tiles or a contrasting color to create accents, patterns, or stripes. Using accurate measurements, make a layout drawing of your walls. The first step is to decide how much of the wall will get tiled and to plan the pattern. You can tile the wall partway up or all the way up to the ceiling. You will also need to choose the proper type and number of special tile trim pieces such as bullnose, cove, and quarter-round edging tiles, as well as accessories like soap dishes, toilet paper holders, and toothbrush holders. If you are unsure of how to figure any of this, a tile store can provide advice and direction. Keep in mind that such things as fancy borders, accent strips, and diagonals can add considerably to the cost of materials and complicate the installation.

Floor preparation is straightforward. It involves removing the toilet and existing tile floor. Since leaks often occur around

the toilet and tub, check these areas carefully for damage, and replace/repair the subfloor as necessary. Repair any damage caused by the removal of the tile. In many cases the easiest course of action is to replace the entire subfloor. The intent is to provide a solid smooth surface for the installation of the new tile.

Wall preparation amounts to providing a solid, smooth surface for the new tile and any paint and wallpaper you might use. This can be as simple as it sounds or, as is more often the case, can involve a lot of heavy, messy work.

It may be necessary to remove the toilet to gain access to the wall behind it. Wallpaper should be completely stripped and the walls washed down to get rid of the glue. Glue may also be softened enough to be scraped off with a wide putty knife. After washing and scraping, let the walls dry and sand off any glue residue with fine sandpaper.

If wall tiles have to be removed, be sure to cover all fixtures to protect them from damage during this stage. Remove the shower door or the rod and curtain. Shut off the water supply, drain the pipes, and remove the faucet handles, shower head, and other trim parts. Wear eye protection and use a hammer and flat pry bar to pop the tiles off. This is a time-consuming operation and may cause a lot of wall material to be broken out in the process. Professionals prefer to cut and pry off the entire wall – tiles and backing – in large sections, right down to the framing. This is quicker and produces less mess. In the case of water damage from a leak, the area around the tub must certainly be stripped to the framing. Its condition should be assessed and any severely rotted studs removed and replaced. Do the same for any wet or torn insulation.

"Sistering" (attaching new material directly to old material to create a composite member) new framing lumber to the old studs should be done wherever necessary to provide solid backing and nailing for the new wallboard. Be very careful when hammering and prying around pipes so that you don't bend them or rupture any soldered joints.

The wall area around the tub should be covered with a fiberglass-reinforced cement board (such as Wonderboard) that will stay intact even when soaking wet. It comes in 3' x 5' sheets, which means that two horizontal layers will cover 6' above the tub. The walls above that height, and any other walls in the bathroom, should be covered with MR (moisture-resistant) sheetrock. Before hanging new backer material, install proper blocking for new accessories (e.g., grab bars, handicapped fixtures).

Joints in the cement board should be covered with mesh tape and filled with thin-set mortar. Joints in the MR board should be taped and finished with joint compound. Any holes or depressions in

the wall surface caused by the removal of the old tiles should be filled and smoothed with patching plaster.

Be sure you lay out accurate horizontal and vertical working lines to help you keep the tiles properly aligned. Use a water-based adhesive and let the tiles set for 24-36 hours before grouting. Fit the tiles close together to make the grout lines as small as possible. This makes cleaning easier and will mean fewer problems in the future. The tub-to-tile joint should be sealed with matching mildew-resistant silicone caulk. Unlike grout, silicone is flexible and will take the expansion and contraction along this joint without breaking the seal. Use the same caulk to seal all openings between fixtures or pipes and the tile.

LEVEL OF DIFFICULTY

Planning the layout is very important, in order to purchase the proper materials in the proper amounts. Beginners should not attempt the project without some self-education or experienced advice. Preparing the walls or floor is not a pleasant task, but calls for more endurance and patience than a high degree of skill. Assessing the condition of existing floor, walls, and framing demands no more than common sense, and you can't go wrong by reinforcing, repairing, or replacing anything that looks doubtful. Installing the tile requires careful planning and the ability to work with a tile cutter. Use of this tool (which can usually be borrowed from the store where you buy the tile) is not difficult to learn. Tiling and grouting is rather

sloppy work, but satisfying because it generally goes quickly and the improvement is immediately noticeable. Be sure to follow all directions regarding the mixing and application of adhesive and grout.

A beginner should have experienced advice throughout all phases of the project, and should at least double the estimated time for all tasks. In a house with one bathroom, homeowners might seriously consider hiring a professional. An intermediate can expect to spend 50% to 100% more time than a pro; the expert, 20% to 30% more.

WHAT TO WATCH OUT FOR

When selecting tile and accessories, care should be taken to make sure all the components are compatible. Some decorative tiles are not always the same dimension as stock accessories. Take care in selecting colors that will be pleasing for a long time. Also, consider purchasing extra materials for future repairs.

Cement board (e.g., Wonderboard) is a product that does its job very well, but it is not very pleasant or easy to work with. It can be cut with a carbide-tipped circular saw, but use an old blade, as this material will generally render the blade useless after cutting, and do it outside if you can. Cutting this material kicks up a lot of dust, so wear safety glasses and a breathing mask. This is a mortar-based material, which means the sheets are heavy and hard. Nails are difficult to start, and nailing close to the edges can cause the material to crumble. The residue is gritty and can scratch the surface of the tub, so be sure such surfaces are adequately protected. If you're tiling the floor as well as the walls, the walls should be done first to avoid messing up your newly-tiled floor.

SUMMARY

Retiling bathroom floors and walls calls for careful planning and quick execution so the bathroom is not out of commission too long. It can involve a series of rather messy tasks, from tearing out the old tiles to grouting in the new, but the satisfaction of seeing the dramatic improvement in the appearance of the room makes the effort worthwhile.

For other options or further details regarding options shown, see

Painting and wallpapering

Tile Walls & Floor

Description	Quantity/ Unit	Labor-Hours	Material
Cement board, taped and finished	60 S.F.	1.8	74.88
Water-resistant 1/2" drywall, taped & finished	32 S.F.	0.5	9.60
Corners, taped & finished	32 L.F.	0.5	2.30
Paint, primer	92 S.F.	0.4	4.42
Paint, 1 coat	92 S.F.	0.6	5.52
Tile for walls, ceramic, 4-1/4" x 4-1/4", thin set, adhesive	140 S.F.	11.8	351.12
Tile for floor, ceramic, 2" x 2", thin set	43.50 S.F.	3.7	207.23
Totals		19.3	$655.07

Project Size	7' x 8'	Contractor's Fee Including Materials	**$1,770**

Key to Abbreviations
C.Y.–cubic yard   Ea.–each   L.F.–linear foot   Pr.–pair   Sq.–square (100 square feet of area)
S.F.–square foot   S.Y.–square yard   V.L.F.–vertical linear foot   M.S.F.–thousand square feet

VENTILATION SYSTEM

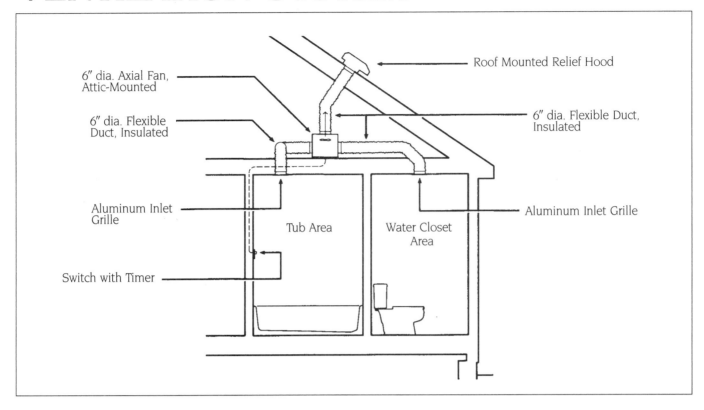

6" dia. Axial Fan, Attic-Mounted

6" dia. Flexible Duct, Insulated

Aluminum Inlet Grille

Switch with Timer

Roof Mounted Relief Hood

6" dia. Flexible Duct, Insulated

Aluminum Inlet Grille

Tub Area

Water Closet Area

Bathrooms generate more humidity than any other room in a house. A typical family's normal schedule of showering and bathing produces large amounts of moist air in a relatively small space. This hot, moisture-laden air penetrates the walls and ceilings, where it can dissolve sheetrock, soak insulation, and rot wood framing, not to mention encourage the growth and spread of mildew.

Most building codes require that bathrooms have either natural or forced ventilation. Natural ventilation means a window, and every bathroom should have one. This provides a source of fresh air and natural light. Mechanical ventilation is a useful adjunct to a window, but it is a strict necessity in a windowless bath. An exhaust fan can ventilate a bathroom much faster than even a wide-open window can, effectively removing moisture and odor-laden air, protecting the room from damage, and making it more pleasant to use.

MATERIALS

Bathroom ventilation fans come in a number of styles: ceiling fan, ceiling fan with light, ceiling fan with infrared heater,

wall fan, and attic-mounted fan. Fan motors are of two types: the axial, or propeller blade, and the centrifugal, or squirrel-cage design.

Whichever style you choose, an exhaust fan must have adequate capacity. The Home Ventilating Institute recommends that the fan be capable of exchanging the air at least eight times per hour. Fans are given cfm (cubic feet per minute) ratings based on the following calculation: assuming an 8' ceiling, multiply the room's length and width by a factor of 1.1. For example, $7 \times 8 \times 1.1 = 61.6$ cfm for a 7' x 8' bathroom. Rounding this number off, your bathroom would require a fan capacity of at least 62 cfm. If your exhaust ducts are long or contain several elbows, the fan will need greater capacity to overcome the added resistance. You must fully assess your specific needs before buying a fan. Experienced or professional advice is recommended.

Fans are also given a noise rating, using a unit of sound called a "sone." One sone is equivalent to the sound level of a quiet refrigerator. Fan ratings range from one sone (the quietest possible) to over four. A lower sone-rated fan generally costs more, but its quiet running may encourage household members to use it

more often and may also indicate a longer life for the motor.

The system illustrated here calls for an attic-mounted, 6" diameter axial fan with an air-flow rating of 270 cfm and a sone rating of 2.5 at the fan itself, in the attic. In the bathroom, the sound is reduced by distance and insulation to not much more than the swish of the moving air.

This is one of the easiest and most adaptable systems to install. It is attached with its mounting bracket anywhere in the attic. It allows for two ceiling air inlets: one over the shower or tub, and the other over the toilet or lavatory. Use 6" insulated flexible duct to connect the fan to the inlet grilles, and to the discharge opening in the relief hood. The insulation will prevent the moisture in the exhaust air from condensing inside the duct and entering the fan housing and motor. Use duct tape at all junctions to get an airtight seal. This will greatly enhance the fan's effectiveness.

The best location for the exhaust outlet is a matter of opinion. Some builders recommend that the fan be mounted so the duct can travel down an inside wall and exit the house just above the foundation. This is an option better suited to new construction. It would be quite difficult, if not impossible, in a remodeling project such as our example. Others feel that a wall unit is better because it goes straight out, and does not require lengths of ductwork. Still others think a spinner-type roof ventilator does the best job. What is *not* a good idea is to exhaust all this warm, moist air into the attic. The resulting condensation can be ruinous. Even directing the air out through the soffit, as is sometimes recommended, will cause the same potentially serious problems when the soffit vents draw the air back up into the attic. In any design, try to keep the ductwork as short and straight as possible, from the intake grilles to the fan, and from the fan to the relief hood.

Electrical connections should be made only by a professional contractor or an expert familiar with local code requirements. Some codes specify that, in a windowless bathroom, a fan must be on the same switch as the light. A fan with an infrared heater should be wired to its own 20-amp circuit. It is fairly common practice to wire the fan to a timer switch. This ensures that the fan will run long enough to clear the bathroom of excess moisture, and also that it will shut off automatically if someone forgets to do it.

A fan with sufficient capacity will ventilate a bathroom rather quickly. To minimize the migration of damp air throughout the house, and reduce heat loss, keep the door closed until the room is clear. You may have to cut a half inch or so off the bottom of the door to allow for enough dry air intake.

LEVEL OF DIFFICULTY

This is not one of the more glamorous home improvement projects. You'll be walking, or more likely, crawling around a dark attic, moving insulation, cutting through sheetrock and sheathing, and raising and breathing dust. Cutting the hole in the roof is not an appropriate task for those who lack the tools and roofing skills, or are uncomfortable climbing ladders and working high off the ground. Designing a venting system that will adequately serve your bathroom may require more figuring than a simple multiplication formula. Fan units come with detailed instructions for installation, but these can never cover every contingency. Often the only way to make things work is with creative problem-solving skills that only experience can provide. A beginner, therefore, is advised to seek experienced help and guidance throughout all phases of the project. Cutting a hole in a roof is major house surgery, and should not be undertaken lightly even by a moderately adept intermediate. Electrical work should be handled only by an expert or a licensed electrician.

Beginners, working with assistance, should add a minimum of 100% to the estimated time for any tasks they are qualified to do. Intermediates should add 50% for the installation, and hire a professional for the wiring and possibly the roof work. Experts can expect to add 20% to the entire job.

WHAT TO WATCH OUT FOR

Unfinished attics often lack both flooring and adequate light. Before lifting a tool, provide yourself with a well-lighted work platform. Lay down lengths of wide lumber (3/4" is safest) across the joists, from the attic opening to the area above the bathroom. Make sure the ends are supported and are not resting on any pipes or wiring that may be damaged. Run an extension cord to the work area. Use a multiple-outlet cord to service lights and any power tools you may use. A clip-on type work light is better than a flashlight, but the latter will be useful for illuminating any hard-to-reach nooks and crannies.

Your main concern, when doing the roof work, should be for your own safety. If your roof is moderately pitched, dry, and swept clean of loose grit, you may be able to walk around on it safely enough in rubber-soled shoes. If the pitch is steep and the surface slick, or if it simply makes you feel more secure, nail on a series of 2' to 3' lengths of 2 x 4 with 12d nails to provide hand and foot holds to move on and work from. When you complete the job, these can be pried off as you descend the roof, and the nail holes filled with roofing tar.

Try to wait for a cool, cloudy day to do the work. Walking on heat-softened shingles can cause damage to them and they get hot enough to burn bare skin. The sun can also turn even a well-ventilated attic into a veritable oven and a miserable place to work.

SUMMARY

Installing a ventilation system such as this is not the most pleasant task, nor one suited to an inexperienced remodeler, but getting the job done can obviate the serious damage condensed moisture can cause, thereby saving you the more expensive cost of future repairs.

Ventilation System

Description	Quantity/ Unit	Labor-Hours	Material
Selective demolition	1 Job	2.0	
Fan, axial, attic-mounted, damper & curb, 6" diameter, 270 CFM	1 Ea.	0.5	268.80
Aluminum air return grilles	2 Ea.	0.6	29.64
Insulated ductwork	20 L.F.	1.2	35.28
Time switch, single pole, single throw, 24 hour dial	1 Ea.	2.0	92.40
Totals		6.3	$426.12

Project Size	Small	Contractor's Fee Including Materials	$910

Key to Abbreviations
C.Y.–cubic yard   Ea.–each   L.F.–linear foot   Pr.–pair   Sq.–square (100 square feet of area)
S.F.–square foot   S.Y.–square yard   V.L.F.–vertical linear foot   M.S.F.–thousand square feet

VANITY AND LAVATORY REPLACEMENT

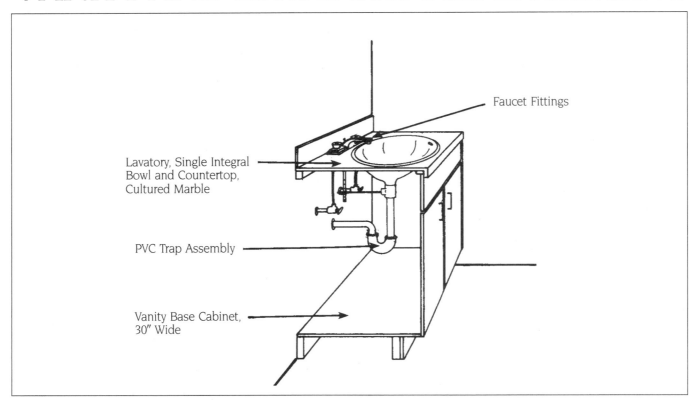

Faucet Fittings

Lavatory, Single Integral
Bowl and Countertop,
Cultured Marble

PVC Trap Assembly

Vanity Base Cabinet,
30" Wide

In most bathrooms the vanity cabinet is the only piece of "furniture" and, as such, offers a certain visual relief from all the fixtures, faucets, and fittings that otherwise dominate the room. Besides hiding the lavatory plumbing, a vanity base cabinet can also provide very useful storage and countertop space in a room where these are always needed and often lacking. The style and design of the vanity, lavatory, and faucet, selected to complement each other and the existing decor, can really help dress up a tired-looking bathroom. Whether you are replacing an old unit or adding one for the first time, the transformation can, in many cases, be achieved quite economically.

MATERIALS

Stock vanities come in two basic depths – 18" or 21" – and in widths from 18" to 72", in 3" or 6" increments. Standard heights vary from 28" to 36". If you are replacing an existing vanity, you can use its dimensions as a reference when planning the new unit. For a first-time installation, you will have to design within the available space. You'll probably want the unit as large as possible to gain

maximum storage and countertop area, but don't crowd the other fixtures. Try to stay at least 2" from the toilet tank, and at least 3" from the side of the tub or shower stall. You will also save a lot of time and money if you locate the new lavatory to within a very few inches of the old sink's centerline. This will allow you to connect to the drain and water lines without having to relocate them in the wall.

Both the vanity cabinet and lavatory get heavy use, so before you buy, carefully consider the construction features and all the pros and cons of the many products available. Base cabinets are made of solid wood, or particleboard with a wood or plastic laminate veneer. Check to see that all joints are solid, secure, and, ideally, glued. Doors and drawers should operate smoothly on good quality hardware. Shelves, racks, towel bars, and the like, if not included, are usually available as modular units that can be purchased separately and installed later.

In most cases, the countertop is also purchased separately from the base cabinet. This gives you some flexibility in mixing and matching materials, colors, and surface textures to achieve the best-looking results. Countertop options

are essentially the same as those offered for kitchens – Formica-type plastic laminates and ceramic tile – as well as polished granite and marble, and synthetic stone materials such as Corian. Wood exposed to water invites problems, so it is best not to use it as a countertop material, no matter how attractive it may look.

Lavatories, like kitchen sinks, are available in a wide array of shapes, sizes, and colors. Common materials are vitreous china, synthetic marble, fiberglass-reinforced plastic, acrylic, enameled cast iron, and enameled steel. These can be purchased in self-rimming, flush, or unrimmed varieties in a deck-mount design, or as an integral bowl-and-countertop unit. A plastic-type countertop and backsplash with a deck-mount, self-rimming sink is generally less expensive than the integral bowl-and-counter combination, but the latter takes less time and effort to install and, being jointless, is completely leakproof. The molded, seamless construction also makes it easier to clean and maintain.

Once the old lavatory and vanity have been removed, cut a hole in the back of the new cabinet to accommodate the water lines and drain pipe. Try to center the new lavatory in the location of the old one to make the plumbing hookups easier. Level the cabinet, front to back and side to side, shimming or trimming the bottom as required, and secure it to the wall with three or four screws driven into the studs. Mount the drain and faucet assembly to the lavatory using plumber's putty to seal the joints between sink and fittings. Installation instructions are provided with the faucet; read and follow them carefully. Lay a bead of construction adhesive on the top edge of the cabinet, and lower the countertop unit into place, centering it and pushing the backsplash tight against the wall. Reassemble the trap, using PVC plastic parts, and add a flexible trap connector if necessary. Attach the faucet to the water lines with flexible risers and new compression fittings. Seek experienced advice if you are unsure of the sizes, lengths, and materials of the plumbing components and how to properly install them.

LEVEL OF DIFFICULTY

This is a project that a competent intermediate should have no great difficulty accomplishing, with minimal disruption to the use of the room. If the floor is level, and the wall plumb, installing the vanity base cabinet should be quite easy. Plumbing hookups require working to very close tolerances, and in a situation like this, where an attempt is being made to line up with existing pipes, careful planning and very accurate

measurements are essential. Be sure you are absolutely certain of the height of the cabinet, the depth of the lavatory bowl, and the on-center location of the drain. You have to make all these variables coordinate to fit the existing plumbing if you are to avoid the necessity of breaking into the wall to rework the water and drain lines. If you have no choice but to move these pipes, the vanity will cover the mess caused by any patch work, but this will add to the time and expense of the job, especially if you need the services of a plumber.

A beginner should try to get experienced help and advice, most of all in the planning and plumbing phases. If a professional carpenter or plumber has to be called in to correct mistakes, it generally ends up costing more than if one had been hired to do the job in the first place. Beginners are advised to proceed with caution, adding at least 100% to the time estimates for the cabinet installation, and 150% for plumbing tasks. Intermediates should add 50%; experts, 20%.

WHAT TO WATCH OUT FOR

Once you have removed the old lavatory and vanity, check the condition of the wallboard and floor. The advent of indoor plumbing, most would agree, was a giant step for mankind, but bringing water inside the house also brought problems.

Leaks have an insidious way of going undetected, sometimes for years, and it is possible for a vanity to hide one of these. Take a careful look for any evidence of dampness or rot. Also check the condition of the shut-off valves and drain pipe. Your water may contain corrosive elements harmless to you, but ruinous to metal fittings. Replace pipes and couplings that look questionable, and valves that don't work smoothly. Use PVC plastic and top-quality metal replacement parts, and be sure they are correctly installed. Now is a good time to add shut-offs for the hot and cold water supplies to simplify future washer replacement and adjustments.

You have to locate studs in the wall to which to attach the vanity cabinet. If there is a baseboard along the wall, it must be removed, and the nails in it should indicate where the studs are. You may have to make a horizontal series of probing holes with a nail to find the studs. Use long drywall screws with finish washers to attach the cabinet to the wall. Just be sure you are hitting wood, not water pipe or wiring. If for any reason studs are not conveniently located, or are inaccessible, you'll have to secure the vanity with wall anchors.

If the wall is irregular or not plumb, scribe and trim the back edges of the cabinet's sides to fit. Some vanities come with "scribing strips" – extra material you can shave down to achieve a perfect fit between cabinet and wall. If your unit doesn't have these strips, you can cover any unsightly gaps with moulding.

Vanity and Lavatory Replacement

Description	Quantity/ Unit	Labor- Hours	Material
Remove existing lavatory	1 Ea.	1.0	
Vanity top, cultured marble, 25" x 32", single bowl	1 Ea.	2.5	204.00
Rough in supply, waste and vent	1 Ea.	7.0	88.80
Piping supply, copper 1/2" Type L	10 L.F.	1.0	18.48
Piping, waste, 4" PVC	7 L.F.	2.3	40.91
Piping, vent, 2" PVC	12 L.F.	3.3	38.16
Vanity base cabinet, two doors, 30" wood laminate	1 Ea.	1.0	222.00
Faucet fittings	1 Set	0.8	128.40
Totals		18.9	$740.75

Project Size	Small		Contractor's Fee Including Materials	$2,012

Key to Abbreviations
C.Y.–cubic yard   Ea.–each   L.F.–linear foot   Pr.–pair   Sq.–square (100 square feet of area)
S.F.–square foot   S.Y.–square yard   V.L.F.–vertical linear foot   M.S.F.–thousand square feet

SUMMARY

The vanity cabinet and lavatory unit is often the main focal point of a bathroom, and is always a very important element in its decor. Lacking a vanity, or having one that is old and worn, can make a bathroom look bland, unfinished, or simply "tired." Installing a new vanity is, in most cases, a weekend job for a homeowner or a short, one-day job for a contractor. Quality products and components are recommended for good looks, durability, and ease of maintenance.

WATER CLOSET (TOILET) REPLACEMENT

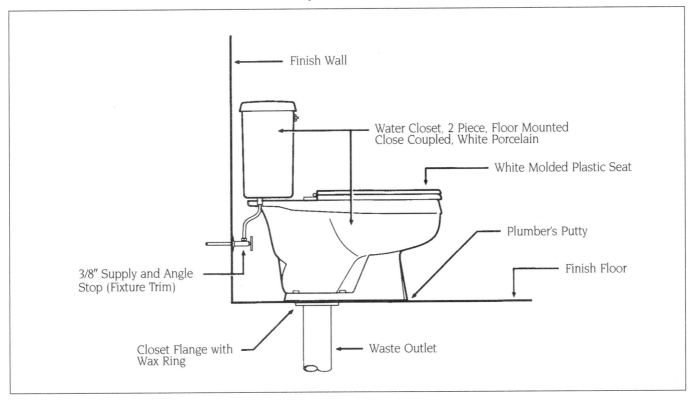

Finish Wall

Water Closet, 2 Piece, Floor Mounted
Close Coupled, White Porcelain

White Molded Plastic Seat

Plumber's Putty

Finish Floor

3/8" Supply and Angle
Stop (Fixture Trim)

Closet Flange with
Wax Ring

Waste Outlet

All the plumbing fixtures in an average household receive a great deal of constant, daily use, but the one most likely to break down is the water closet – or as it is more commonly and less genteely known – the toilet. There are three reasons for this. First, it is a receptacle for solid waste, and its drain is prone to clogging. Second, it is mechanical, with a number of moving parts that can break or malfunction. And third, a typical toilet has as many as five places – seals and valves – where leaks can develop, not including cracks in the bowl or tank.

Most homeowners have performed minor repairs or installed new parts on an old toilet, perhaps never considering a full replacement. It might be a good idea to rid yourself of a cranky, worn-out toilet and put in a new one – an operation that is usually neither difficult nor expensive.

MATERIALS

Porcelain toilets come in a variety of styles, sizes, shapes, and colors. They also can vary in terms of installation and flushing action. Unless you are prepared for the extra expense and labor involved

in moving the toilet to a new location, it is best to choose a replacement unit that will line up with your existing drain and water supply.

One-piece fixtures are available in floor-mount and wall-hung designs. Their low-profile look and ease of installation make them an increasingly popular choice. Two-piece toilets are perhaps the most familiar type. They are available in standard floor-mount, corner-tank, wall-hung, and back-outlet styles. Another version, in which the tank is raised high on the wall, is available for people who want a Victorian motif. Again, the type of unit you are replacing will largely determine what choices you should consider.

Tank widths vary from 20" to 24". Toilets with the tank mounted high on the wall are around 15" wide. Toilets project out from the wall about 26" to 30", requiring a room depth of 44" to 48".

In an attempt to follow, or initiate, decorating trends, and thereby encourage sales, many manufacturers are now offering bathroom fixtures in a veritable rainbow of colors. Unless you are attempting to match the color of the other components of your bathroom decor, it is not a bad idea to stick to plain

white. Nothing goes out of date faster than trendy colors, with the possible exception of popular songs. A color that is "au courant" today can look sadly unstylish in five or six years, and leave you wondering how you could ever have chosen it. A conservative, neutral color is a safer long-term choice.

Also, be wary of the latest, fashionably chic, imported products. Above all else, a toilet should provide years of dependable service, have proven, reliable industry support in terms of guarantees and parts, and enjoy wide acceptance by those who know them best – professional plumbers. A fancy "designer" toilet from Europe will lose a lot of its appeal when it needs fixing and your plumber can't get parts for it.

Perhaps the most important consideration of all in choosing a new toilet is the question of flushing action and the amount of water it uses. The need to conserve water has produced legislation and building codes that require new and replacement toilets to be ultra-low-flush (ULF) models, which use no more than 1.6 gallons (6 liters) of water per flush. Conventional toilets use five gallons, and "water-saver" models use only about one-third less than that. Massachusetts,

Los Angeles, and other areas have mandated the use of ULF toilets, and they are likely to be required nationwide within a few years. Water-conserving toilets cost a bit more than the older five-gallon units but the money you save in water use will eventually make up for the difference. A couple of stabs on a calculator will show that vast quantities of water can indeed by saved by these new flush designs.

Check with your local building inspector to see if the plumbing code requires a ULF toilet; then visit a well-stocked bath showroom, compare what is available, and ask which models the professional plumbing contractors prefer.

The toilet seat must be purchased separately. Materials and colors vary and, as with most products, price is a factor of quality.

LEVEL OF DIFFICULTY

Under normal circumstances, to install a toilet on the existing supply and drain lines of the old unit is quite easy, and it should take no more than a few hours, using common household tools, to do the entire job. The project becomes more complex if you have to run water and drain pipes to a new location, and unless you are an expert, you should hire a licensed plumber to do this kind of work. Care must be taken when handling both the old and the new fixtures, to avoid messy accidents, expensive damage, or possible injury. A porcelain toilet is odd-shaped, somewhat heavy and slippery, and is an awkward thing to move around. If dropped, it can crack or, worse, shatter into sharp, jagged pieces that can easily gash an arm or hand.

Assuming no replumbing is necessary, a beginner should increase the time estimate by 100%; an intermediate, by 20%, and an expert, by 10%.

WHAT TO WATCH OUT FOR

The one crucial dimension you need to determine for your new toilet is its roughing-in size; that is, the distance from the wall to the center of the drainpipe. You can usually determine this before removing the old toilet by simply measuring from the wall to one of the hold-down bolts that secure the unit to the floor. If the bowl has four bolts, measure to one of the rear bolts. The new toilet's roughing-in size can be less than that of the old fixture, but if it is greater, the new toilet won't fit.

When removing the old toilet, turn off the water supply valve, flush the toilet to empty the bowl and tank, then sponge out any remaining water. After disconnecting the water supply line, you may want to remove the tank to make the old toilet easier to carry out of the house. Unscrew the nuts from the hold-down bolts and gently rock the bowl from side to side to break the seal. Lift the bowl straight up and keep it level so any remaining water doesn't spill from the trap. Stuff a rag in the open drainpipe to block sewer gas and to keep debris from falling into the drain. Clean off the residue of the old wax ring from the floor flange and scrape the remains of the putty from the floor.

When installing the new toilet, position the new wax ring over the toilet horn on the bottom of the bowl and lay a strip of plumber's putty around the bottom edge of the bowl base. Remove the rag from the drainpipe and carefully lower the bowl into place over the flange, using the bolts as guides. Form a seal by pressing the bowl down firmly while twisting it slightly. Level the bowl in both directions, shimming it with brass washers if necessary. Be careful not to overtighten the hold-down nuts at the base of the bowl, lest you crack it. Seal the joint between the bowl's base and the floor with a bead of caulk. Attach the tank, connect the water supply, turn it on, and check for leaks. Any minor leak may be the result of not tightening all bolts equally; minor adjustments should solve this problem.

SUMMARY

Whatever the reasons for replacing a toilet, it is a job that a homeowner can accomplish, as long as the existing water supply and drain lines can be used. The main thing you want a toilet to provide, above all else, is long, dependable service. Seek the advice of qualified experts – plumbing fixtures dealers and installers – to help you come to an informed decision.

Water Closet Replacement

Description	Quantity/ Unit	Labor- Hours	Material
Remove existing water closet	2 Ea.	1.0	
Water closet, close-coupled, standard 2 piece	1 Ea.	3.0	168.00
Seat, plastic	1 Ea.	0.3	31.80
Closet flange and ring	1 Ea.	0.9	51.60
Wax ring	1 Ea.	0.1	1.49
Totals		5.3	$252.89

Project Size	Small	Contractor's Fee Including Materials	$635

Key to Abbreviations
C.Y.–cubic yard Ea.–each L.F.–linear foot Pr.–pair Sq.–square (100 square feet of area)
S.F.–square foot S.Y.–square yard V.L.F.–vertical linear foot M.S.F.–thousand square feet

Section Three
CLOSETS AND STORAGE

If you can afford the space, the cost of creating new storage areas or closets is minimal when compared to almost any other home improvement and, dollar for dollar, probably adds more to the value of your home than many more elaborate and costly projects. Following are some tips regarding the addition of storage in your home.

- Building a closet is a good project for a beginner who wants to develop skills because it entails framing, drywall hanging and taping, and trim carpentry, all on a small scale.

- Think about your future remodeling plans when planning the placement of a closet or other storage centers. For example, would the area someday be a good location for a skylight, chimney, or second stairway?

- Whenever you plan to cut into a wall, first determine if there are any obstructions in the form of water pipes, electrical wires, heating ducts, etc.

- Buy the closet door before you begin framing so you can follow the manufacturer's directions regarding the opening size for the door.

- If you plan to use a closet for long-term storage, consider finishing the interior with cedar.

- Consider making a portion of the wall or ceiling in the closet removable to provide access to pipes or wires.

- Prime, stain and/or paint the first coat on all trim prior to installing it; then all you need to do is touch-up work in awkward areas like the interior of the closet.

- Consider installing a fluorescent light fixture in the new closet. Incandescent fixtures are a fire hazard and frequently not allowed in building codes.

- Closet organizers can easily double your closet storage capacity. There are many different styles of organizers available at home improvement centers and department stores.

HOME ENTERTAINMENT CENTER

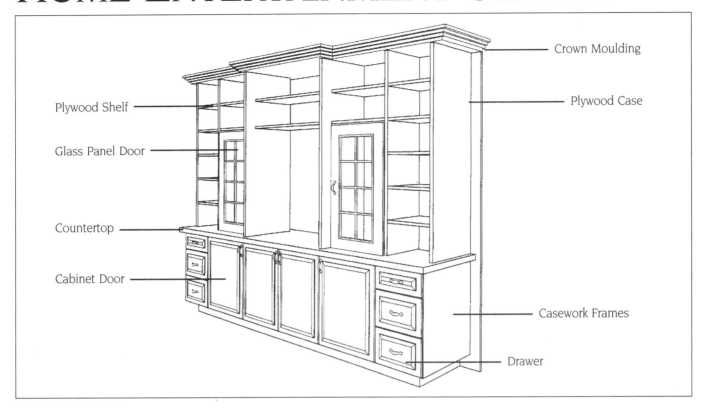

Crown Moulding

Plywood Case

Plywood Shelf

Glass Panel Door

Countertop

Cabinet Door

Casework Frames

Drawer

Years ago, the typical family owned a radio and clustered around it in the evenings to listen to favorite programs, such as *The Green Hornet, The Shadow* or a baseball game. For pictures, a trip to the local theater or ballpark was necessary. Over the years, our electronic options have increased to include TVs, VCRs, CD players, tape decks, cable boxes, and other components. The family room has become a center of activity – more like a theater than a playroom. And, as the sophistication and quality of our audio and video equipment has increased, many have found the need to organize it into a home entertainment center. Building this unit can be a complex and costly project. To get maximum enjoyment, components should be selected and installed with advice from an expert.

MATERIALS

Once the components have been selected, you will need to find a good location in your home. A room of at least 10' x 12', off the major circulation routes, will be best. The larger the TV screen, the larger the room should be. Once you have identified a suitable location, casework should be designed to contain your system. It should be of a style and quality consistent with the rest of your home. It should be designed to include all the equipment you own or are likely to purchase within the next few years. Thought should be given to a flexible design, so that new pieces of equipment can be incorporated without having to rebuild the casework.

The largest component will probably be the TV. Televisions have become increasingly larger over the years, and this trend will probably continue. Surrounding the TV will be the other components of the system, which might include the AV receiver, VCR, DVD player, CD player, tape deck, cable box and/or satellite receiver, turntable, speakers, and speakers switcher. In addition, there are all the associated parts and pieces – CDs, videotapes, and so forth. Plenty of storage space is a necessity. If your room is large, you may wish to install a combined system, and using the TV for watching the news and sitcoms, but add an overhead video projector for watching movies and sports. In this event, a projector would be included on the ceiling, 10' to 14' away from the cabinet, and a pull-down or motorized movie screen would be built into the cabinet or into the ceiling just in front of the TV. All this equipment will take up space, and a wall space of about 10' wide will be needed to neatly contain everything.

Wire management, support, equipment placement, and ventilation are critical to the success of cabinet design. It will be difficult to meet the demands of a home entertainment center and its electronics by using stock cabinets. If you are advanced in cabinetmaking, your skills will pay off in the production of your home entertainment center. If you are less skilled, you may be able to use 30" high base units, but you should definitely plan to build adjustable shelves above. Make sure the shelf material you choose and the support system are able to support the weight of all the components. Again, be sure to allow room for future equipment.

The first task in design is to determine the placement of the components. At the center will be the TV. The minimum practical size will be a 27" diagonal, with an overall size of 24" high by 28" wide and 20" deep. More typical would be a 36" TV, with dimensions of 30" high by 34" wide by 24" deep. A surround-sound system usually has front speakers to the left, center, and right of the TV, and effects speakers either on right and left walls or on the wall opposite the TV to the left and right, depending on whether they are directly radiating or dipolar. The front speakers are typically bookshelf speakers built into the cabinet in the top left and right sectors. They are often concealed by acoustically transparent cloth. The center channel speaker is usually placed horizontally on top of the TV or on a shelf above the TV.

Other components are placed around the TV. The industry standard dimensions for components are 17-1/4" wide by 14" to 18" deep. Their exact locations will be determined by user preference and accessibility requirements. For instance, a turntable has a dustcover that swings up, and will need to either have sufficient height above, or be on a pull-out shelf to change records. Similarly, a CD player should be placed so that you don't have to stand on a step ladder to see what CDs are on the player! Work all of this out

first. The storage you allow should provide easy organization of CDs, records and tapes, the video camera, and more.

The electronic equipment has its own special demands. First among them is wire management. You need to include sufficient outlets for all the components. Design the casework in such a way that you can connect the components without a lot of visible wiring. Allow for access points between the cabinet boxes for running wire between components. Secondly, electronic equipment can produce a surprising amount of heat. Ventilation is very important. In most cases, a flow-through cabinet design will be fine, but be prepared to add a fan if needed, especially in the case of powerful amplifiers.

For our home entertainment center, we have used stock 30" high hardwood base cabinets with a wood countertop. This can be built of either solid wood or veneer plywood with solid wood band edging. Above this we have placed a box frame

with wood upright dividers and adjustable metal shelf standards. The uprights are held 3" off the back wall to provide ventilation. A 1 x 6 pine board is used as a backer on each upright to hide wiring behind. The shelves are constructed of 3/4" plywood with a hardwood edge band.

LEVEL OF DIFFICULTY

This is a project that will require expert advice in the planning and design stages. Homeowners should seek advice from a custom AV installer on the individual requirements of components, as improper installation can harm the equipment. Once the dimensions and configuration are determined, the construction of the casework should not be beyond the skills of the expert or advanced intermediate. Even beginners can complete some aspects of the project. All do-it-yourselfers should proceed slowly and seek advice before starting unfamiliar tasks.

WHAT TO WATCH OUT FOR

Be sure that you know the dimensions of your components before you lay out your casework. Allow room for additional components, or for a larger TV in the future. It is wise to plan on the obsolescence of your current equipment, and anticipate the next generation of electronics. In other words, be sure that you allow plenty of space and flexibility in your design. Invest in high quality adjustable shelf standards. They will pay for themselves many times over as your system evolves. Also, make sure to include enough capacity in your electrical supply so that you don't overload your outlets in the future. If necessary, add new circuits. Consider installing a new phone line, especially if your computer will be included as part of the home entertainment center.

SUMMARY

A home entertainment center can become the focus of many happy hours for your family. It must be designed to accommodate specific components now, but be flexible enough to grow with the changes that will occur in home electronics over the years. If it is well designed and carefully constructed, your home entertainment center will become an important furnishing in your home, and add to its value.

Home Entertainment Center

Description	Quantity/ Unit	Labor-Hours	Material
Casework frames, 2-bay, 36" wide x 36" high	3 Ea.	10.4	475.20
Cabinet doors, raised panel, 21" wide x 30" high	4 Ea.	2.3	196.80
Cabinet doors, glass panel, 21" wide x 36" high	2 Ea.	0.6	68.40
Drawer, hardwood front, 15" wide, 6" high	2 Ea.	1.1	55.20
Drawer, hardwood front, 15" wide, 9" high	4 Ea.	2.3	122.40
Shelves, 3/4" plywood with edge band	33 L.F.	3.5	49.10
Shelf standards and clips, adjustable metal	90 L.F.	0.5	32.40
Drawer/door pulls	12 Ea.	1.4	108.00
Drawer glides	6 Pr.	1.5	71.64
Cabinet hinges	6 Pr.	0.5	18.86
Plywood, 3/4", wood veneer, AA grade	96 S.F.	4.8	286.85
Backer strip, 1 x 6 pine	20 L.F.	0.7	28.56
Countertop, 24" x 3/4" plywood shelf with edge band	10 L.F.	1.1	28.20
Stain casework	192 S.F.	2.4	11.52
Varnish casework, 3 coats, brushwork	192 S.F.	4.7	39.17
Crown molding, stock pine	10 L.F.	0.4	23.04
Face band, 1 x 4, stock pine	10 L.F.	0.3	10.32
Face band, 1 x 2, stock pine	20 L.F.	0.6	9.84
Totals		39.1	$1,635.50

Project Size	2' x 10'	Contractor's Fee Including Materials	$4,261

Key to Abbreviations
C.Y.–cubic yard   Ea.–each   L.F.–linear foot   Pr.–pair   Sq.–square (100 square feet of area)
S.F.–square foot   S.Y.–square yard   V.L.F.–vertical linear foot   M.S.F.–thousand square feet

STANDARD CLOTHES CLOSET

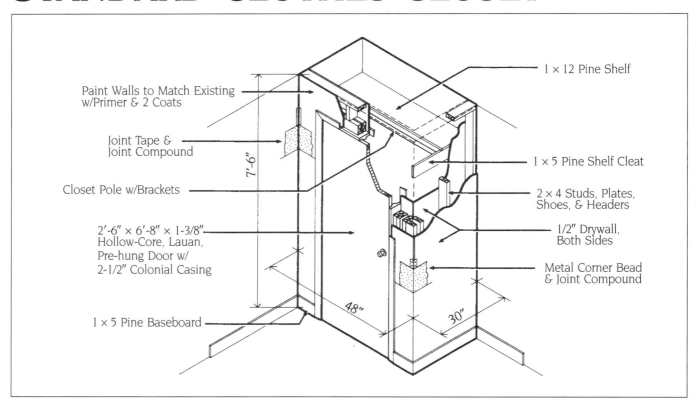

Paint Walls to Match Existing w/Primer & 2 Coats

Joint Tape & Joint Compound

Closet Pole w/Brackets

2'-6" × 6'-8" × 1-3/8" Hollow-Core, Lauan, Pre-hung Door w/ 2-1/2" Colonial Casing

1 × 5 Pine Baseboard

7'-6"

1 × 12 Pine Shelf

1 × 5 Pine Shelf Cleat

2 × 4 Studs, Plates, Shoes, & Headers

1/2" Drywall, Both Sides

Metal Corner Bead & Joint Compound

48"

30"

The project described here is a typical clothes closet formed by squaring off the inside corner of a bedroom. The basic principles involved can be adapted in a variety of ways depending on the availability and location of an appropriate space. Increasing its size is an obvious option.

Customized shelving and built-in or prefab storage systems should also be considered, as well as interior fluorescent lighting. In short, use some creative planning to adapt this project to fit the specific needs of your home and family.

MATERIALS

Framing consists of 2 x 4s, although if space is very limited, the walls may be constructed of 2 x 3 stock; every inch counts. Walls are of 1/2" sheetrock. The shelf is standard-grade 1 x 12 pine on 1 x 5 cleats, which also provide support for the closet pole. There are several choices for the door. Bi-fold doors, either panelled or louvered, allow maximum access to the interior but may not be in keeping with the look of the room. In that case, a prehung, hollow-core, flush door would be an inexpensive option. In an

older house a solid, panelled door might be the only good-looking choice even though it costs more.

All finish trim, such as baseboard, base moulding, and door casing, should match the existing trim in the room. This can be difficult, if not impossible, in the case of antique or custom-made houses. All you can do is come as close as possible with the standard stock carried by your local lumberyard. If your aesthetic sense demands a perfect match, you may be able to locate a mill or craftsperson who can duplicate your mouldings, but be prepared to pay a premium price.

If the floor is carpeted, carefully cut it on the layout lines with a sharp utility knife. You will need to trim back more for the sheetrock and baseboard, but it's safer to do that later to ensure a good final fit.

Locate ceiling joists and wall studs to use as nailing points for the new walls. Do this carefully – you may encounter water pipes and/or electrical wires. A stud finder device can be purchased at your local lumber supply or hardware store. This is a handy addition to any homeowner's tool kit. An old fashioned way to find a stud is to locate the nails in baseboards or male mouldings; this generally indicates the location of solid

nailing. Chances are there won't be nailers in every place they are needed so you will have to install them by cutting access holes with a drywall saw and attaching pieces of lumber to existing framing members. This will call for some creative improvisation. Keep in mind, however, that the walls you are building are simply partitions and there is a good deal of stability inherent in the right angle they form. You don't need a lot of heavy duty nailers; just a few along the walls and ceiling will stiffen up the framing.

In the absence of nailers, you may choose to use any of several construction adhesives on the market today. Most lumberyards can make recommendations for alternate fastening systems. If you are including a light fixture in the closet, the wiring for the fixture and the switch should be run when the framing is completed.

Install the sheetrock, inside and out. Start with a full width sheet across the top of the wall. This will put a tapered, factory edge tight to the ceiling and make it easier to produce a good-looking finished corner joint. Attach a metal outside corner bead, tape all seams and joints, and finish with joint compound.

Mark a level line at the desired shelf height (around 64″ to 66″ is standard) for the 1 x 5 cleats on the back and side walls of the closet's interior.

If your room has a unique ceiling finish or a wall finish you do not wish to disturb (e.g., wallpaper), wall and ceiling mouldings might be used to cover the area where the room wall (or ceiling) meets the new closet exterior wall, thereby eliminating the need to disturb the existing finish. This will also cut in half the amount of necessary taping.

Construction is completed with the hanging of the door and the installation of casing and baseboards.

LEVEL OF DIFFICULTY

Because its location at an inside corner of a room requires the building of only two walls, this is a fairly simple project. Hanging the door will be a challenge for a beginner. If a bi-fold door is used, be sure the finished opening is square and plumb, and follow the manufacturer's instructions carefully. If you choose a standard swing door, buying it prehung will simplify matters considerably. Take extra care to cut the door to fit.

Take the time to lay out and frame everything square, plumb, and level. The look of the finished closet is predicated on the quality of the framing, and no amount of fudging on the finish will completely camouflage earlier errors.

A smooth finish on the walls, without the need for a lot of sanding, will be a challenge for anyone not experienced in the application of joint compound with broadknife and corner tool.

A beginner should add 100% to all times, and even more for the finish work. The intermediate and the expert should add 30% and 10%, respectively, except for the finish work, for which an increase of 50% and 20% would be appropriate.

WHAT TO WATCH OUT FOR

The trickiest part of this project is getting a smooth finish with the joint compound. Use sticky mesh tape, and be sure to put it on flat and tight. Any lumps, bumps, or ragged, stringy edges will come back to haunt you later. You will need a minimum of three, if not four, coats to achieve a professional-looking surface. Each succeeding layer should feather out slightly beyond the one beneath. For the final coat or two, thin out the compound with water, mixing it to a smooth, soft ice cream consistency. This will enable you to lay on a very thin coat and feather it out to nothing.

A good outside corner is fairly easy to achieve if the metal corner bead was carefully nailed on so as to avoid nicks and dents. Simply use the edge of the metal bead as a guide for your broadknife as you smooth the compound with a few long up-and-down strokes. Inside corners are more difficult, and this small project gives you 12 of them, including six that intersect. This is where you will need the corner tool to build up each layer, feathering out the edges on the final coat with the broadknife. Final smoothing with fine sandpaper can be done after the last coat is dry.

SUMMARY

Homeowners often "live around" the problem of lack of closet space by substituting furniture such as trunks, armoires, chests, and extra dressers. The more permanent solution of building a closet is often avoided because it is seen as costing too much in both floor space and money. A closet, however, eliminates the need for those extra pieces of furniture and, if properly sized, will hold a greater variety of clothing in a more accessible space. Keep in mind that high on the list of a home's desirable features, usually right after "modern spacious kitchen," is "ample closet space."

For other options or further details regarding options shown, see

Interior doors

Understairs closet

Walk-in cedar closet

Standard Clothes Closet

Description	Quantity/ Unit	Labor- Hours	Material
Wood framing partitions, 2 x 4	12 L.F.	1.9	50.69
Blocking, misc. to wood construction	40 L.F.	1.3	17.28
Drywall, 1/2″ thick, plasterboard, taped and finished, 4′ x 8′ sheets	128 S.F.	2.1	38.40
Corner bead, galvanized steel, 1″ x 1″	15 L.F.	0.3	1.80
Door flush, HC, lauan, 1-3/8″ x 2′-6″ x 6′-8″, prehung with casing	1 Ea.	0.8	195.60
Passage set, non-keyed	1 Ea.	0.7	45.60
Casing, 2-1/2″ wide	1 Set	0.5	34.20
Trim for baseboard	22 L.F.	0.7	30.10
Shelving, 1 x 12	4 L.F.	0.3	20.06
Closet pole	4 L.F.	0.2	3.84
Paint, walls and ceiling, primer	100 S.F.	0.4	4.80
Paint, walls and ceiling, 2 coats	100 S.F.	1.0	10.80
Paint, trim and baseboard, primer	100 S.F.	1.2	2.40
Paint, trim and baseboard, 1 coat incl. puttying	100 S.F.	1.2	3.60
Totals		12.6	$459.17

Project Size	48″ x 30″	Contractor's Fee Including Materials	**$1,273**

Key to Abbreviations
C.Y.–cubic yard   Ea.–each   L.F.–linear foot   Pr.–pair   Sq.–square (100 square feet of area)
S.F.–square foot   S.Y.–square yard   V.L.F.–vertical linear foot   M.S.F.–thousand square feet

UNDERSTAIRS CLOSET

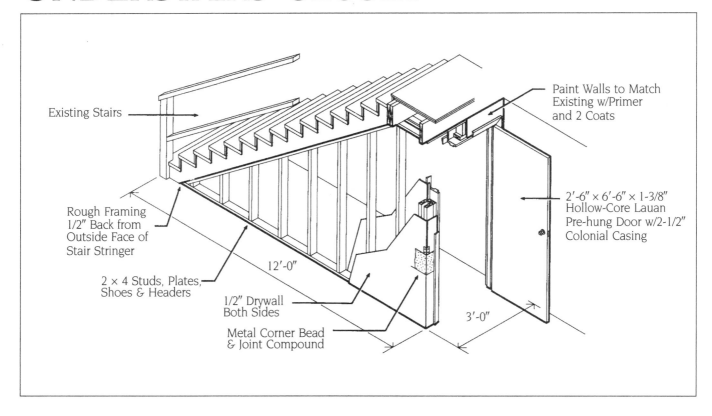

Existing Stairs

Paint Walls to Match Existing w/Primer and 2 Coats

Rough Framing 1/2″ Back from Outside Face of Stair Stringer

2 × 4 Studs, Plates, Shoes & Headers

12′-0″

1/2″ Drywall Both Sides

Metal Corner Bead & Joint Compound

2′-6″ × 6′-6″ × 1-3/8″ Hollow-Core Lauan Pre-hung Door w/2-1/2″ Colonial Casing

3′-0″

Most houses never seem to have enough closet space, but often do have numerous nooks and crannies, which, with a little imagination and design creativity, can yield a significant amount of useful storage area. A good example is the space below the basement stairs. The lower portion of its triangular shape provides somewhat difficult access and might therefore suggest long-term storage use. The area closest to the door could be fitted out in a number of ways with shelves, racks, poles, hooks, or prefabricated closet systems to suit your needs. Refer to the *Standard Clothes Closet* and *Walk-in Cedar Closet* projects for tips on finishing the interior.

Be sure this area of the basement is dry, or take steps to make it so. Any moisture will soon be absorbed into the framing lumber, causing rot as well as encouraging mildew that can develop on the walls and contents of the closet because of the damp air trapped within.

MATERIALS

Framing is standard 2 x 4 lumber covered with 1/2″ sheetrock. Other types of wall material could be used, but sheetrock, unlike some cheaper but less attractive alternatives, would not have to be replaced should you ever decide to finish off the basement. If you are a beginner, it provides you with the opportunity to learn how to work with this very common building material.

The top and bottom plates require an angle cut where they come together at the apex of the triangle. Use a bevel gauge to take this angle off the stair stringer and the floor. The top plate should be nailed 1/2″ in from the edge of the stringer, and the bottom plate plumbed to line up with the top. This will allow the sheetrock to be laid flush to the outside of the stringer. This seam can later be covered with a length of decorative molding such as flat astragal, panel strip, or other. The bottom plate should be anchored to the floor with fluted masonry nails, driven by hand with a small sledge hammer. These nails are brittle and chip easily, so wear safety glasses.

The studs have an angle cut at the top and are toenailed in on the bottom. Use a level and a stud to transfer each layout line from the bottom plate to the top, and at the same time mark the angle on each stud.

Frame the rough opening for the door; square and plumb it in both directions. Use a framing square and a straightedge to mark a line on the floor to ensure that the bottom of the frame lines up across the opening. Attach a stud to the foundation wall with masonry nails. If your stairs are not against the wall as shown, but are in the middle of the basement, you will have to build another partition on the other side of the stairs.

It is not necessary to do so, but you may want to cover the inside of the closet with some sort of sheet material such as plywood or cedar particleboard. If so, this should be done prior to hanging the sheetrock on the outside. After the sheetrock is up, attach the metal outside corner bead. Do this carefully so as not to nick or dent it. Tape the seams, and finish with joint compound. Plan on at least three or more coats to get a nice, smooth surface. You may need to sand the final coat.

Hanging the door and installing the passage set (door handle with no locking action) is next. Have a supply of shim shingles handy. Keep the jamb square and plumb, checking as you go along to make sure the door opens and closes freely, without binding or rubbing. When the passage set is in, the door should close with a solid click and no rattle.

Finally, mark a 1/4" reveal around the outside of the jamb, and cut and nail up the casing. Set the nails, and fill the holes with vinyl spackle.

LEVEL OF DIFFICULTY

This is an excellent beginner's project. Probably the worst consequence of a mistake would be the door not working too well. It is a good opportunity to learn to work with the various layout and measuring tools – level, square, chalkline, and bevel gauge. There is rough carpentry involving angle cuts, joint compound to be applied and smoothed, a door to be hung, and finish trim to be cut with 45° miters. And, best of all, any mess you make is in the basement.

A beginner would be encouraged to undertake this job, but should plan on adding 100% to the listed time it takes to complete. The intermediate and expert should add 20% and 10%, respectively.

WHAT TO WATCH OUT FOR

Basement stairs are often subjected to a good deal of heavy use and are not always well built. Over time, the components – stringers, risers, treads, and railings – can become damaged, loosened, and worn to the point where they are actually dangerous. This project provides a good opportunity to make whatever repairs are necessary. This may be something as simple as renailing the treads and risers, or it may involve a more complicated task, such as replacing broken treads or reinforcing split stringers.

Railings are intended to provide support and protection. They themselves become a hazard when broken, loose, and wobbly. If yours are in this condition, now is the time to replace them.

The framing of this closet will form a solid unit with the stairs, further stiffening and strengthening them. Everyone using the stairs will appreciate the safe and solid feeling under foot that replaces the old spring and sway.

Understairs Closet

Description	Quantity/Unit	Labor-Hours	Material
Plates, untreated, 2 x 4	28 L.F.	0.6	12.10
Studs, 8' high, 2 x 4	80 L.F.	1.2	34.56
1/2" thick sheetrock	64 S.F.	0.5	16.13
Tape and finish	64 S.F.	0.5	16.90
Cornerbead, galv. 1-1/4" x 1-1/4"	8 L.F.	0.2	0.86
Hollow-core door, 1-3/8" x 6'-6" x 2'-6", pre-hung	1 Ea.	0.8	135.60
Casing, 2-1/2" Colonial, 1 head, 2 sides	1 Set	1.4	15.24
Passage set, no key	1 Set	0.7	45.60
Paint, wall, primer	48 S.F.	0.2	2.30
Paint, wall, 2 coats	48 S.F.	0.5	5.18
Paint, trim, primer	1 Face	1.2	2.40
Paint, trim, 1 coat, incl. puttying	1 Face	1.2	3.60
Totals		9.0	$290.47

Project Size 3' x 12'

Contractor's Fee Including Materials $847

Key to Abbreviations
C.Y.–cubic yard   Ea.–each   L.F.–linear foot   Pr.–pair   Sq.–square (100 square feet of area)
S.F.–square foot   S.Y.–square yard   V.L.F.–vertical linear foot   M.S.F.–thousand square feet

SUMMARY

Creating a clean, useful storage space in an otherwise wasted, odd-shaped corner is a satisfying accomplishment. This is especially true if, at the same time, you are acquiring or refining building skills that you may later apply to more ambitious and rewarding home improvement projects.

For other options or further details regarding options shown, see

Interior doors

Linen closet

Standard clothes closet

Standard stairway

Walk-in cedar closet

WALK-IN CEDAR CLOSET

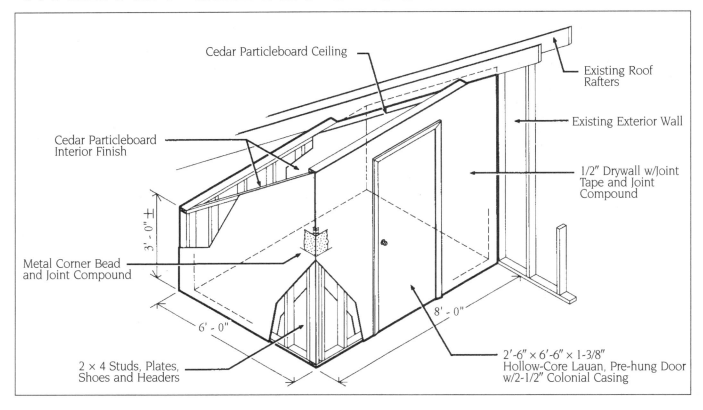

Cedar Particleboard Ceiling

Cedar Particleboard Interior Finish

Existing Roof Rafters

Existing Exterior Wall

1/2″ Drywall w/Joint Tape and Joint Compound

3′ - 0″ ±

Metal Corner Bead and Joint Compound

6′ - 0″

8′ - 0″

2 × 4 Studs, Plates, Shoes and Headers

2′-6″ × 6′-6″ × 1-3/8″ Hollow-Core Lauan, Pre-hung Door w/2-1/2″ Colonial Casing

A walk-in cedar closet provides a simple and not too costly solution to a number of very common long-term storage problems: seasonal clothing and sports equipment; holiday decorations and costumes; heirloom linens and seasonal bedclothes; sewing fabrics; legal documents and financial records; papers, magazines, and books; in short, any item of value, intermittently used, that is prey to moths, mites, and other hungry bugs. Relocating these items to a cedar closet protects them from the depredations of the insect world, while freeing up the space they occupy throughout the rest of the house.

An attic location has been chosen because in most houses attic space is under-utilized, reasonably accessible, and, most importantly, dry. An unfinished basement is an acceptable alternative to an attic location provided it, too, is dry. Cedar can repel bugs, but it cannot prevent mildew.

Economizing on material costs does not adversely affect this closet's use or effectiveness. This is a strictly utilitarian storage area in an unseen and unvisited part of the house, and thus it would be hard to justify spending extra for such niceties as solid brass door hardware and fancy mouldings.

MATERIALS

Framing begins with a 3′ high kneewall along the eaves. This wall could be somewhat higher or lower depending on the pitch of the roof and the desired storage layout, but keep in mind that a very low wall makes building and future access difficult, while a higher wall reduces useful space.

The main wall is framed to provide adequate headroom – around 7′-6″ – with a rough opening for the door. As always, when performing renovations, be sure to check for electrical wiring, water and vent pipes, and heating ducts before doing any nailing. These items are generally easy to spot in the walls or under the roof of an unfinished attic, but the flooring could conceal a potentially unpleasant surprise.

Unless this closet is to be incorporated into a finished attic (living space) at a later date, there is no need to insulate the walls or ceilings. (If insulation is to be installed, be sure the soffits are vented and attach styrofoam vent channels in the roof bays from the eaves to the ridge vent before laying up the fiberglass.) The wiring for the light fixture and switch is run when the framing is complete. It may

be advantageous to install a vapor barrier (4 mil poly film) prior to hanging wallboard should you decide to heat the closet at a future date.

The closet's exterior is covered with 1/2″ sheetrock, taped and finished. Looks are not very important here, but you may want to spend the time practicing how to produce a smooth finish wall so you'll be able to do it later for a more visible project.

If the interior is to be covered with cedar particleboard sheets, these are simply cut and attached to the walls and roof framing through the drywall. Tongue-and-groove cedar strips can be installed vertically or horizontally. If you want to put them up vertically, the walls and ceiling must be furred with 1 x 3 strapping, 16″ on center, to which the cedar is attached through the drywall with 8d finish nails. Tongue-and-groove strips are attractive, but they cost more and take more time to install. Sheets don't look as good but are cheaper and can be installed much more quickly. Both keep the moths at bay, so the choice is based on aesthetics and economics.

The door is installed and trimmed out along with the light fixture, shelving, and closet poles to complete the project.

LEVEL OF DIFFICULTY

A rank beginner would need advice and guidance during some phases of this job: plumbing and squaring the main wall; measuring and cutting the angles on the studs; and hanging the door. But this is a very forgiving project. Even if the wall leans a bit, the joint tape shows through, and the door rattles, the closet will still function.

The intermediate and expert would find this an easy job to complete over a couple of weekends working at a comfortable pace.

The major problem to be dealt with is getting the materials to the site. Attic stairs are almost always narrow and steep and usually located in a hallway with very little elbow room. The framing lumber and strapping, assuming 8' lengths, should go up easily enough. If necessary, they could be brought in through a window. Tongue-and-groove cedar strips are packed in small, light bundles. Particleboard in 4' x 8' sheets are thin, light, and somewhat flexible and may bend enough for two people to manipulate them around a tight corner. Not so the sheetrock. Even in the roomiest of situations, sheetrock panels are just big and heavy enough to be an awkward burden. If they won't go up, they must be cut in half. This will mean more joints to tape and finish, but you'll get more practice with the "mud."

Beginners should add 100%, intermediates 50%, and experts 20% to the estimated time for each task.

WHAT TO WATCH OUT FOR

An attic is a long way from where most tools and supplies are kept. Plan and prepare so that once the materials are in place you can work without a lot of annoying and time-consuming trips to the basement workshop for this tool or that fastener. Clear enough space for the materials to be stacked out of the way and arrange them so they will be accessible in the order they are needed. This may require a major, but temporary, rearrangement of the items stored in the attic. Take the time to do it so as to leave yourself plenty of room to work. The job will be safer and more pleasant if things are not constantly in the way and underfoot. If extra work space is simply not available, you may have to store and cut the building materials elsewhere and bring them up for installation a few at a time. This will certainly add to the time required to complete the project, and appropriate allowances should be made.

Aromatic cedar is highly allergenic and its fumes and dust can irritate skin, eyes, nose, throat, and lungs (that's why bugs don't like it). You'll be working in a confined space, so be sure to provide yourself with as much ventilation as possible. Establish fresh air circulation by opening all available doors and windows, and by setting up an exhaust fan or two. Wear a dust mask when making cuts and sweep up the sawdust periodically.

SUMMARY

A walk-in cedar closet is a pleasant and useful amenity in any home that has the space to accommodate it. Older homes frequently lack bedroom closet space, but often have roomy attics perfectly suited to this type of improvement. Considering that this closet is really nothing more than an enclosure of unfinished attic space, the level of skill and quality of materials required need be no more than adequate. As such, it represents a good low-risk project for a beginner to undertake, and the minimal cost incurred is more than offset by the value derived.

For other options or further details regarding options shown, see

> *Full attic*
> *Half attic*
> *Interior doors*
> *Standard clothes closet*
> *Understairs closet*

Walk-In Cedar Closet

Description	Quantity/ Unit	Labor- Hours	Material
Wood framing partitions	216 L.F.	3.1	93.31
Blocking, misc., wood construction	40 L.F.	1.3	17.28
Drywall, 1/2" thick, taped and finished	128 S.F.	2.1	38.40
Drywall, 1/2" thick, no finish (for closet interior)	320 S.F.	2.6	80.64
Aromatic cedar, 4' x 8' x 1/4" thick particle board	320 S.F.	12.8	326.40
Door, flush lauan hollow-core, 1-3/8" x 2'-6" x 6'-6" prehung	1 Ea.	0.8	195.60
Casing, 2-1/2" Colonial, wide trim	1 Set	0.5	34.20
Closet pole	12 L.F.	0.5	11.52
Light fixture, fluorescent, interior surface, 32-watt and 40-watt	1 Ea.	0.4	90.00
Light switch	1 Ea.	0.5	7.80
Totals		24.6	$895.15

Project Size	6' x 8'	Contractor's Fee Including Materials	$2,515

Key to Abbreviations
C.Y.–cubic yard Ea.–each L.F.–linear foot Pr.–pair Sq.–square (100 square feet of area)
S.F.–square foot S.Y.–square yard V.L.F.–vertical linear foot M.S.F.–thousand square feet

LINEN CLOSET

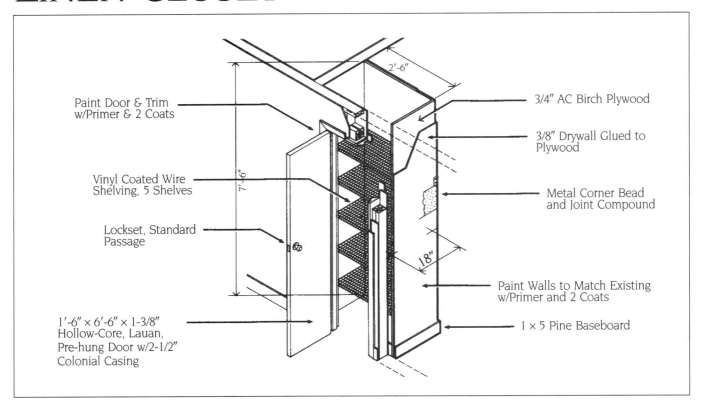

Paint Door & Trim w/Primer & 2 Coats

Vinyl Coated Wire Shelving, 5 Shelves

Lockset, Standard Passage

1'-6" × 6'-6" × 1-3/8" Hollow-Core, Lauan, Pre-hung Door w/2-1/2" Colonial Casing

3/4" AC Birch Plywood

3/8" Drywall Glued to Plywood

Metal Corner Bead and Joint Compound

Paint Walls to Match Existing w/Primer and 2 Coats

1 × 5 Pine Baseboard

2'-6"

7'-6"

18"

The design of most modern houses usually includes a linen closet in the upstairs hallway near bedrooms and bath. Older houses often lack this amenity (and closet space in general) because bedclothes and other linens were often stored in furniture such as blanket chests, or in shelved cupboards known as linen presses. Even in new houses, builders sometimes cut expenses by eliminating certain features that homeowners often take for granted. Building a linen closet along the lines suggested here is a fairly easy and economical way to add one.

MATERIALS

As with any remodeling project of this type, you must first deal with the question of location. The typical placement of a linen closet is in the upstairs hall near bedrooms and bathrooms where the contents of the closet – sheets, blankets, and towels – are used. Your hallway may contain a corner or alcove that would make a suitable closet location, or the hall passageway may be spacious enough to accommodate a cabinet-type installation against a wall. If such is the case in your house, the dimensions will be dictated by the available space. The

inside dimensions of the closet illustrated here, 18" x 30", represent close to a minimum size. If circumstances require it, you could reduce these dimensions an inch or so but, by making the closet much smaller than this, you run the risk of building a storage space too small to be practical. In that case it would be better to seek another location, perhaps on the first floor near the laundry center, rather than compromise too much on the size. In a typical household, a linen closet is not used so often that locating it a distance from bedrooms and bath would cause much, if any, inconvenience.

This plan calls for breaking through a hallway wall into a bedroom, sacrificing just a little over four square feet of the bedroom floor area. By utilizing the room's inside corner, you need only build two walls, thereby saving a good deal in both materials and labor.

To keep the loss of bedroom space to a minimum, this closet partition is not framed in the usual way with 2 x 4s and drywall. Using 3/4" plywood covered by 3/8" sheetrock on the bedroom side, you can build a wall that is stiff and sturdy enough to support the relatively light load of the shelf system and its contents. Positioning the "good" side of the AC

plywood on the inside of the closet gives you a finished wall surface that needs only to be painted. Of course, if you have plenty of space, you can use conventional framing.

Having removed any baseboards and moldings on both sides of the wall, cut and frame a rough opening for the door. The plywood partitions must be attached at several points to the existing bedroom walls, ceiling, and floor. If you are lucky, you may locate studs in the walls, and joists or furring strips in the ceiling to which you can secure the new walls, or you could install surface nailers. The simplest method, however, might be to attach a few angle irons to the tops and sides of the plywood panels and use wall anchors to secure them in place. This is not the fine woodworking you read about in magazines, but this is, after all, only a light-duty closet. If you join the plywood panels with glue and screws, reinforced with three or four angle irons, you have built a right-angled partition that has a lot of inherent stability. Once this is anchored – sides, top, and bottom – it is not going to be moved without a great deal of violent force pushing from the inside, and it's unlikely that sheets and towels will generate any of that.

74

The 3/8″ sheetrock can be glued to the plywood using construction adhesive. If you prefer to use drywall screws, be sure they are not longer than 1″, and don't countersink them too much or the points will protrude through the plywood on the inside of the closet. For the same reason, be careful when attaching the metal corner bead.

Wallpaper should be stripped back to allow the joint compound to adhere directly to the wallboard or plaster. A textured ceiling is always very difficult, if not impossible, to blend into. In such a case, the easy way out would be to nail up two pieces of narrow crown moulding to hide the ceiling joints.

Most ready-to-assemble, modular closet systems use three basic components: a wardrobe shelf with clothes-hanging capacity, a linen rack, and wire baskets in a frame. Linen racks come in three-, four-, eight-, and twelve-foot lengths, with nine-, twelve-, sixteen-, or twenty-inch depths. These can be cut with a hacksaw, allowing you a great degree of flexibility in customizing your own system. If your closet is large enough, you could design a system using all three of these components to maximize its storage potential.

LEVEL OF DIFFICULTY

The most difficult aspect of this job might well be finding a suitable location. Once that has been done, the mechanics of building are fairly straightforward. Cutting and framing the rough opening for the door can be a bit tricky. Framing stock has to be fitted and attached between the layers of the wall. In very confined spaces like this, long drywall screws – 3″ or so – can often be used where you can't reach with a hammer to drive a nail. There will also be some improvised floor patching and matching required where the sole plate has to be cut out of the door opening. All other building and installation tasks are well within the abilities of an intermediate. Getting a smooth finish on the walls with the joint compound will probably call for a good deal more troweling and sanding than a professional would require, but elbow grease is free, as long as it's your own.

A beginner will need advice in planning this job and should have some experienced help in carrying it out, adding 100% to the time it takes. An intermediate might find this project challenging in certain phases – especially the planning

and framing – but not beyond his or her capabilities. Intermediates should figure on 50% more time; experts, 10% more.

WHAT TO WATCH OUT FOR

Unless you are absolutely positive that no obstructions are present, make your initial cuts with a keyhole or drywall saw. That way, if you do encounter a pipe or wire, you will be able to feel it before any damage is done to it, or any harm done to you. If you blindly plunge in with a power saw, such an encounter could result in a very unpleasant, if not shocking, surprise.

Installing a typical modular shelf system can be done with a few common household tools. Normally, 1/4″ holes are drilled for J-hook wall anchors to be screwed into the wallboard. The plywood wall construction of this closet eliminates the need for such holes. The fin-back ends of the hooks can be clipped off and the hooks screwed directly into the plywood. Just be sure the screws provided for the hooks are short enough so they don't protrude through the sheetrock on the other side.

SUMMARY

If your house lacks a linen closet, one can be built as long as a suitable location can be found. The floor plan and layout of your home might be such that the usual second-floor hall location is not a viable option, in which case a first-floor alternative must be planned. The exact location of this closet is not important. Depending on your circumstances, a cupboard or cabinet-type unit built against a wall or in a small alcove might be the best solution. The closet design illustrated here is of minimal size and requires only a small sacrifice of bedroom space.

For other options or further details regarding options shown, see

Cedar closet

Interior doors

Standard clothes closet

Understairs closet

Linen Closet

Description	Quantity/ Unit	Labor- Hours	Material
3/4″ AC birch plywood, 4′ x 8′	32 S.F.	1.6	71.81
Gypsum plasterboard, 3/8″ thick, 4′ x 8′, taped and finished	32 S.F.	0.3	7.30
Corner bead, galvanized steel, 1″ x 1″	7.50 L.F.	0.2	0.90
#2 pine, 1 x 5 baseboard	6 L.F.	0.2	6.26
Crown molding, pine	5 L.F.	0.2	9.12
Door trim set, 1 head, 2 sides, pine, 2-1/2″ wide	1 Set	1.4	15.24
Lauan hollow-core door, 1′-6″ x 6′-6″ x 1-3/8″, pre-hung	1 Ea.	0.8	135.60
Lockset, standard duty, non-keyed passage	1 Ea.	0.7	45.60
Vinyl-coated wire closet system, 5 shelves	1 Ea.	0.5	39.75
Paint, walls, primer	71 S.F.	0.3	3.41
Paint, walls, 2 coats	71 S.F.	0.7	7.67
Paint, door and frame, primer	2 Faces	0.8	2.46
Paint, door and frame, 1 coat	2 Faces	0.8	2.36
Totals		8.5	$347.48

Project Size	18″ x 30″	Contractor's Fee Including Materials	$905

Key to Abbreviations
C.Y.–cubic yard Ea.–each L.F.–linear foot Pr.–pair Sq.–square (100 square feet of area)
S.F.–square foot S.Y.–square yard V.L.F.–vertical linear foot M.S.F.–thousand square feet

Laundry Center

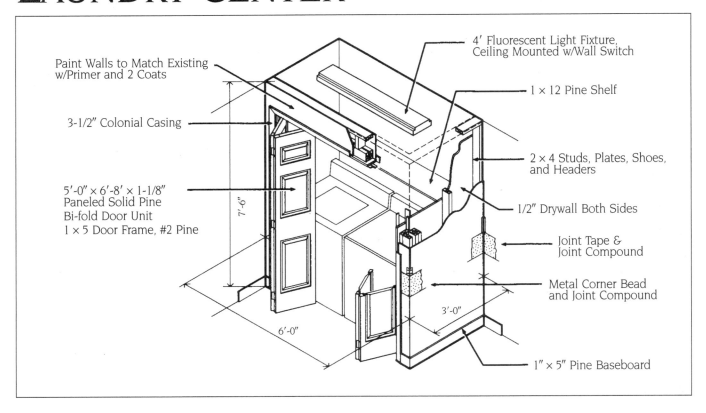

Paint Walls to Match Existing w/Primer and 2 Coats

3-1/2" Colonial Casing

5'-0" × 6'-8' × 1-1/8" Paneled Solid Pine Bi-fold Door Unit 1 × 5 Door Frame, #2 Pine

4' Fluorescent Light Fixture, Ceiling Mounted w/Wall Switch

1 × 12 Pine Shelf

2 × 4 Studs, Plates, Shoes, and Headers

1/2" Drywall Both Sides

Joint Tape & Joint Compound

Metal Corner Bead and Joint Compound

1" × 5" Pine Baseboard

7'-6"

6'-0"

3'-0"

Today's washers and dryers operate with virtually no wet mess. Recognizing this, many designers and architects place the laundry center near the kitchen or, increasingly, upstairs near the bedrooms, where clothes and bed linens are stored and changed, rather than in inconvenient places like basements and mud rooms. If your washer and dryer are currently banished to some out-of-the-way spot, it may be time to bring them closer to the center of your home's daily activities.

MATERIALS

As with any project of this type, you are forced to work within the constraints of the general layout of your house, in addition to the limitations imposed by the size and style of your existing washer/dryer set. You could, of course, buy new appliances of smaller dimensions, or of stacked design, which would allow you greater flexibility in planning and locating this closet. You may, in fact, discover that you have no other choice. The cost of the project would, naturally, increase dramatically.

The closet layout given here presupposes the standard appliance size of 29" wide

by 26" deep. Thus, the minimum inside dimensions of the closet must be 32" x 66-1/2" with a 5' door opening, to provide enough space for washer hoses, a dryer ventilation duct, and user access. It is constructed by squaring off the inside corner of a room or hallway. There exist, no doubt, other possible locations in your house – some obvious, and others that may call for a bit of imagination. Instead of creating a new space, you may be able to convert and adapt an existing one, such as a walk-in closet.

Another consideration is the ease with which water lines, a drain pipe, and a vent stack can be installed. Virtually anything is possible for a price, but if your budget is not unlimited, you would do well to map out your existing plumbing and try to locate the laundry closet near a "wet wall" (one that contains a drain and vertical runs of pipe), or near a "chase" (an avenue specially built for running pipes and ducts). This will allow you to tap into existing feed and drain lines, which is generally more economical than running new ones.

Electrical wiring is needed for the washer, the light fixture, and, on its own circuit, the dryer. The flexibility of wires makes them easier to run than rigid plumbing,

so bringing new circuits to the laundry center directly from the main service panel ("home runs" in electrician's jargon) might actually be more economical than trying to extend existing circuits. Electric dryers require a 220V power cable and breaker. A do-it-yourselfer should not attempt wiring of this type.

The basic framing and construction of this closet is essentially the same as that described in the *Standard Clothes Closet*. If you have the space, you may want to include shelves and racks or a cupboard for linen or laundry supply storage. For linen storage, you might consider lining the inside with cedar (see *Walk-in Cedar Closet*) in addition to drywall. Western red cedar resists moisture, and its fragrance and good looks make for a very pleasant interior.

A 5' bi-fold door allows full access to the machines, besides providing the necessary space for installing them, and for removing them when servicing is needed. Paneled doors have a more substantial look and are sturdier than the common louvered design. The dryer must be vented to the outside, and because neither it nor the washer generates any significant amount of interior moisture, the ventilation provided by louvers is not strictly necessary. The doors can be closed while the washer is operating, to help reduce noise, but they should be left open while the dryer is running to prevent overheating.

LEVEL OF DIFFICULTY

As is the case with many home improvements, this one calls for the skills of the three main trades – carpentry, plumbing, and electrical. The framing and finish work are quite basic, and installing the bi-fold doors is really a matter of reading, understanding, and following directions. A closet this size or larger, equipped with a stacked washer/dryer unit, would allow you the space to add a number of shelves, racks, and hangers, as well as such amenities as a rollout hamper or a fold-down ironing board. This kind of fitting-out can be done very easily using modular, prefabricated closet systems, or can be built from scratch, according to the level of your woodworking skills.

Electrical and plumbing work should be performed only by professionals or experts. Even if you manage to find the ideal location, near existing pipes and wiring, hookups must be made in conformity with local codes and must pass inspection. The one-time expense of hiring professionals to do these jobs is a small price to pay for the safety and well-being of your home and family.

Beginners should, as always, try to get experienced help and advice for the carpentry, and should add 100% to the time estimated below. The intermediate should be able to build this closet easily, adding 50% to the time. Both should hire contractors for the wiring and plumbing. An expert should figure on 20% more time for all tasks.

WHAT TO WATCH OUT FOR

In many locations a pan with floor drain must be placed under a washer if it is located on the second floor. Washers do, on occasion, overflow or leak, and this precaution could prevent expensive repairs later on. Even if it's not required, it's a good investment.

Bi-fold doors come with installation instructions that are quite easy to understand. These are predicated, however, on a properly built finished opening, which, in, turn, presupposes a properly framed rough opening. There is an old saying about one good framing carpenter being worth two finish carpenters, and there is more than a little truth to it. Good finish work is all but impossible to produce over poor framing. The fudging that has to be done to compensate for badly framed work always makes things look a little "off," lacking the crisp, clean lines associated with fine craftsmanship. Take pains to frame the rough opening plumb, square, and level, to the dimensions called for by the manufacturer. When buying your framing lumber, spend a little time inspecting it, and don't take any that is warped or bowed. Put the straightest 2 x 4s aside for the door opening. Also, remember that you can't do good work with bad tools. An accurate measuring tape, level, and framing square are basic necessities for any carpentry work. You should own them and use them.

SUMMARY

Creating a laundry center close to the active living space of the house is an improvement that can eliminate a lot of arduous stair climbing with heavy loads of clothes, and this convenience can add value to your home. Finding a suitable location may take a little extra imagination and work, but in most houses an appropriate space can usually be created or converted. Construction can be accomplished by a competent intermediate, with contractors being hired for the plumbing and wiring. A centrally located laundry center makes these essential appliances easier and more pleasant to use, while still keeping them out of sight.

Laundry Center

Description	Quantity/ Unit	Labor- Hours	Material
Wall studs, 2 x 4, 16″ O.C., 8′ long	15 Ea.	1.8	51.84
Drywall, 1/2″, taped and finished	128 S.F.	2.1	38.40
Corner bead, galvanized steel, 1″ x 1″	30 L.F.	0.6	3.60
Door, bi-fold w/hardware, raised panel, pine, 1-1/8″ x 5′ x 6′-8″	1 Ea.	1.5	303.60
Doorjamb, 1 x 5, #2 pine	20 L.F.	0.8	20.88
Door casing, 11/16″ x 3-1/2″, Colonial	40 L.F.	1.5	71.52
Baseboard, pine, 1 x 5	12 L.F.	0.4	16.42
Light fixture, fluorescent, C.W. lamps, surface-mounted	1 Ea.	1.1	85.20
Light switch	1 Ea.	0.5	7.80
Shelf, 1 x 12 pine	6 L.F.	0.5	30.10
Paint, walls, primer	150 S.F.	0.6	7.20
Paint, walls, 2 coats	150 S.F.	1.5	16.20
Paint, door and frame, primer	2 Faces	1.3	2.11
Paint, door and frame, 1 coat	2 Faces	1.3	2.36
Totals		15.5	$657.23

Project Size 3′ x 6′

Contractor's Fee Including Materials **$1,701**

Key to Abbreviations
C.Y.–cubic yard   Ea.–each   L.F.–linear foot   Pr.–pair   Sq.–square (100 square feet of area)
S.F.–square foot   S.Y.–square yard   V.L.F.–vertical linear foot   M.S.F.–thousand square feet

For other options or further details regarding options shown, see

Electrical system: *Light fixtures*
Electrical system: *Receptacles*
Interior doors

KNEEWALL STORAGE UNIT

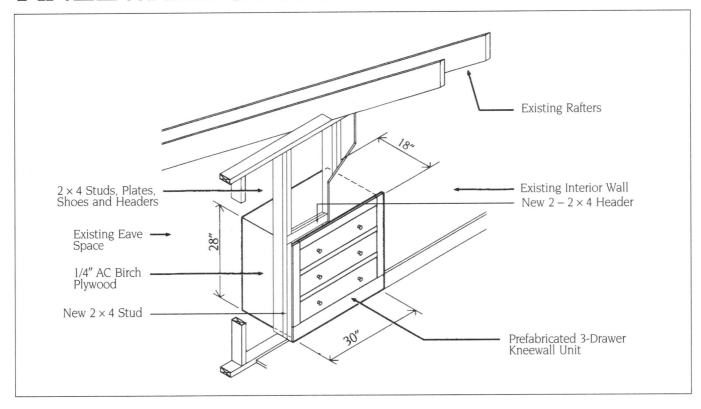

Existing Rafters

2 × 4 Studs, Plates,
Shoes and Headers

Existing Eave
Space

1/4″ AC Birch
Plywood

New 2 × 4 Stud

28″

18″

30″

Existing Interior Wall
New 2 – 2 × 4 Header

Prefabricated 3-Drawer
Kneewall Unit

The never-ending search for storage space has frequently inspired homeowners and professional remodelers to take advantage of the various, and sometimes curious, nooks, corners, and crannies that can be found in almost any house. Once such a space has been discovered, it becomes a test of one's cleverness and creativity to suitably adapt it for storage while, at the same time, blending the new construction into the old so that the end result does not look like a patch job. Some house designs, such as the traditional Cape, as well as most houses with finished attics, have short walls running up to the slope of the roof and parallel to the ridge, called "kneewalls." Behind these walls is a space, triangular in cross-section and generally running the length of the house, that often contains only the cold air needed to ventilate the roof. Some houses have a small door that allows a degree of access to the area, but the severely restricted headroom means crawling on hands and knees to get in, and thus most of the items stored there are clustered very near the door, leaving the inner recesses quite inaccessible and empty. Installing one or more built-in cabinets, bookcases,

or drawer units is a very effective way to utilize this particular cranny.

MATERIALS

The dimensions of the frame of any storage unit depend on the height of the kneewall and the amount of unobstructed lateral wall space available. Remember that the slope of the roof, which forms the hypotenuse of the cross-section triangle, somewhat limits both the height and the depth of the unit. This should present few problems, however, because bookcases should be no more than 12″ deep, and cabinets and drawers are generally 24″ or less.

Kneewalls are usually framed like regular stud walls, with the top plate notched into the rafters. Some kneewalls are framed without a top plate, the studs being simply nailed alongside the rafters, but this makes for a weak, unbacked sheetrock joint where the wall and ceiling intersect. If you are framing a new wall, it's best if all the framing members – floor joists, kneewall studs, and roof rafters – line up. The rough opening for the storage unit should be cut out and framed as you would for a window, with a rough sill a few inches above the floor and double

side studs and header. A kneewall generally provides only auxiliary support to the rafters and is not, by strict definition, a load-bearing wall; the double-stud framework simply provides adequate and convenient nailers.

You can buy a prefabricated storage unit, have one custom made, or build one yourself. Many home centers and unfinished millwork shops sell a pine, three-drawer "kneewall unit" about 28″ high, 30″ wide, and 18″ deep. These are often of stick carcass construction, with a face frame of 1 x 3 top and sides and a 1 x 5 bottom. It's a good idea to reinforce the stapled joints of the frame with 1-1/4″ drywall screws. For additional strength, and for protecting the drawers' contents from dust and helping to seal out cold air, you can sheathe the framework with 1/4″ AC birch or lauan plywood, attaching it with glue and 1″ screws.

Building your own cabinet or bookcase unit is not very difficult, even if you are an intermediate with no specialized cabinet-making skills or tools. You need only build a five-sided open box of appropriate dimensions, using 3/4″ AC plywood held together with glue and 2″ drywall screws. Shelves can be added by installing adjustable metal brackets, drilling a series of holes for shelf clips or dowels, or simply gluing and screwing the shelves permanently in place. A face frame of #1 clear pine is glued and nailed to the carcass with 6d finish nails. The base of the frame should be of wider stock than the top and sides, as in the illustration. For a more finished look, band moulding with mitered corner joints can be attached to the frame with glue and 4d finish nails. Doors can be mounted to the frame with appropriate hinges after the unit has been installed in the rough opening. Paint, or stain and polyurethane, and the installation of knobs or pulls complete the job.

LEVEL OF DIFFICULTY

This is a small project that any homeowner with a basic tool kit and fundamental carpentry skills can accomplish. If you install a prefabricated or custom-made unit, the most difficult task will be the cutting and framing of the rough opening. Take careful measurements, and allow yourself about 1/4" all around. The unit is held in place by nailing the face frame to the backing studs, so you don't have to worry about

a very snug side-to-side fit. Cutting and framing is always somewhat messy, but the opening is fairly small and the dust and dirt should be easy enough to contain. Building your own plywood box carcass is not hard. Straight cuts are important to ensure that everything is square when assembled, and these cuts are best made on a table saw. If you don't have a table saw, cutting carefully with a hand-held circular saw will produce perfectly acceptable joints. If you are planning to stain the piece, your joints, good or bad, will show; if you paint, you can hide any gaps with caulk. If you lack the tools or skill to build the type of doors you want, you can probably get them custom made of pine or poplar for a reasonable price.

A beginner should add about 100% to the time estimates; an intermediate, 50%; and an expert, 10%.

WHAT TO WATCH OUT FOR

An existing kneewall probably has electrical wires running through it. Locate these before you start cutting for the rough opening. If you are lucky, there may

be enough slack in the wire to run it under the unit; if not, you'll have to install a longer length of new wire or, if circumstances allow, add another receptacle and from it, continue the circuit with new wire to the next outlet.

The top of the rough sill for the storage unit should be high enough above the floor to allow the bottom rail of the face frame to line up with the baseboard, which is typically 1 x 5, or 4-1/2" wide. This will give the piece a more built-in look. If the baseboard is of a different width, you may want to design or adapt your trim to accommodate it. If you can match the existing base moulding, you could continue it up and around the face frame to help visually tie the new unit into the existing woodwork. Also remember that you must build a support for the back of the carcass the same height as the rough sill, which supports its front.

Be sure to seal all the gaps in and around the new storage unit with foam insulation or caulk to prevent the infiltration of cold air from the soffit vents and to minimize heat loss in winter.

SUMMARY

Finding and creating new storage space in your house can be a challenge to your inventiveness. The low, open area behind the kneewalls in a Cape-style house or in a finished attic is one that is often underutilized or completely overlooked. The odd shape and lack of headroom in this space make it quite inconvenient to get to through a door, but it lends itself very well to the installation of various types of storage units. As with any cabinet installation, careful planning, accurate measurements, and precise cuts will produce the best results, and provide you with a little more of that which few homes have in sufficiency – useful storage space.

For other options or further details regarding options shown, see

Understairs closet

Kneewall Storage Unit

Description	Quantity/Unit	Labor-Hours	Material
Selective demolition, cut opening for storage space	1 Opening	2.0	
Kneewall studs, 2 x 4, 4' long	24 L.F.	0.4	10.37
Plywood, 1/4" AC Birch, 4' x 8'	32 S.F.	1.0	28.42
Custom drawer unit	1 Ea.	3.0	113.40
Drawer pulls	6 Ea.	0.5	23.18
Totals		6.9	$175.37

Project Size	28" x 30"	Contractor's Fee Including Materials	$566

Key to Abbreviations
C.Y.–cubic yard   Ea.–each   L.F.–linear foot   Pr.–pair   Sq.–square (100 square feet of area)
S.F.–square foot   S.Y.–square yard   V.L.F.–vertical linear foot   M.S.F.–thousand square feet

Section Four
ELECTRICAL SYSTEMS AND FIXTURES

The most important thing to remember when doing electrical work is to turn off power to the area or device you are working on. Always confirm that the power is off by testing the circuit with a neon circuit tester. Restore power only when your work is completed. Following are some additional tips for doing electrical work in your home.

- Always make sure the circuits are dead before starting any work. Let everyone in the house know what you are doing.

- Your municipality probably requires any new electrical work in your home to conform to the National Electrical Code. Most bookstores and home centers carry a simplified guidebook. Be sure to ask your local building inspection department about permit requirements.

- Light fixtures should not only illuminate a room but also enhance the basic architectural and decorative style of the room. Be sure the lighting fixtures you have in mind complement the room and your home in general. For example, contemporary fixtures in a colonial home may look out of place.

- When choosing a light fixture, consider the weight of the fixture, height of the fixture above the floor, location of the fixture, and placement of furniture in the room.

- Some hanging light fixtures are extremely heavy – too heavy for one person to install. Be sure to get help if you need it.

- Remember this rule of thumb: If a light fixture is high, it will light a wide area. If it is low, it will light a smaller area.

- If a new light fixture is the only planned improvement, bear in mind that the added light may draw attention to previously unnoticed blemishes in the wall, floor, and ceiling finishes. Thus this project may lead to other renovations.

- Before spending the time and money for electrical add-ons, be sure your existing electrical service can handle the extra load. If not, you will need to upgrade the service. The minimum upgrade for a single-family residence is 100 amps. It may, however, be more effective to upgrade to a higher amperage (125, 150, or 200) to allow for future power requirements.

- Consider purchasing long bits and a cordless drill before attempting to run wires. You may have to shut off the power to move a wire or do other electrical work, and often stock bits are not quite long enough to penetrate several members of a wall or ceiling structure.

- Running cables in finished walls is usually easier if you work with a helper.

- Never use oil or petroleum jelly as a lubricant for pulling cables; they can damage the cable's thermoplastic sheathing.

ELECTRICAL SYSTEM: LIGHT FIXTURES

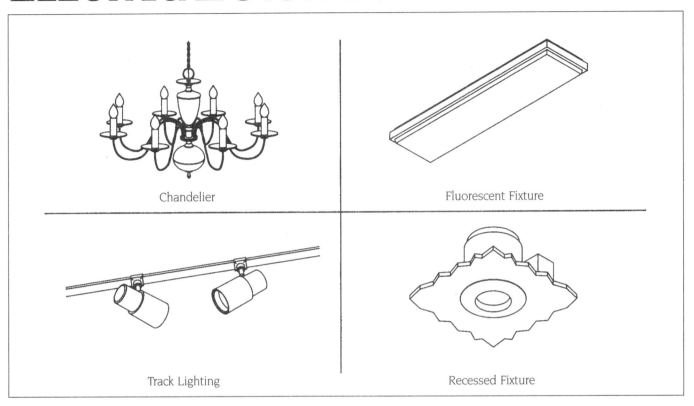

Chandelier

Fluorescent Fixture

Track Lighting

Recessed Fixture

Although lighting is an important feature in most home improvements, it is often overlooked in the planning stages. It is unwise to wait until the project is nearly complete to select a fixture, even though it is one of the last pieces of a project to be installed. Important considerations that can affect other parts of the project include the weight of the fixture, height of the fixture above the finished floor, location of the fixture, and future placement of furniture in the room.

MATERIALS

Selecting light fixtures can be compared to choosing wallpaper because there are so many choices. Lighting needs can be broken down into three categories: general lighting, task lighting, and accent lighting. The primary sources of general lighting are windows and overhead, ceiling-mounted fixtures. Ideally, general lighting provides bright, even illumination to the entire room. Surface-mounted fixtures are usually located in the center of the room. Recessed lights should be placed throughout the room in such a way that their beams overlap. Switches should be located at doorways, in conformity with local codes.

Task lighting is placed above work areas such as counters, desks, sinks, stoves, sewing tables, and the like. Track lights, recessed fixtures, and undercabinet strip units are typical, and should be located so as to produce soft, even illumination without harsh, work-surface glare.

Accent lights are more a decorative option than a functional necessity. They are designed to illuminate and draw attention to elements of a room for dramatic effect. The same type of fixtures used for task lighting can serve this purpose. A little accent lighting can attract a great deal of attention, so be careful not to overdo it.

Lamps for residential lighting fixtures can be incandescent, fluorescent, or high-intensity discharge (HID). HID lamps include mercury vapor, metal halide, high-pressure sodium, or low-pressure sodium. Low-pressure sodium lamps are used principally for outdoor applications because of their strong yellow color.

Incandescent lamps, in which a filament is heated to incandescence by an electric current, are the least efficient lamps, but offer a low initial cost. Their warm (high red content) color spectrum is pleasing for many applications, and they light up instantly. In halogen lamps,

an incandescent contains a halogen gas that recycles tungsten back onto the filament surface. Halogen lamps are expensive, but the high initial cost is offset by increased energy savings.

Fluorescent lamps are based on an electrical discharge of ultraviolet energy, which excites a fluorescent coating and transforms some of that energy to visible light. The initial cost of fluorescent lamps is higher than the cost of incandescent lighting fixtures. Various fluorescent lamps are available, with trade-offs in efficiency, color spectrum, and lamp cost.

HID lamps generally are more efficient than fluorescent, but do not generate the same broad light spectrums, and require several minutes to warm up before full output is reached. They also have a higher initial cost.

For wiring of lighting fixtures, you can use armored BX cable or nonmetallic sheathed cable; electrical metallic tube (EMT) is not often used because of the inflexibility of the EMT wiring system.

LEVEL OF DIFFICULTY

Installing new light fixtures in an existing, finished space is not an easy task and requires substantial skill, knowledge, and experience with electrical installations. Lacking these qualifications and knowledge of local codes and requirements, the homeowner should hire an electrician. Homeowners may be able to replace existing lamps or light fixtures if they consult with professionals.

Depending on the structure of the house, there may be complications in routing the wires for the fixture or switches. Experienced electricians can provide creative solutions to such problems.

Upgrade Lighting Fixtures

Description	Quantity/ Unit	Labor-Hours	Material
Light, chandelier	1 Ea.	1.3	348.00
Light, recessed, high hat can, round reflector, 150 Watt	1 Ea.	1.0	97.20
Light, fluorescent, surface-mounted, 2' x 2', two 40 Watt lamps	1 Ea.	1.1	103.20
Track lighting, 4' section of track	1 Ea.	1.2	57.00
Track fixtures, low voltage, 25/50 Watt	3 Ea.	1.5	349.20
Switch device, quiet type	3 Ea.	0.6	16.92
Dimmer type, for use with pendent light	1 Ea.	0.5	12.96
Totals		7.2	$984.48

Contractor's Fee Including Materials	$1,800

Key to Abbreviations
C.Y.–cubic yard   Ea.–each   L.F.–linear foot   Pr.–pair   Sq.–square (100 square feet of area)
S.F.–square foot   S.Y.–square yard   V.L.F.–vertical linear foot   M.S.F.–thousand square feet

WHAT TO WATCH OUT FOR

Be sure to turn off the power at the circuit breaker or fuse box before you begin any electrical work. The appropriate branch circuit must be identified and the corresponding breaker thrown.

If a new light fixture is the only planned improvement and will add light to the area, it may also draw attention to previously unnoticed blemishes in the wall, floor, and ceiling finishes. Thus, as is often the case in home improvement, this project may lead to one or more other renovations.

SUMMARY

Adding a new light fixture to a room, whether for cosmetic reasons only or simply to brighten a previously dark room, can make a significant difference in the appearance of the space. For this reason, you should give serious thought to the type and style of fixture, as well as the placement of the fixture in the room.

For other options or further details regarding options shown, see

Electrical system: Receptacles

ELECTRICAL SYSTEM: RECEPTACLES

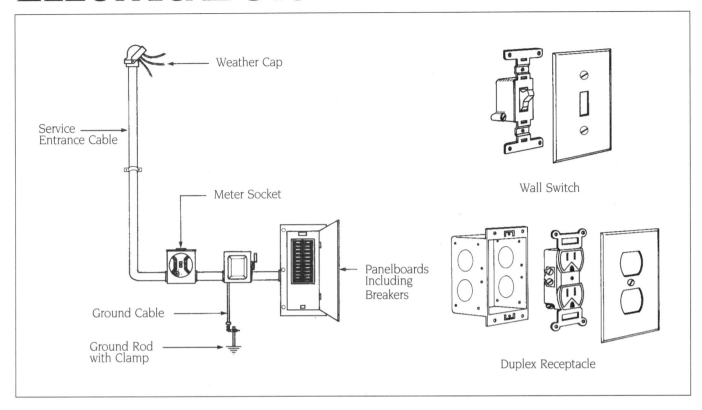

Weather Cap

Service Entrance Cable

Meter Socket

Ground Cable

Ground Rod with Clamp

Panelboards Including Breakers

Wall Switch

Duplex Receptacle

The two most common electrical deficiencies of any home are insufficient lighting and a lack of wall receptacles. Light fixtures are the topic of the previous project; upgrading the rough wiring system and installation of additional receptacles are covered here. This project describes, in general terms, a few basic electrical necessities and amenities that you may consider as improvements or additions to your home's electrical system. Homeowners may do their own electrical work, but the work must be inspected for conformity to local codes.

MATERIALS

Before adding any electrical components to your home, you must first determine whether your existing electrical service can handle the increased load. If your house is more then 15 years old, it may well be underpowered and unable to safely handle the demands of more fixtures and appliances. If your fuses blow (or circuit breakers are thrown) frequently, it could be an indication that your system is not adequate for your *present* needs. If your house is not new, and the electrical system is original, an upgrade should be considered and may be necessary before

any more circuits or outlets are added to it.

A service upgrade involves increasing the total number of amps – the rate at which electricity is delivered – to meet or exceed your home's electrical demand. To measure your existing power needs and estimate future ones is a matter of analyzing and evaluating your entire system, from the main service panel to each outlet and appliance. Because of the variety of equipment and the different ways it may have been installed, it is best to have this assessment performed by a licensed electrician.

Some houses may have an electrical system as low as 30 amps, far below the needs of today's average household. Typical residential service today is rated at 100, 125, 150, or 200 amps. Current federal guidelines and national codes require a minimum upgrade of 100 amps for a single-family residence. Such an upgrade is not inexpensive, but it may be necessary before further electrical work can be done. Consider this as buying insurance for your home and family; an overworked electrical system is a serious potential fire hazard.

Locating new boxes in the room will depend on your needs, but in general

the easiest installations are: below an existing switch; in a common stud space with an existing box offset behind it on the opposite wall surface; and converting from a single- to a double-gang box. Before starting the job, be sure to count the number of outlets and fixtures that are already on the circuit you plan to add to. You should have no more than a total of 8 outlets and fixtures on a 15-amp circuit, and no more than 10 on a 20-amp circuit.

Any receptacle located in a potentially wet area, such as a bathroom, kitchen, laundry room, or basement (consult local codes), must have a ground fault interrupter (GFI). This type of receptacle is designed to cut power instantly when it senses a false ground – your wet hand, for example. A GFI receptacle is more expensive than the standard type, but the protection it affords is worth the extra cost.

LEVEL OF DIFFICULTY

Analyzing and assessing the capacity and condition of your electrical service is a job for an expert or a professional contractor. Installing a new service panel is a task that should be undertaken only by a licensed electrician. Local codes specify who can legally perform this kind of work. Some will allow a homeowner to do the work as long as the proper permits are acquired and the installation is inspected. But remember that whenever you work with electricity there is a potential for injury, or even death. In this case you are dealing with the total incoming electrical power to your home. Prudence would dictate that you hire a professional.

Laying out the lighting in a room to achieve optimum illumination calls for experience, and a knowledge of the pros and cons of the various types and designs of lighting fixtures. An expert can run wiring and install lights and receptacles. Surface-mounted fixtures are easier to install than recessed units. Running the wiring to outlets and switches can be fairly easy, for example, in a second-floor room with attic access; or it may be very difficult in an insulated, exterior wall containing fire stops.

Both the beginner and the intermediate would be well advised not to undertake any electrical task except with the direct aid and assistance of an expert or professional. No one – expert, intermediate, or beginner – should attempt even the smallest electrical task unless in possession of a thorough understanding of electrical circuitry, the proper UL-rated materials, and the tools and skill to properly do the work. The fact that a bulb lights up, or that a receptacle carries current, does not mean that the outlet is wired correctly. Understand that the safety of your home and family is at stake, and be cautious.

Beginners, intermediates, and experts should add at least 100%, 60%, and 20%, respectively, to all tasks they do perform.

WHAT TO WATCH OUT FOR

As part of the evaluation of your electrical system, you should map all your circuits to determine where they are and what outlets they serve. The electrician you hire will have to do this, so you can save him the time and yourself the money by doing it yourself. Using a separate sheet of paper for each floor of your house (including the attic, basement, and garage), draw a complete floor plan, and indicate the location of every outlet – that is, light fixtures, switches, and receptacles. Shut off one circuit at the main panel and systematically flip each light switch, and plug a small hand tool or radio into each receptacle to determine which outlets are dead and, therefore, on that circuit. Do the same for each circuit until you have accounted for every outlet and switch in the house. Number and color-code the circuits on your map for easy identification, and label them on the main breaker panel or fuse box.

SUMMARY

Before spending the time and money for electrical add-ons, be sure your existing electrical service has the capacity to handle the extra load. If not, upgrading the service is required. The minimum upgrade for a single-family residence is 100 amps. It may, however, be more cost effective to upgrade to a higher amperage (125, 150, or 200) to allow for possible future power requirements.

Ignorance in performing electrical work can be, literally, fatal, and overloaded or improperly wired circuits can cause fires. It cannot be too strongly recommended that this type of upgrade and improvement be done only by an expert or, preferably, a licensed electrician. Your peace of mind in knowing that the work was done right will contribute to your enjoyment of these added amenities.

For other options or further details regarding options shown, see

Electrical system: Light fixtures

Upgrade Electrical System

Description	Quantity/ Unit	Labor-Hours	Material
200 AMP service			
Weather cap	1 Ea.	1.0	21.96
Service entrance cable	10 L.F.	1.1	63.00
Meter socket	1 Ea.	4.2	50.40
Ground rod with clamp	1 Ea.	1.8	35.40
Ground cable	10 L.F.	5.0	146.40
3/4" EMT	5 L.F.	0.3	3.42
Panel board, 24 circuit	1 Ea.	12.3	540.00
GFI with 12/2 Type NM cable	2 Ea.	1.5	81.60
Outlet box	12 Ea.	4.8	21.31
Non-metallic sheathed cables, #12 2-wire	120 L.F.	4.4	25.92
Total		36.4	$989.41

Contractor's Fee Including Materials	**$3,301**

Key to Abbreviations
C.Y.–cubic yard   Ea.–each   L.F.–linear foot   Pr.–pair   Sq.–square (100 square feet of area)
S.F.–square foot   S.Y.–square yard   V.L.F.–vertical linear foot   M.S.F.–thousand square feet

WHOLE-HOUSE FAN

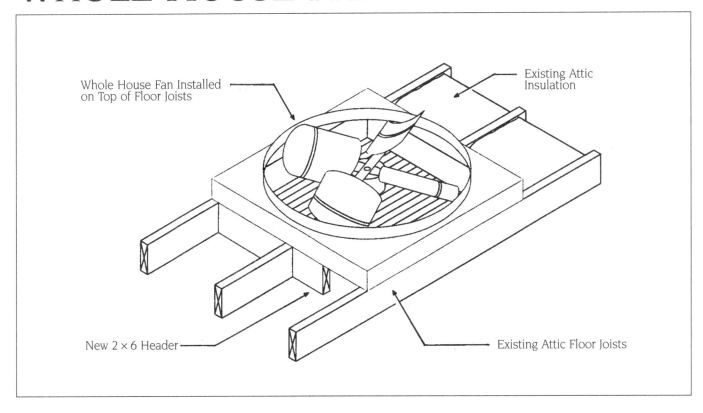

Whole House Fan Installed on Top of Floor Joists

Existing Attic Insulation

New 2 × 6 Header

Existing Attic Floor Joists

A properly ventilated attic is an essential element in a well-built house. Inadequate ventilation, and the condensed moisture that attends it, can be the cause of a number of problems, some more serious than others, but all unnecessary. One problem is heat buildup. Roofs in northern climates are generally dark colored and soak up the sun's rays like heat-absorbing sponges. Much of the heat is radiated into the attic, and from there passes through the ceilings to the second-floor living areas, usually bedrooms. Thus a hot attic above makes for uncomfortably hot rooms below. Ventilation keeps cooler outside air circulating, in effect making the attic a buffer zone between the intense heat on the roof of the house and the inhabitants within. A whole-house fan mounted in the attic floor greatly increases this circulation by drawing air up through the house from the cooler regions of the ground floor and basement.

MATERIALS

If you plan to install a whole-house fan, you must first evaluate the capacity and condition of your existing attic ventilation system. This fan can do its job only if the air can escape the attic fast enough through the vent openings. The rule of thumb states that a free-air area (total area of vent openings) be at least 1/300 of the ceiling area. With the addition of a whole-house fan, the free area will generally have to be substantially increased. In a typical pitched-roof house, this might mean increasing the size of the gable-end vents, enlarging the soffit vents, and even adding a ridge vent. A more elegant and possibly less costly solution to this problem would be to install a roof-mounted ventilator with a motor-driven fan wired to a thermostat. When the hot air drawn into the attic by the whole-house fan reaches a given temperature, the roof exhaust fan kicks in and blows the hot air out. Designing this fan system to adequately service your house demands professional know-how. Get the advice of an experienced electrician or heating contractor.

A multispeed fan costs more than other types, but at slow speed the motor will last longer and run more quietly. You will greatly appreciate the latter feature on hot summer nights.

A whole-house fan is normally located in a hallway ceiling near the stairwell. This placement helps to induce an updraft from the lower, cooler levels of the house.

If the attic has flooring and insulation, portions of these materials must be removed to expose the joists in the area where the fan is to be located. Following the manufacturer's specific directions, a section of ceiling is removed, the joists are cut back, and headers installed to frame the opening. The fan is mounted directly to the framing. This is a two-person job; one supports the unit from below, and the other is in the attic securing it in place. Be sure the installation is squared off to the hallway walls. Shims or spacers can be used to make any necessary adjustments. Also, be sure everything you install, from the framing to the fan itself, is tight. Looseness means movement; movement means noise.

The fan must be hard-wired and a switch installed in the hall below. In an attic it is usually not too difficult to locate a feed and fish a wire down a convenient wall for a switch box. Cut the hole for the box very carefully. A switch plate doesn't cover very much, and though walls can be patched, damage to wallpaper or a discontinued paint color can be impossible to hide.

Finish trim is either integral to the unit or provided along with it. If any other trim is required, make sure it is as flat and innocuous as possible, and installed so as not to interfere with the working of the fan louvers.

LEVEL OF DIFFICULTY

A high degree of expertise is required in the planning stages to ensure that the attic ventilation system has the capacity to exhaust the volume of hot air the whole-house fan will push into it. A design failure here means the entire system will not work, and the time and money spent on it will have been wasted. It is recommended that you seek the advice of several qualified experts and, if possible, talk to other homeowners who have had similar systems installed.

The actual installation involves fairly routine framing and finishing for someone familiar with the tools and materials required. Electrical work should be done only by an expert or a professional contractor. Wiring mistakes are easy to make, and results can be disastrous. A fixture could work, yet still be wired wrong, and the error could be discovered only later when someone smells something burning.

A beginner should increase the time estimate by 100% for any tasks he or she is qualified to do. An intermediate should add 50% to the mechanical installation and hire out the electrical work. An expert should add 15% to all tasks.

WHAT TO WATCH OUT FOR

Detailed installation instructions for the whole-house fan are provided by the manufacturer. Read and follow them carefully. Be sure the layout lines you draw on the ceiling are accurate and square. Before taking a saw in hand, make sure your cuts will not weaken any structural members. Temporary bracing may be required while creating the opening. Mistakes can be patched up, but you'll kick yourself if you end up having to refinish or repaint the entire hallway ceiling just because you misinterpreted a direction or misread a measuring tape.

As with any ceiling work, be prepared for a mess. Cover the floor with canvas or plastic, and isolate the area as much as possible from adjoining rooms. If there is loose insulation in the ceiling, clear it well back from the planned opening. Cutting out the ceiling and joists with a reciprocating saw will cause a lot of vibration, which will shake down any debris and dust left near the hole and deposit it all over everything, including you. Wear a dust mask.

Remember that headers are made up of two layers of the framing lumber. For nominal 2″ stock, that means cutting the joists back 3″ from the inside dimension of the rough opening. Be sure not to cut the ceiling back that much or you'll be faced with a major patch-and-finish job such as the one mentioned above.

SUMMARY

A whole-house fan cannot cool your home in the same way or to the same degree as air conditioning. A fan draws air from the lower, and generally cooler, portions of the house, up into the attic where it is exhausted through the vents. It cannot lower the temperature of the house below the temperature of the ambient air, but the air it moves does produce a cooling effect. Even if it requires the addition of a roof-mounted exhaust fan or other attic ventilation, a fan of this type can be a very efficient alternative to air-conditioning, especially in northern climates.

Whole-House Fan

Description	Quantity/ Unit	Labor- Hours	Material
Selective demolition, ceiling cut-out	9 S.F.	2.0	
Framing through ceiling 2 x 6	12 L.F.	0.4	8.21
Whole house exhaust fan, ceiling mounted, 36″, variable speed,			
remote switch, incl. shutters, 20 amp- 1 pole circuit breaker,			
30′ of 12/2 type NM cable	1 Ea.	2.3	708.00
Totals		4.7	$716.21

Project Size	3′ x 3′	Contractor's Fee Including Materials	$1,242

Key to Abbreviations
C.Y.–cubic yard   Ea.–each   L.F.–linear foot   Pr.–pair   Sq.–square (100 square feet of area)
S.F.–square foot   S.Y.–square yard   V.L.F.–vertical linear foot   M.S.F.–thousand square feet

Section Five
FIREPLACES

Installing a new fireplace is a challenging project for any do-it-yourselfer. Attractive brickwork on the fireplace hearth and facing are important, so even expert do-it-yourselfers may consider hiring a professional mason. If you decide to do all or part of the work yourself, the following tips will be useful.

- Ask your masonry supplier about the best cleaners for washing down the brick after the job is done.

- Protect existing surfaces before starting your masonry work. Mortar and cleaning solutions can stain floors and walls.

- If you are adding decorative detail you may need to use a bonding agent to ensure that the new materials bond correctly and securely. Check with your masonry supplier.

- You can purchase an inexpensive set of mason's tools that includes a trowel, pointers, mortar box, brush and lines, and brick chisel.

- Masonry work is generally viewed as *quality* when it "looks good." Maintain even spacing and joints. Keep the tops of the units in line and maintain a smooth face. Masonry units installed poorly can cast shadows and give the impression that quality workmanship was not important. Beginners should practice before starting. Try a test panel.

- Fireplace installation requires a building permit and inspection, and some aspects of the work must be done by a professional – particularly gas piping and venting.

BUILT-IN FIREPLACE

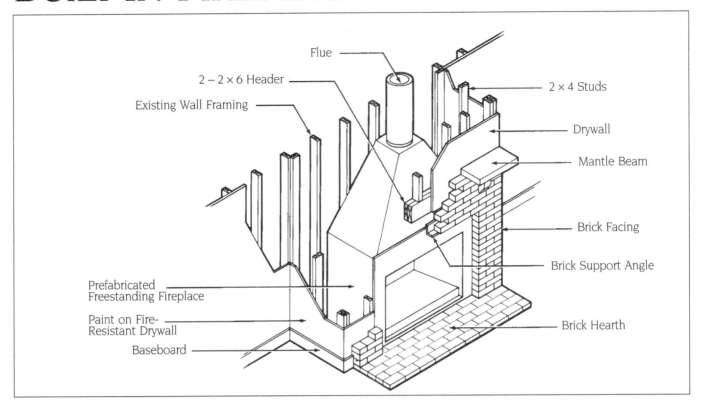

Flue

2 – 2 × 6 Header

Existing Wall Framing

2 × 4 Studs

Drywall

Mantle Beam

Brick Facing

Brick Support Angle

Prefabricated
Freestanding Fireplace

Paint on Fire-
Resistant Drywall

Baseboard

Brick Hearth

With careful planning and appropriate design, a new fireplace can make a drab room come alive with atmosphere. Conventional fireplaces made of brick and/or other masonry materials are difficult to install in an existing house, but other options are available that involve less work and expense. One of the most popular alternatives is a prefabricated fireplace unit that comes from the supplier ready for installation, complete with flue pipe and accessories. Before placing the unit, the area where it is to be located has to be framed and prepared to receive it. The extent of preliminary work depends on many variables, such as the floor level where it is to be placed, whether it is to be installed on an interior or exterior wall, and, especially in old houses, the amount of leveling and reconditioning required for the support floor surface. If the floor is not level, or its support system is inadequate, the materials cost estimate of this project will be higher.

MATERIALS

The materials required to complete the installation include the prefabricated freestanding fireplace unit, its flue, related accessories, a simulated-brick chimney top, brick and masonry materials for the facing, and conventional framing and wall supplies. The heart of the project is the prefabricated fireplace, which is an efficient and economical alternative to standard brick and mortar installations. Many different manufacturers produce these units, usually from sheet steel. The structure may look heavy, but it is lightweight enough to be carried and placed by two people. Also, most prefabricated fireplaces do not require additional floor support, as they weigh about as much as a heavy piece of furniture. Make sure you purchase a unit that complies with the standards of stove and freestanding fireplace approval agencies. Also, carefully follow the manufacturer's instructions and precautions during installation of the fireplace and its flue assembly, and closely adhere to the requirements of the local building code.

Before the unit is placed, the cavity or wall area where it will sit permanently has to be framed with 2 x 4s. The cost of the framing materials for this operation will depend on the particular location of the fireplace. For example, if the unit is to extend outside of the house, the exterior wall will have to be built, at more expense, to enclose the fireplace unit and the flue. If the installation is to be located inside the room (as in this example) and backed to an existing partition or wall, the materials cost will be lower. Remember that if the fireplace unit is located in the room, it will take up considerable floor space. The exterior placement may cost you more, but the extra expense may be worth it to save floor space. After the fireplace has been put into position and the flue pipe safely tied in, the new framing should be covered with 1/2″ fire-resistant drywall that is then taped, finished, and painted or wallpapered to suit the room's decor.

The brick hearth and facing recommended in this plan can enhance the fireplace opening. Other noncombustible veneers provide viable options at varying costs. The laminated hardwood mantel is one of many alternatives available for dressing up the opening. The design and choice of materials for the facing, hearth, and mantel should be given careful consideration before the selections are made. Brick, marble, and stone are among the materials that can be used. Some extra expense may be worthwhile, as the aesthetic appeal of the fireplace is important.

Take care in cutting the openings for the flue pipe and be sure to use high-quality fittings and flashing where necessary. Joist shields should also be installed where needed for safety. A simulated brick chimney top is a feature that adds to the appearance of the job by hiding the top of the insulated piping. If the pipe extends from a rear-facing or hidden section of your roof or if you are not fussy about the exposed pipe, then omit the simulated chimney as a decorative extra and deduct its cost from the project.

LEVEL OF DIFFICULTY

Although the installation of a prefabricated fireplace may appear to be a major undertaking, with the right conditions it can be a fairly uncomplicated project for the capable do-it-yourselfer.

The masonry tasks demand specialized skills and should be completed by a professional if you have not had experience in laying stone or brick. Most of the other tasks, including cutting the ceiling and roof openings, can be completed by intermediates and experts. Get some advice from a knowledgeable person if you have not cut ceilings or roof holes before. Beginners should limit themselves to the finish work, the framing, and drywalling. Experts should add about 10% to the labor-hour estimates and more for the masonry operations. Intermediates should tack on about 40% to the time for the framing, drywalling, and finish work, excluding the masonry. They should increase the time for hole-cutting and fireplace and flue installation by 100% if they have not had prior experience with these tasks. Beginners should double the time estimate for any jobs they intend to complete.

WHAT TO WATCH OUT FOR

Two aspects of this project should be noted as critical, one affecting materials and labor costs and the other, safety. The expense for both materials and labor will vary significantly from the plan's estimate if the prefabricated fireplace unit will be used as a replacement for an existing fireplace facility or if extensive demolition and alteration of the site are required before the new fireplace is installed. The materials cost will also be increased considerably if an exterior chase has to be built to house the fireplace and flue pipe, especially if the fireplace is to be located on the first floor of a two-story house. Safety, another aspect of this project, has to be monitored carefully throughout the installation process. Follow the manufacturer's safety recommendations closely and observe local building codes that deal with fireplace or woodburning stove regulations. Most communities require approval of all new fireplace and stove installations, so arrange to have the inspection before you drywall the fireplace enclosure.

Built-In Fireplace

Description	Quantity/ Unit	Labor-Hours	Material
Framing, fireplace enclosure, 2 x 4 studs and plates, 16" O.C.	14 L.F.	2.8	72.41
Header over opening, 2 x 6	12 L.F.	0.5	8.21
Cut ceiling and roof opening for flue	2 Ea.	3.2	
Frame through ceiling and roof, 2 x 6	16 L.F.	0.2	10.94
Fittings, ceiling support, 10" diam.	1 Ea.	0.7	142.80
Fittings, joist shield, 10" diam.	1 Ea.	0.7	79.80
Drywall, 1/2" on walls, fire-resistant, taped & finished, 4' x 8'	128 S.F.	2.1	39.94
Fireplace, built-in, 42" hearth, radiant, recirculating, small fan	1 Ea.	8.9	1,020.00
Facing, 6' x 5', standard size brick	30 S.F.	10.8	92.34
Hearth, 3' x 6', standard size brick	1 Ea.	8.0	168.00
Mantel beam, laminated hardwood, 2-1/4" x 10-1/2" x 6'	1 Ea.	1.6	114.60
Simulated brick chimney top, 3' high, 16" x 16"	1 Ea.	0.8	202.80
Fittings, roof flashing, 6" diam.	1 Ea.	0.5	58.80
Trim, baseboard, 9/16" x 3-1/2"	10 L.F.	0.3	13.68
Paint, walls and trim, primer	90 S.F.	0.4	4.32
Paint, walls and trim, 1 coat	90 S.F.	0.6	5.40
Totals		42.1	$2,034.04

Project Size	5' x 6' x 13'	Contractor's Fee Including Materials	**$4,906**

Key to Abbreviations
C.Y.–cubic yard   Ea.–each   L.F.–linear foot   Pr.–pair   Sq.–square (100 square feet of area)
S.F.–square foot   S.Y.–square yard   V.L.F.–vertical linear foot   M.S.F.–thousand square feet

FIREPLACE FACING OPTIONS
Cost per Square Foot, Installed

Description	
Red Brick	$11.15
Field Stone	$27.00
Marble	$37.59
Ceramic Tile	$14.40

SUMMARY

A new fireplace is one way to give your family room, living room, or den new life and atmosphere and some extra heat during the cooler months. With the installation of a prefabricated fireplace unit, you can save on material costs and labor by doing much of the work on your own.

For other options or further details regarding options shown, see

Masonry fireplace

Gas Fireplace

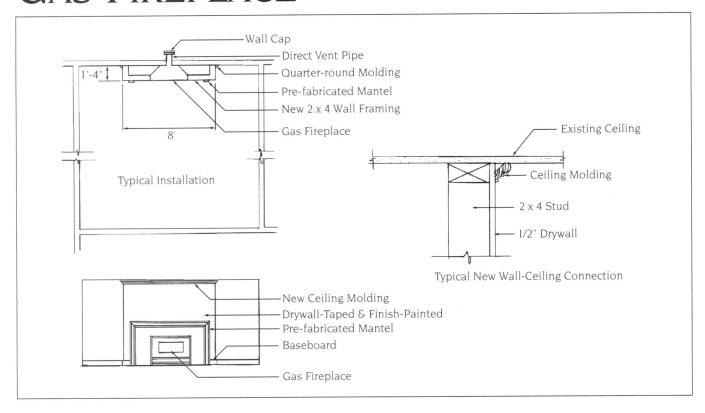

Wall Cap
Direct Vent Pipe
Quarter-round Molding
Pre-fabricated Mantel
New 2 x 4 Wall Framing
Gas Fireplace

1'-4"

8'

Typical Installation

Existing Ceiling
Ceiling Molding
2 x 4 Stud
1/2" Drywall

Typical New Wall-Ceiling Connection

New Ceiling Molding
Drywall-Taped & Finish-Painted
Pre-fabricated Mantel
Baseboard
Gas Fireplace

Fireplaces continue to be one of the best home improvement investments in many regions. They add atmosphere and enjoyment to your home and often enhance its resale value. While wood-burning fireplaces are still the first choice for many homeowners for their authenticity, gas fireplaces and stoves are gaining popularity for their convenience and the increasingly natural look of their fire and glow. Gas fireplaces can be controlled instantly by wall-mounted thermostats or remote control.

There is a gas fireplace or stove and venting technique for just about any application, and a wide range of fireplace styles, decorative log sets, and surrounds to suit any decor. Many models do not need a chimney (they can be vented through the wall or roof), nor do they require electricity to light or operate, making them available for use during power failures.

Selecting a particular model will involve several criteria: how much space the fireplace or stove will occupy in a room, the fireplace's appearance and suitability to the room decor, and your requirement for heat in that room. Models with high heat output may overwhelm a small room, heating it up quickly, but having

to be turned off often. Basements may require a model that delivers more heat.

In choosing a fireplace, you'll also be comparing heat value in BTUs and the Maximum Steady State Efficiency and Annual Fuel Utilization Efficiency of various models. (An efficient natural gas fireplace can operate for one hour at a cost of approximately 10 cents.) Many fireplaces have blowers that deliver warm air from the fireplace body (behind the wall) into the room, adding to the radiant heat from the front of the fireplace.

MATERIALS

Gas fireplaces and stoves are sold through dealers who offer installation services or recommend professional installers. Many models burn either natural gas or propane. If an existing chimney is used, it may need to be lined with a stainless steel liner to meet a specific fireplace's venting requirements.

For this project we are installing a gas fireplace unit into a traditional-looking fireplace surround, including a mantel. The surround is constructed using standard wood framing techniques and minor finish woodwork. The surround is not structural, that is to say, it does

not support any significant load. The new construction is blended with the old using wall and ceiling moldings, with no requirement to repaint the existing ceiling or wall surface. The fireplace will be vented directly through the wall. A major focus is the proper venting of the fireplace and connection of the gas (and electricity, if the model has a blower) to the unit. This project assumes that the home is already supplied with natural gas for heating, and the plumber will tie into an existing gas line in the basement. The thermostat purchased with the unit will be installed by an electrician or fireplace installer.

LEVEL OF DIFFICULTY

Homeowners can perform some of the preparation and finish work for this project. Professional installers are generally called upon for the stove or fireplace and vent/chimney work—per building department and safety requirements, and gas connections must be made by a licensed plumber or a professional gas fireplace installer. Novice remodelers clearly would not undertake gas or electrical connections, but could prepare for the installation by opening up the wall, re-framing for the exterior vent cap, and creating a basic surround using a prefabricated mantel and moldings. Beginners should consult with a professional before undertaking any of these tasks, particularly if the exterior wall is masonry or stucco. It is a good idea to talk to your plumber beforehand to verify the appropriate location for wall openings, and the timing and sequence of tasks. The novice's tasks may take 100% longer than the same work done by a professional. Intermediate and advanced do-it-yourselfers may, in addition to the prep and finish work, want to design and fabricate their own mantels and trim. With a prefabricated mantel, the work may take 50% longer than a professional; with their own mantel construction, 100%.

Gas Fireplace

Description	Quantity/Unit	Labor-Hours	Material
Framing, fireplace enclosure, 2 x 4 studs and plates, 16" O.C.	80 L.F.	0.4	10.34
Header over opening, 2 x 6	12 L.F.	0.1	1.37
Cut wall opening for vent	1 Ea.	2.3	
Frame through wall 2 x 4 x 8	8 L.F.	0.1	0.86
Drywall, 1/2" on walls, fire-resistant, taped & finished, 4' x 8'	96 S.F.	0.1	0.62
Direct vent gas fireplace, including vent kit	1 Ea.	17.8	2,040.00
Fireplace mantel, 6" molding, 6' x 3'-6" opening, prefab. pine, colonial	1 Ea.	5.3	614.40
Moldings, base, stock pine, 9/16" x 3-1/2"	10 L.F.	0.1	2.74
Moldings, ceiling, stock pine, 9/16" x 1-3/4"	10 L.F.	0.1	1.51
Moldings, trim, quarter round, stock pine, 3/4" x 3/4"	16 L.F.	0.1	1.01
Paint, walls and trim, primer	100 S.F.	0.1	0.10
Paint, walls and trim, 1 coat	100 S.F.	0.1	0.12
Totals		26.6	$2,673.07

Project Size 5' x 6' x 13'

Contractor's Fee Including Materials $5,111

Key to Abbreviations
C.Y.–cubic yard   Ea.–each   L.F.–linear foot   Pr.–pair   Sq.–square (100 square feet of area)
S.F.–square foot   S.Y.–square yard   V.L.F.–vertical linear foot   M.S.F.–thousand square feet

WHAT TO WATCH OUT FOR

Check with your local building department about permit requirements, including manufacturer's specifications or other information you may need to submit along with your permit application. Check to see if there are requirements for annual follow-up inspections or service.

Try to position the fireplace so that the vent pipe is in the space between two studs, so that no re-framing is required. Allow at least 48" of clearance between the front of the fireplace and any objects, such as furniture.

Take care to avoid damaging walls, ceilings, and floors as you construct the surround and position the unit. Keep in mind that it is generally easier to handle longer stock and awkward-shaped equipment with two people. Read the manufacturer's specifications carefully to avoid difficulties.

SUMMARY

Gas fireplaces provide instant heat and a relaxing atmosphere in our time-starved world. Their various venting options make it possible to install them in just about any room in a house. While some parts of this project must be performed by professionals, homeowners can contribute by doing preparation and finish work, and can take satisfaction in the thought that their costs for material and professional installation will be a good investment.

For other options or further details regarding options shown, see

> *Elevated deck*
> *Elevated L-shaped deck*
> *Entryway steps*
> *Redwood hot tub*

MASONRY FIREPLACE

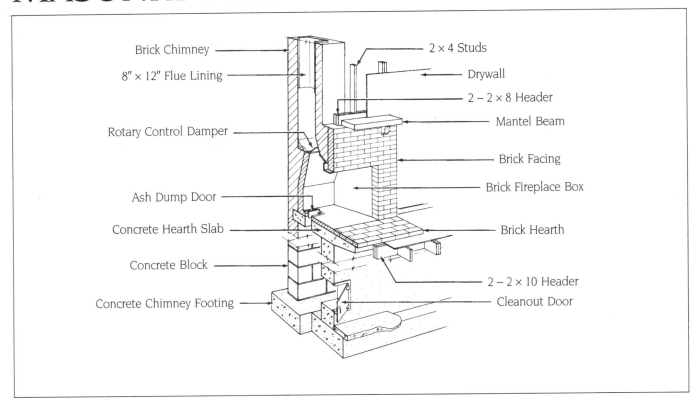

Brick Chimney

8" × 12" Flue Lining

Rotary Control Damper

Ash Dump Door

Concrete Hearth Slab

Concrete Block

Concrete Chimney Footing

2 × 4 Studs

Drywall

2 – 2 × 8 Header

Mantel Beam

Brick Facing

Brick Fireplace Box

Brick Hearth

2 – 2 × 10 Header

Cleanout Door

Although the construction of a masonry fireplace involves considerable time, inconvenience, and expense, it is a durable fixture that contributes to the appearance and atmosphere of a room. Masonry work does require some specialized skills and tools, however, and do-it-yourselfers should be cautioned about tackling this project if they do not have any experience in this area. The site preparation, actual construction of the fireplace and chimney, and inside and outside finishing operations should all be carefully planned in the proper sequence so that inconvenience can be kept to a minimum. If you have not done masonry work before, get advice from a qualified person before you start, and do some reading on the subject. When you begin the project, plan to work slowly and patiently with a gradual increase in productivity as you acquire skill in the masonry operations. The rewards of personal satisfaction, savings on construction costs, and added value to your home can make your efforts worthwhile.

MATERIALS

The materials used in the installation of a masonry fireplace include conventional masonry products as well as framing lumber for alterations that need to be made to accommodate the new structure. There are also some incidental materials associated with the fireplace, chimney, and flue. The total cost of these items will vary with each installation, depending on the amount of required site preparation, the height of the chimney, and the extent of finish work needed to receive the new chimney on the side and roof of the house. Costs for materials will also vary with the type of masonry veneer selected for the facing of the fireplace, hearth, and chimney. Before the fireplace and chimney can be put into position, extensive site work is required on the inside and, usually, on the outside of the house. The estimates given in this plan will help you to determine the cost for this work. On the inside of the house, framing materials are required to maintain the floor and exterior wall support system when the openings are cut for the hearth and fireplace. Before cutting the wall opening and installing the headers, be sure to provide temporary

ceiling support, as the outside wall may be a load-bearing structure. The same precaution should be taken before cutting the floor joists for the hearth. The materials for these temporary 2 x 4 support partitions have not been included in the estimate, but the same lumber used for the ceiling support can be reused for floor support and, possibly, elsewhere in the installation where 2 x 4s are required. If you have not performed framing and support operations before, get some advice before you tackle them, as mistakes and poor workmanship can cause serious structural damage to the house.

Outside the house, the site must be prepared for the construction of a foundation to bear the weight of the chimney. Because this foundation should extend to the full depth of the house's foundation, a considerable amount of excavating has to be done.

This excavation usually has to be done by hand because of its proximity to the structure and finished landscaping. Machine excavation can sometimes be done at extra cost. After the excavation has been completed, a footing is placed; later, concrete blocks are laid to serve as the foundation for the fireplace and chimney. Additional labor costs also have to be included for removing the siding and preparing the rake boards or eaves to receive the new chimney.

The masonry materials used in this plan include bricks, concrete blocks, flue liner, and mortar. These items are readily available from most building and masonry supply outlets, but it pays to shop around for them. Here are a few suggestions to assist you in the purchase. First, try to buy all of the materials from the same retailer. By consolidating your order, you will get a better price. Second, have the materials delivered to your property, with the delivery charge included in the bill at the time of purchase. Unless the materials have to be transported an excessive distance, most retailers will not charge you for delivery of an order of this size. Third, select the veneer material carefully, as conventional red brick is only one of many options. Field stone, quarry-prepared stone, and cobblestone are other choices that function well and give a different appearance. The cost for these alternative materials will vary depending on the required preparation of the stone, its quality, and its availability. Concrete block with stucco veneer is an attractive yet economical option for chimney design.

The appropriate fittings, prefabricated firebox and roof flashing should be selected with the aid of a mason and/or a reputable retailer if you are unfamiliar with these specialty items. Make sure, for example, that you match the size of the cast-iron damper unit and firebox and that you install appropriately sized clean-out fittings. The flashing required to join chimney and roof must be carefully installed to assure that the structure is weathertight. If the new chimney is placed on the gutter side of the roof, a cricket (a sort of diverter) will have to be constructed at extra cost. Additional expense will also be incurred if a chimney cap is needed to avert downdrafts in your fireplace. The rental of staging or scaffolding is one more cost to figure in for work on the chimney construction.

LEVEL OF DIFFICULTY

A significant amount of skill and experience in masonry work is required to take on the building of this fireplace. If you are accomplished in the use of tools, but have not worked with masonry materials before, you should do some reading on the subject, seek advice from a mason, and even complete a trial project like a small brick-and-block retaining wall. Experts and some experienced intermediates should be able to complete the framing tasks and the excavation, and leave the remainder of the project to a professional. Intermediates and experts should add about 60% and 30%, respectively, to the labor-hour estimates for the framing and site preparation operations. If you plan to do the masonry work and are inexperienced in bricklaying, then double or even triple the professional time, even if you are an expert in other remodeling skills. Plan to work very slowly at first, with a gradual increase in productivity as you gain experience and the project progresses.

WHAT TO WATCH OUT FOR

One of the most important elements of any fireplace is the firebox. Inaccurate alignment or incorrect sizing of this component can adversely affect the fireplace's operation by causing inefficient draft. Seek advice from a mason before you construct the firebox, as his or her experience and knowledge can help prevent the incorrect or ineffective placement of this vital fireplace component. It may be worth the expense of hiring a mason for this task if you do not have the skill and know-how to complete it on your own.

SUMMARY

Installing a conventional masonry fireplace in your home is a major remodeling project, but one that can be tackled, at least in part, by do-it-yourselfers. It is a fairly involved project, but will result in a pleasing and practical addition to your home.

For other options or further details regarding options shown, see

Built-in fireplace

Masonry Fireplace

Description	Quantity/ Unit	Labor-Hours	Material
Excavating by hand, heavy soil, for chimney foundation	9 C.Y.	18.0	
Concrete chimney footing	0.50 C.Y.	2.4	97.80
Concrete block, 8" thick, for foundation wall, 32" x 60" x 8'	85 S.F.	8.0	171.36
Concrete block, 8" thick, for foundation wall, 32" x 60" x 8'	70 S.F.	6.6	141.12
Framing joists, 2 x 10, fireplace foundation opening	30 L.F.	0.5	44.64
Cleanout door & frame, cast iron, 8" x 8"	1 Ea.	0.7	28.80
Framing, walls, header over fireplace opening, 2 x 8 & 2 x 4	20 L.F.	0.9	20.88
Fireplace box, complete	1 Ea.	8.0	150.00
Damper, rotary control, cast iron, 30" opening	1 Ea.	1.3	85.20
Facing, 6' x 5', standard size brick	30 S.F.	5.5	96.84
Hearth, 3' x 6', standard size brick	1 Ea.	8.0	168.00
Mantel beam, rough texture wood, 4 x 8 x 6'	6 L.F.	1.3	30.53
Demolition, remove siding, cut roof	1 Job	8.0	
Chimney, 16" x 20" x 12", standard size brick	12 V.L.F.	12.0	324.00
Flashing at siding and roof	30 S.F.	1.6	157.32
Totals		82.8	$1,516.49

Project Size	6' x 6' x 20'	Contractor's Fee Including Materials	$5,810

Key to Abbreviations
C.Y.–cubic yard   Ea.–each   L.F.–linear foot   Pr.–pair   Sq.–square (100 square feet of area)
S.F.–square foot   S.Y.–square yard   V.L.F.–vertical linear foot   M.S.F.–thousand square feet

Section Six
KITCHEN REMODELING

Start designing your kitchen by planning the location of the sink, stove, and refrigerator, preferably in a working triangle, with a perimeter of 12-24'. Following are additional tips (based on National Kitchen & Bath Association recommendations).

Countertops:

- A work area should have at least 36" of continuous countertop.

- At least one work area should be next to the sink.

- There should be at least 9" of countertop on one side of the cooktop, 15" minimum on the other side.

- There should be at least 15" of counter space next to or across from (no more than 48" away) the refrigerator.

- Natural stone, such as marble, granite or slate, are costly and can take two weeks or more to fabricate. You can reduce the cost by using stone for an island top, and laminate on the other countertops.

- Laminates come in a nearly endless range of colors and patterns, in widths from 18-60". Wood or solid surface edging can provide a custom look, with an additional cost of between $40-$80 per five-foot run of countertop.

- Solid surfacing offers durability, seamless construction, and an increasing variety of color and pattern. The cost is significantly higher than laminate, and this material is fabricated and installed by the distributor.

Flooring:

- Vinyl and linoleum are the most economical, though it pays to purchase the best grade your budget allows, due to the high traffic on kitchen floors. If you choose a wood or stone pattern, look for as little repeat as possible.

- If you choose ceramic tile or stone, look for a slip-resistant finish.

- Wood floors offer warmth and charm, but require regular sweeping and will not tolerate spills. New laminate "floating" floors resemble wood, offer resistance to wear, and can be installed over existing floors.

Cabinets:

- It's a good idea to shop around to get a feel for price, appearance and features as you formulate choices and priorities within your budget. Visit kitchen dealers, home centers and lumberyards. Determine whether you want stock, semi-custom, or custom cabinets. Kitchen dealers offer more personal attention and usually a design service as part of their overall fee. Home centers may provide some design assistance as part of your purchase.

- Crown mouldings and trim pieces can provide a custom look. Select a cabinet and moulding style that harmonizes with the rest of your home.

- Consider accessories for maximizing storage space. Lazy Susans, pull-out shelves, and appliance garages are among the options.

Sinks & Appliances:

- Sinks should be durable and easily maintained, whether inexpensive stainless or solid surfacing. Extra depth may be an important consideration and is often lacking in less expensive sinks. Thicker gauge stainless steel is better, and less noisy. The lower the number, the heavier the gauge.

- Be wary of plastic faucet parts. Solid brass is most corrosion-resistant.

- Compare appliance prices at department and discount stores, and home centers. Get a guarantee with any discontinued model.

CABINET/COUNTERTOP UPGRADE

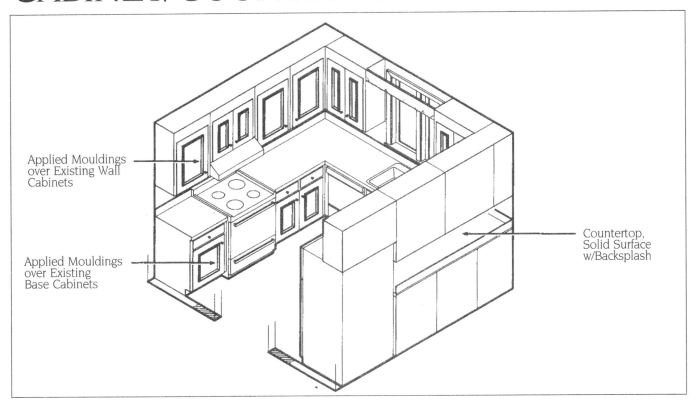

Applied Mouldings over Existing Wall Cabinets

Applied Mouldings over Existing Base Cabinets

Countertop, Solid Surface w/Backsplash

Most kitchens serve as all-purpose utility rooms for a sometimes bewildering variety of tasks, all of which take their toll in spills, burns, scrapes, and scratches. Years of heavy use and constant traffic produce a worn appearance on even the most durable of finishes. Very few kitchens wear their age well, and it is a rare house that has never required a full kitchen renovation. If your kitchen layout is satisfactory and the major appliances do not need replacing, however, you can achieve the look of a new kitchen by simply dressing up the existing cabinets and, if necessary, installing a new countertop.

MATERIALS

There are a number of options for upgrading kitchen cabinets:

- The easiest and most obvious sort of facelift calls for repainting or refinishing, and possibly installing new hardware – hinges, door handles, and drawer pulls. As mentioned above, the heavy use to which kitchen surfaces are subjected dictates the use of top quality materials. Use high or semi-gloss paint, and take the time to properly

prepare the cabinets and drawers with adequate sanding, cleaning, and priming to ensure that the paint will adhere well. If you are replacing hardware, be sure to fill all holes carefully before sanding. Apply two thin coats with a good quality brush so as to avoid brush marks, drips, and puddling. When the paint is thoroughly dry, the new hardware can be installed. Be careful in mounting the hinges so that the doors line up vertically and horizontally.

- If the old surface is badly damaged, cracked, or peeling, sanding alone may not produce an acceptably smooth surface. You may have no choice but to use chemicals to strip the paint to the bare wood. You may then prime and paint, or stain and varnish, depending on the look you want. Paint strippers are hazardous, so follow all the safety precautions recommended by the manufacturer.

- If your cabinet doors are flush, you may want to dress them up by adding decorative moulding. There are a variety of moulding profiles available. Cut mitered corners and attach them in whatever rectangular pattern you choose, using brads and glue. Fill the

brad holes with vinyl spackle and cover any open seams with a thin bead of caulk.

- Old doors and drawer fronts might also be replaced. This is easy enough if the cabinets are painted. If they are stained, matching the wood type, grain, and color could prove to be difficult, if not impossible.

The other major component in a kitchen facelift is the countertop. The most typical countertop material found in modern kitchens is plastic laminate, and though it is very durable, it can become faded, burned, chipped, or delaminated. Some damage can be repaired, but there comes a time when the only option is replacement.

Plastic laminate is economical, easy to maintain, and resists moisture superbly. Hundreds of colors and patterns are available. Alternative countertop materials include solid wood, ceramic tile, Corian and stone. Kitchen design centers can provide information about durability, ease of maintenance, availability, cost, and other variables.

To take out the old countertop, the sink must be disconnected and removed. Be sure the water shut-off valves are closed and the circuit breaker is tripped for the garbage disposal. Locate and remove any screws or brackets holding the top to the base cabinets. The sink hole is a good spot to cut the countertop in two. This will greatly facilitate removal. If there is a tile backsplash it will probably have to be removed to get the old countertop out and the new one in. If you can leave it, be careful not to chip any tiles. It is possible to install new Formica over old. The sink must be lifted out and the old surface thoroughly sanded and cleaned to prepare it for the contact cement. The cement is applied to the top of the old and the underside of the new Formica, which has been precut to fit, with the outside edges and the edges of the sink opening left slightly larger to be trimmed with a router after installation. The counter edge can be lined with an oak strip, sanded and oil-finished. Working with contact cement is extremely difficult. You get only one chance to make a perfect fit – it grabs instantly and allows for no adjustment. If you decide to follow this option, leave it to a professional.

LEVEL OF DIFFICULTY

Producing a good paint job is not as easy as it looks. A professional painter is a skilled tradesperson, but sanding, stripping, staining, painting, and varnishing are all tasks that can, to some degree, be handled by a beginner willing to spend the time to do a thorough and careful job.

Mounting new hardware on new or old cabinet fronts requires minimal tool handling skills, but it does involve precise measuring to ensure that the doors hang straight and line up horizontally, and that the handles and drawer pulls are properly located.

Cutting mouldings and achieving tight miter joints calls for more skill but, again, with enough care (and extra material for mistakes) it can be undertaken by a non-expert.

Only an expert or professional should attempt a countertop replacement. There is virtually no room for error in this task, from taking the initial measurements to the final installation. Even the cheapest countertop is expensive and must be custom-made to fit precisely. Installing it can be extremely difficult, depending on its size and shape, access to the kitchen, and the presence of any immovable obstructions. It is not a one-person job. Two or more people will be required to lift and maneuver it into position, while taking great care not to chip the edges or corners. Solid surfacing materials and stone are highly durable, premium countertop surfaces. Both require expertise in fabrication that should be left to a professional.

The expert should add 30% to the time for installing the countertop. The intermediate and beginner are advised not to attempt it unaided. They should add 50% and 100%, respectively, to the cabinet-finishing tasks; the expert, 10%.

WHAT TO WATCH OUT FOR

This project will disrupt the family's use of the kitchen. The cabinet facelift portion poses no more than a minor inconvenience, but removing and replacing a countertop can easily go beyond one day if unanticipated problems arise.

This is a good time to add undercabinet light fixtures, outlets, and switches. Also, you might consider having a new sink and faucet or garbage disposal installed. If these jobs are to be done by contractors, be sure to coordinate their schedules with yours so the kitchen doesn't stay out of commission beyond an acceptable length of time.

Plastic laminate must be handled carefully. When making cuts such as that for the sink, measure precisely and score the line with a parrot-beak linoleum knife prior to cutting with a saber saw. To prevent the countertop being scratched, tape a piece of cardboard or felt to the sole of the saw. Attach a piece of scrap lumber across the underside of the sink opening, or have a helper below, to support the waste piece so that at the end of the cut it doesn't drop and tear a piece out of the working surface.

SUMMARY

A complete kitchen renovation project is a major undertaking in terms of time and money. If you analyze your kitchen and your family's needs, you may find that the only objectionable feature of your present set-up is the color or the condition of the cabinet and countertop surfaces. An upgrade project can provide you with the look of a new kitchen.

For other options or further details regarding options shown, see

> *Electrical system: Light fixtures*
> *Electrical system: Receptacles*
> *Island unit*
> *Range hood*
> *Sink replacement*

Cabinet/Countertop Upgrade

Description	Quantity/Unit	Labor-Hours	Material
Custom pine molding, 1 x 1	220 L.F.	6.5	191.40
Paint, cabinets, prime	135 S.F.	1.7	8.10
Paint, cabinets, 1 coat	135 S.F.	2.7	19.44
Countertop, 24" wide, solid surface	21 L.F.	12.0	1,398.60
Countertop cutout for sink	1 Ea.	0.7	57.60
Cabinet hardware, pulls	22 Ea.	1.9	85.01
Concealed hinges, avg.	22 Pair	2.6	170.28
Drawer pulls	6 Ea.	0.3	11.30
Totals		28.4	$1,941.73

Contractor's Fee Including Materials	$4,152

Key to Abbreviations
C.Y.–cubic yard   Ea.–each   L.F.–linear foot   Pr.–pair   Sq.–square (100 square feet of area)
S.F.–square foot   S.Y.–square yard   V.L.F.–vertical linear foot   M.S.F.–thousand square feet

STANDARD STRAIGHT-WALL KITCHEN

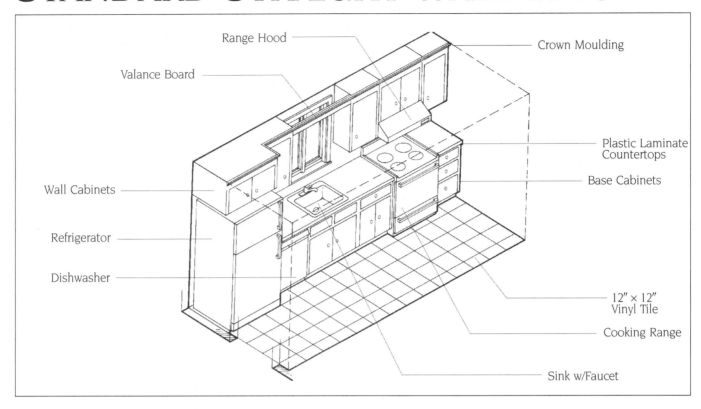

Range Hood — Crown Moulding

Valance Board —

Plastic Laminate Countertops

Wall Cabinets —

Base Cabinets

Refrigerator —

Dishwasher —

12″ × 12″ Vinyl Tile

Cooking Range

Sink w/Faucet

Because of limited floor or wall area, particularly in smaller houses or apartments, kitchen facilities must sometimes be confined to a single wall. Often, as with corridor-style kitchens, the components must be cleverly arranged to meet a pressing need for space. It is important in these cases, therefore, that you select the cabinets, appliances, and other materials with efficient use of space as one of the primary guidelines. Even a relatively basic kitchen renovation like this straight-wall facility requires careful planning and selection of materials to fit the particular area, food preparation needs, and your budget.

MATERIALS

When compared to larger facilities, this standard straight-wall kitchen plan is limited in floor, counter, and cabinet area, but it maintains an efficient arrangement nevertheless. The use of standard-grade products keeps the materials cost to a modest level, yet the result of the project should be an attractive new facility that will function well for many years. For most do-it-yourselfers, this project is challenging, but more manageable than larger kitchen renovations. If you choose

to do some of the work on your own, the cost for this installation can be reduced considerably.

Although the standard-grade cabinets included in this plan lack the deluxe paneled features of more expensive models, they are comparable in all other aspects. They are made of quality materials, include durable hardware, and come completely finished from the factory or shop. Before you purchase the cabinets, do some research, shop extensively, and know precisely what you are buying. Be wary of products that are suspiciously low in price; examine sample units to determine their materials and construction, and find out if auxiliary items like sufficient shelving and durable hardware are included in the quoted price. If you plan to install the cabinets on your own, get some advice before you start the job and work slowly. Beginners should leave the cabinet installation to a professional, but experts and moderately skilled intermediates should be able to handle this single-wall placement if they are given some professional advice before starting.

The surface coverings used in this kitchen project are economical materials that have proven to be attractive, durable, and

easily maintained in kitchen use. You can purchase the plastic laminate countertop as a prefabricated unit from the cabinet retailer or from suppliers who specialize in countertop manufacture. If you decide to do the countertop on your own, hold up on cutting the opening for the sink until you have consulted your plumber. He or she may prefer to do the entire sink installation, including the cutting of the opening, at one time. The vinyl flooring material is also a proven product for kitchen use. Many decorative options are available in both sheet and tile format. Sheet goods cost slightly more; and because they must be laid in single or several large pieces, they require more skill to install than tiles. The materials are comparable in durability, but the fact that sheet flooring is seamless makes it less prone to water penetration. Experts can lay the floor in this kitchen, and skilled intermediates can do the same if they are given professional guidance.

The appliances for this standard straight-wall kitchen have been selected with economy in mind. If space is not a problem, a standard-sized refrigerator and range should be placed at a workable distance from one another along the run of the facility. Undersized appliances are available for situations where space is a critical concern, but they are usually special-order units and cost more. If you can fit standard-sized appliances into the new facility, you should do so whenever possible. Be sure to allow for the extra cost of new electrical or plumbing rough-in necessitated by relocation or reconditioning of the old lines. The estimate for this plan includes plumbing rough-in for a double-bowl sink, but it does not allow for additional plumbing for relocation of the fixture. A smaller single-bowl sink can be installed in place of the double-bowl unit. By making this change in the plan, you would add space to the counter area while reducing the cost.

LEVEL OF DIFFICULTY

This kitchen remodeling project can be accomplished by all experts and those intermediates who are advanced in their building skills. Many different construction operations are required to complete the project, and all do-it-yourselfers should proceed slowly and seek advice before starting unfamiliar tasks. The installation of cabinets, countertop, and flooring, as well as plumbing and electrical work, requires knowledge, special skills and tools, and careful work. Taking on more than you can handle may result in costly damage to materials and the additional expense of hiring a contractor to correct faulty installations. Beginners should not attempt the cabinet or countertop placement, but they may be able to handle the other jobs, including the flooring, if they seek guidance along the way. They should add 100% to the professional time for the basic tasks, and more for the flooring work. Experts and intermediates should add 20% and 50%, respectively, to the time estimates for all tasks except the cabinet placement. They should add 50% and 100%, respectively, for that operation to allow for slower, more careful work.

WHAT TO WATCH OUT FOR

One way to save money on this and other kitchen remodeling projects is to remove the old facility on your own. The job may appear to be routine, but it does require skill in the use of tools and knowledge of fundamental building practices. Be especially careful in dismantling the old cabinets, and avoid doing careless damage to the walls and ceiling in the process. Take the necessary precautions of shutting off water and electrical supply lines to the kitchen before removing the sink, range hood, and other appliances. If you are unfamiliar with plumbing and electrical work, call a professional to do the disconnecting before you start the removal.

Standard Straight-Wall Kitchen

Description	Quantity/Unit	Labor-Hours	Material
Blocking, for mounting cabinets, 2 x 4	18 L.F.	0.6	7.78
Plywood, underlayment-grade, 3/8" thick, 4' x 8' sheets	128 S.F.	1.4	73.73
Flooring, vinyl composition tile, 12" x 12" x 1/16", plain	65 S.F.	1.0	78.00
Wall cabinets, 12" deep, 36" wide, economy	3 Ea.	2.1	450.00
Wall cabinet, 2-door, 12" high, 30" wide	1 Ea.	0.7	142.80
Base cabinets, 24" deep, 36" wide, economy	2 Ea.	1.6	508.80
Countertop, 1-1/4" thick Formica, w/backsplash	8.50 L.F.	2.3	78.03
Valance board over sink	4 L.F.	0.2	37.20
Trim, crown molding, stock pine, 9/16" x 3-5/8"	28 L.F.	0.9	51.07
Paint, ceiling & walls, primer	285 S.F.	1.1	13.68
Paint, ceiling & walls, 1 coat	285 S.F.	1.8	17.10
Paint, cornice trim, simple design, incl. puttying, primer	28 L.F.	0.3	0.67
Paint, cornice trim, simple design, one coat	28 L.F.	0.3	1.01
Cooking range, freestanding, 30" wide, one oven	1 Ea.	1.6	304.80
Hood for range, 2-speed, 30" wide	1 Ea.	2.0	42.60
Refrigerator, no frost, 19 C.F.	1 Ea.	2.0	600.00
Dishwasher	1 Ea.	2.5	304.80
Sink w/faucets & drain, SS, self-rimming, 30" x 21", single bowl	1 Ea.	2.9	296.40
Rough-in, supply, waste & vent for sink	1 Ea.	7.5	102.00
Totals		32.8	$3,110.47

Project Size	6'-6" x 14'	Contractor's Fee Including Materials	**$6,018**

Key to Abbreviations
C.Y.–cubic yard   Ea.–each   L.F.–linear foot   Pr.–pair   Sq.–square (100 square feet of area)
S.F.–square foot   S.Y.–square yard   V.L.F.–vertical linear foot   M.S.F.–thousand square feet

FLOOR COVERING OPTIONS
Cost per Square Foot, Installed

Description	
Oak Strip, prefinished	$9.70
Floating "Wood"	$6.05
Sheet Vinyl	$4.16
Quarry Tile	$8.25
Vinyl Tile	$2.40

SUMMARY

This remodeling project demonstrates how an efficient and economical kitchen facility can be installed on a single wall. Its basic, space-saving layout and modest cost make it a practical design for smaller homes and an approachable undertaking for do-it-yourselfers.

For other options or further details regarding options shown, see

> Deluxe straight-wall kitchen
>
> Electrical system: Light fixtures
>
> Electrical system: Receptacles
>
> Flooring
>
> Painting and wallpapering
>
> Standard window installation*

* In Exterior Home Improvement Costs

DELUXE STRAIGHT-WALL KITCHEN

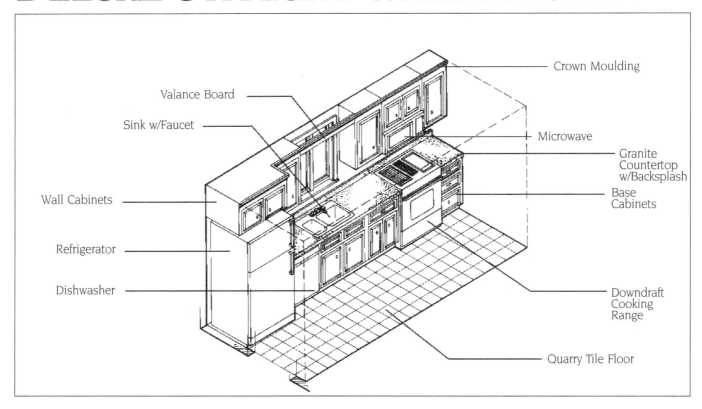

Crown Moulding

Valance Board

Sink w/Faucet

Microwave

Granite Countertop w/Backsplash

Base Cabinets

Wall Cabinets

Refrigerator

Dishwasher

Downdraft Cooking Range

Quarry Tile Floor

The deluxe features of this straight-wall kitchen facility enhance its appearance and convenience while maintaining the efficiency and durability standards required of all kitchens. Although this plan presents a small and relatively simple renovation design, its materials are expensive, and do-it-yourselfers should use discretion in determining which tasks they will undertake.

MATERIALS

The materials used in this deluxe kitchen design are top-of-the-line products that will enhance the appearance of the facility and provide many years of reliable service. Some substitutions can be used, but these modifications should be made with discretion and thought given to their aesthetic impact on the kitchen's appearance. Remember that kitchen products are in place for a long time, and economizing with lower-grade substitutes may lead to problems in the future. If you feel that you need guidance, consult a professional kitchen planner. The cost of such a service may save you unnecessary expense in materials and improve the efficiency of the facility.

The cabinets in this plan are prefabricated hardwood units with paneled cabinet doors. Because the cabinets establish the decor of the room, they are its main feature and are worthy of some advance planning and research. If you are installing the cabinets on your own, follow accepted carpentry methods and work slowly and carefully. Rushing this operation can result in misaligned components and, worse, damage to expensive cabinets.

The granite used for the countertop and the quarry tile used for the floor provide durable, easily maintained, and attractive surfaces. Other countertop options include hardwood, ceramic tile, and solid surfacing materials like Corian.

Quarry tiles come in rectangular shapes in sizes from around 4" per side up to 12". They are also available in hexagons, octagons, and other shapes – in a range of sizes. Colors run from earth tones to various grays. Remember that the flooring should be installed after all other kitchen components, except the refrigerator and freestanding stove, have been placed. The tiles should cover the floor surface under the plug-in appliances, but not the area under the cabinets.

The appliances for this straight-wall facility should be placed in locations that will make them conveniently accessible during their use. Before purchasing the refrigerator, for example, determine which design features of the appliance will best suit its location. Size, left- or right-hand door opening, and side-by-side or upper and lower freezer/refrigerator layout are all important considerations, especially in smaller kitchens. A space-saver microwave may fit above the range. If a built-in range is utilized, special cabinet and countertop planning is required, whereas a freestanding unit requires only a standard opening, so plan accordingly. A change from built-in to freestanding range or vice versa after the cabinets have been purchased will require a costly alteration. If you plan to install the downdraft-type range, you will need to arrange for the required venting to the exterior, either directly through the wall or up through the framing to the roof. This is not always simple. Be sure you can accomplish the modification before you invest in this type of range. A dishwasher is a modern convenience, but in a single-wall kitchen it will take up 30" of already precious base cabinet space. Therefore, you may wish to forgo this feature. If you do intend to install the dishwasher as a new or relocated facility, remember that it requires both plumbing and electrical rough-in and hook-ups at an increase in the cost of the installation.

LEVEL OF DIFFICULTY

Although the undertaking is challenging because of the wide variety of skills required to complete the installation, this kitchen's single-wall layout and small area make it a reasonable project for experts and competent intermediates. Remember that precise measurement, tight fits, and accurate alignment are required in all interior finish operations,

especially cabinet, countertop, and flooring placement. Also, bear in mind that you are working with deluxe materials that are expensive to replace or repair if you damage them with poor workmanship and carelessness. As a general rule, beginners should not attempt the cabinet installation and should double the estimated professional time for basic tasks. Experts and intermediates should add 20% and 50%, respectively, to the labor-hour estimates for the basic operation, and 50% and 100% for the cabinet placement. Specialized jobs like the tiling, electrical installation, and plumbing should be attempted only by individuals experienced in these areas of construction, or by accomplished do-it-yourselfers.

WHAT TO WATCH OUT FOR

Because the cabinet system for this kitchen project is basic in its single-wall design and limited in the number of units, its installation is a manageable undertaking for skilled do-it-yourselfers. If you follow a few basic guidelines, work patiently, and seek advice before you start and when any problems develop, your efforts can produce professional results and considerable savings in labor costs.

Here are a few suggestions to get you started. Thoroughly prepare the walls and floor in the cabinet area by making sure that they are solid, free from voids and obstructions, and reasonably plumb and level. Install the wall units first, making all fastenings temporary as you keep a constant check on tight abutments and the alignment of facings. Maintain plumb and level as you proceed, using shims and spacers where needed. Follow the same process for the base cabinets, giving special attention to irregularities or slope of the floor. Make all permanent fastenings with heavy wood screws or lag bolts after the entire system has been placed. Install the valance and trim after the wall units have been permanently placed. At least two workers are needed for the cabinet installation, so plan accordingly.

Deluxe Straight-Wall Kitchen

Description	Quantity/ Unit	Labor-Hours	Material
Blocking, for mounting cabinets, 2 x 4	18 L.F.	0.6	7.78
Plywood, underlayment-grade, 3/8" thick, 4' x 8' sheets	128 S.F.	1.4	73.73
Flooring, quarry tile 6" x 6", mud set	65 S.F.	7.4	226.98
Wall cabinets, 12" deep, 36" wide, hardwood	3 Ea.	3.0	841.68
Base cabinets, hardwood, 24" deep, 18" wide	1 Ea.	1.0	448.56
24" wide	1 Ea.	1.0	376.32
36" wide	1 Ea.	1.1	409.92
Granite countertop, average, 1-1/4" thick	8.50 L.F.	5.2	943.50
Valance board over sink	4 L.F.	0.2	37.20
Trim, crown molding, stock pine 9/16" x 3-5/8"	28 L.F.	0.9	51.07
Paint, primer, ceiling & walls	185 S.F.	0.7	8.88
Paint, 1 coat, ceiling & walls	185 S.F.	1.1	11.10
Paint, cornice trim, simple design, incl. puttying, primer	28 L.F.	0.3	0.67
Paint, cornice trim, simple design, one coat	28 L.F.	0.3	1.01
Cooking range, built-in, 30" downdraft type with grille	1 Ea.	1.6	840.00
Microwave	1 Ea.	1.0	492.00
Hood for range, 2-speed, 30" wide	1 Ea.	3.3	540.00
Refrigerator, no frost, side-by-side with ice/water dispenser	1 Ea.	5.3	3,330.00
Sink w/faucets & drain, SS, self-rimming, 33" x 22", dbl. bowl	1 Ea.	3.3	534.00
Dishwasher, built-in, 5 cycle	1 Ea.	1.5	308.40
Rough-in, supply, waste & vent for sink & dishwasher	2 Ea.	15.0	204.00
Totals		55.2	$9,686.80

Project Size	6'-6" x 14'	Contractor's Fee Including Materials	$16,371

Key to Abbreviations
C.Y.–cubic yard   Ea.–each   L.F.–linear foot   Pr.–pair   Sq.–square (100 square feet of area)
S.F.–square foot   S.Y.–square yard   V.L.F.–vertical linear foot   M.S.F.–thousand square feet

SUMMARY

This straight-wall kitchen plan demonstrates how a small facility can take on an elegant appearance as deluxe products are incorporated into its construction. The additional expense of its top-quality materials will be returned in many years of pleasant, practical, and dependable use.

For other options or further details regarding options shown, see

> *Electrical system: Light fixtures*
>
> *Electrical system: Receptacles*
>
> *Flooring*
>
> *Painting and wallpapering*
>
> *Standard straight-wall kitchen*
>
> *Standard window installation**

> ** In Exterior Home Improvement Costs*

STANDARD CORRIDOR KITCHEN

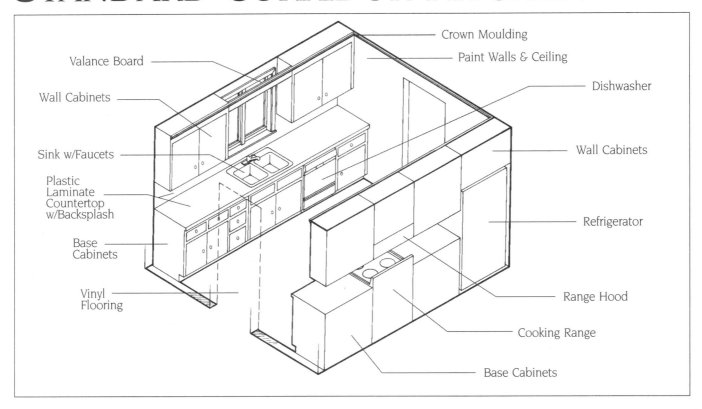

Crown Moulding

Paint Walls & Ceiling

Dishwasher

Wall Cabinets

Refrigerator

Range Hood

Cooking Range

Base Cabinets

Valance Board

Wall Cabinets

Sink w/Faucets

Plastic Laminate Countertop w/Backsplash

Base Cabinets

Vinyl Flooring

With the specialized materials and products that are available, an efficient kitchen facility can be installed in a remarkably small area. This standard corridor kitchen project demonstrates how limitations of size can be overcome to create an efficient and attractive facility at a reasonable cost. Carpentry, electrical, and plumbing tasks are all involved in this renovation. Also, because the kitchen area of any home performs a vital function, expediency in completing the project is an important consideration. If you plan to complete the project on your own, you might want to set aside a block of time to complete most of the work in a concentrated week's effort rather than spreading it out over several weekends.

MATERIALS

The materials used in this kitchen plan are standard-grade items, readily available from kitchen suppliers and building supply retailers. To remodel an existing kitchen completely, you may have to modify some of the suggested cabinet and appliance selections to suit your space limitations. Also, your choice of cabinets and appliances should be carefully determined, since the major

costs for any kitchen facility are focused on these components. Because the corridor kitchen layout allows for only two walls of work space, the design and arrangement of the cabinet system must be efficient. This plan recommends standard-sized wall and base units with depths of 12″ and 24″, respectively. These depths should be maintained wherever possible because other kitchen plumbing fixtures and appliances are designed around these standard dimensions. Altering the base cabinet depth, particularly, will cost you more money for custom or special order units and can cause additional expense and aggravation in selecting appliances, sinks, and other kitchen fixtures. If your kitchen area is too narrow to accommodate the 24″-deep base cabinets on both walls, seek advice and suggestions from a professional contractor before you begin the project. When making the choice between giving up floor space and giving up cabinet space, in most cases you will be better off with diminished floor space and standard-sized cabinets. The countertop, like the cabinets, should also be designed with efficiency in mind so as to produce a maximum amount of surface area on the two walls. Plastic laminate counter coverings are durable, easy to

clean, and economical, but other more expensive materials like ceramic tile can serve as attractive and functional substitutes. The trim work for the cabinets should be selected on the recommendation of the cabinet manufacturer.

Before the cabinets are installed, the ceiling, floor, and walls of the kitchen area have to be reconditioned and prepared. The expense incurred in this step of the project will depend largely on the amount of fundamental restoration required. Tearing out old cabinets and stripping the floors of old linoleum or tile are difficult, in that these tasks take time and require some know-how. Be sure to correct any structural or cosmetic deficiencies in the walls and floor. Any straightening or smoothing of existing rough ceiling, wall, or floor surfaces done at this stage of the project will expedite the rest of the installation. Remember, too, that if you plan to continue to use your kitchen during the period of the remodeling work, you will have to move appliances and temporarily rearrange the facility as the job progresses. If you are doing work on your own, plan to have someone available to lend assistance. Like cabinet systems, appliances vary significantly in quality and cost. The price estimates provided in this plan are based on medium-priced appliances with a limited number of operational accessories and convenience options. Because the recommended range, microwave, and refrigerator are freestanding models, just about any standard-sized unit can be installed. The two-speed, 30"-wide hood and exhaust fan unit is also a common item that can be purchased at most kitchen, electric, or building supply outlets. If possible, duct the exhaust vent to the outside. Remember that the cost for appliance installation will vary if the new unit is to be placed in a different location. Electric or gas ranges, for example, require supply connections that should be installed by qualified professionals.

LEVEL OF DIFFICULTY

Kitchen remodeling demands knowledge of and skill in many different building operations. Cabinet installation, plumbing, electrical work, and flooring placement are just some of the tasks that the intermediate has to contend with during the renovation process. Most beginners can handle the basic jobs like the ceiling, wall and floor preparation, and the painting and staining; but they should leave the specialized work to the carpenter, plumber, or electrician. Even moderately skilled intermediates should hire a professional to install the cabinets and countertop. Because kitchen cabinets involve substantial expense and specialized knowledge of installation procedures, even expert do-it-yourselfers might consider hiring a professional

for this job. Intermediates and experts should add about 50% and 20%, respectively, to the professional time estimate for all tasks, except the cabinet installation and plumbing work. If they intend to perform these operations, they should add 50 to 100 percent to the labor-hours. Beginners should attempt only those tasks mentioned earlier and plan to double the professional time for their work.

WHAT TO WATCH OUT FOR

If new underlayment cannot be placed directly on the existing floor, the difficult chore of stripping the old resilient floor will have to be accomplished. You can use several methods to perform this task, depending on the amount, type, and condition of the covering to be removed. If the area is relatively small, the old material can be peeled and chipped by hand with a flat bar or other blade-type instrument. If the old material has a backing, and some of it stays on the floor, hand or power sanding may be needed to get it up. Larger floors can be stripped in the same way, but you might look into renting a floor grinder if the job looks too big to do by hand. Before the new covering is laid, be sure that the old underlayment is fastened thoroughly and that all seams and voids have been filled with floor leveler and sanded smooth.

SUMMARY

Corridor kitchens, because of their two-wall cabinet layout, require both imagination and careful planning in a remodeling design. With the thoughtful arrangement of cabinets and appliances, a stylish and efficient facility can be created.

For other options or further details regarding options shown, see

> *Deluxe corridor kitchen*
> *Electrical system: Light fixtures*
> *Electrical system: Receptacles*
> *Flooring*
> *Painting and wallpapering*
> *Standard window installation**

* In Exterior Home Improvement Costs

Standard Corridor Kitchen

Description	Quantity/ Unit	Labor- Hours	Material
Blocking, for mounting cabinets, 2 x 4	38 L.F.	1.2	16.42
Plywood, underlayment-grade, 1/2" thick, 4' x 8' sheets	128 S.F.	1.4	98.30
Flooring, vinyl sheet goods, 0.125" thick	96 S.F.	3.8	411.26
Wall cabinets, 12" deep, 36" wide, 24" high, economy	6 Ea.	4.2	900.00
Base cabinets, 24" deep, 36" wide, economy	6 Ea.	4.7	1,526.40
Countertop, 1-1/4" thick, plastic laminate, w/backsplash	18 L.F.	4.8	165.24
Valance board over sink	4 L.F.	0.2	37.20
Trim, crown molding, stock pine, 9/16" x 3-5/8"	42 L.F.	1.3	76.61
Paint, ceiling & walls, primer	425 S.F.	1.7	20.40
Paint, ceiling & walls, 1 coat	425 S.F.	2.6	25.50
Paint, cornice trim, simple design, incl. puttying, primer	42 L.F.	0.5	1.01
Paint, cornice trim, simple design, one coat	42 L.F.	0.5	1.51
Cooking range, freestanding, 30" wide, one oven	1 Ea.	1.6	304.80
Microwave	1 Ea.	2.0	105.60
Hood for range, 2-speed, 30" wide	1 Ea.	2.0	42.60
Refrigerator, no frost, 19 C.F.	1 Ea.	2.0	600.00
Dishwasher	1 Ea.	5.0	338.40
Sink w/faucets & drain, SS, self-rimming, 33" x 22", dbl. bowl	1 Ea.	3.3	534.00
Rough-in, supply, waste & vent for sink	1 Ea.	7.5	102.00
Totals		50.3	$5,307.25

Project Size	11' x 11'-6"	Contractor's Fee Including Materials	$10,041

Key to Abbreviations
C.Y.–cubic yard   Ea.–each   L.F.–linear foot   Pr.–pair   Sq.–square (100 square feet of area)
S.F.–square foot   S.Y.–square yard   V.L.F.–vertical linear foot   M.S.F.–thousand square feet

DELUXE CORRIDOR KITCHEN

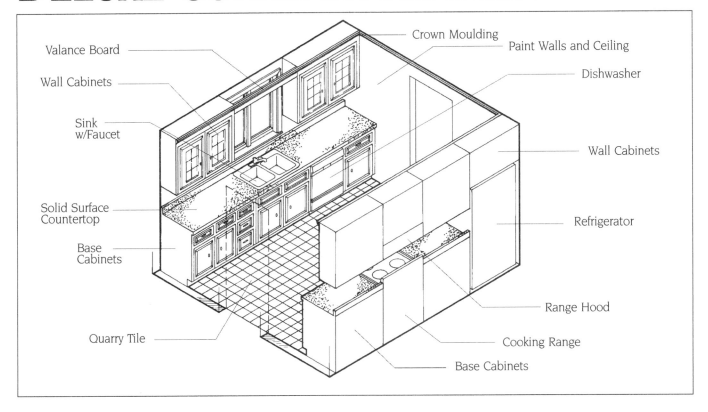

Valance Board — Wall Cabinets — Sink w/Faucet — Solid Surface Countertop — Base Cabinets — Quarry Tile — Crown Moulding — Paint Walls and Ceiling — Dishwasher — Wall Cabinets — Refrigerator — Range Hood — Cooking Range — Base Cabinets

The arrangement of walk-through or corridor kitchens provides the opportunity for the homeowner to install deluxe cabinets, appliances, and finish materials in a two-wall format. This remodeling plan utilizes upgraded appliances, decorative cabinet work, and other amenities, like a solid surface countertop, to enhance the appearance and overall quality of the kitchen. This renovation can make better use of the given space while adding value to your home.

MATERIALS

The materials used in the deluxe corridor kitchen plan consist of deluxe models of the fundamental appliances, cabinet system, plumbing fixtures, and finish materials found in most kitchen installations. As in the standard corridor kitchen plan, this two-wall kitchen design presents the challenge of limited space to the homeowner. For this reason, the materials have to be wisely chosen to ensure efficient use of both floor and work space.

The cabinets are the single most important item installed in any kitchen. Those chosen for this plan are top-quality

units with raised panel doors on both the wall and base systems. In addition to the superior-quality hardwood used in their manufacture, this deluxe cabinet system features roller drawer slides and other durable hardware fittings. The cost of this cabinet system will vary according to the choice of wood and modifications in its layout. If you are doing the job on your own, double-check all of your measurements before you order the cabinets, as miscalculations when dealing with expensive hardwood units like these can be disastrous. In fact, unless you are a skilled do-it-yourselfer, hire a professional to do the entire cabinet installation, including the preorder measuring.

The countertop and backsplash for this kitchen are made of solid surface material instead of the more commonly used plastic laminate. Other options include ceramic tile and stone.

In preparing the ceiling, floor, and walls to receive the new materials, follow accepted construction techniques and avoid any tendency to cut corners. Check the subfloor and its support system and the condition of the ceiling and walls after the old cabinets, appliances, and

flooring have been removed. If you see the need to recondition the existing ceiling, floor, or walls, make the repairs – even if it means adding an unexpected expense for drywall or plywood. Doing the job thoroughly may boost the cost a bit, but will save you money and headaches in the long run. Plan to paint the ceiling and walls as soon as possible after the old cabinets and appliances have been removed. The painting will go faster if it is done before the new appliances, cabinets, and floor are installed.

The appliances included in this kitchen plan complement the high quality cabinets, countertop, and floor covering. Two of the appliances – the range and the dishwasher – require special provision in the cabinet design because they are built-in units. The installation of both of these appliances should be performed by professionals, and extra costs should be figured in for the additional hookups. The range hood is another item that should be professionally installed. The expense for rough-in of the double-bowl sink has been included in the plan, but this cost will exceed the estimate if you plan to change the sink location.

LEVEL OF DIFFICULTY

The remodeling of a deluxe kitchen demands the same levels of expertise and versatility as those involved in a standard kitchen project, but the installer has the added responsibility and risk of working with more expensive materials. Tearing out the old facility, preparing the ceiling, floor, and walls, and painting are all tasks within the reach of beginners and intermediates. Removing the old cabinets and countertop can pose a challenge if these fixtures are tied into the walls, ceiling, or floor. The quarry tile flooring and routine plumbing and electrical work can be accomplished by experts who are familiar with the correct procedures. The cabinet installation should be undertaken only by a carpenter or an expert do-it-yourselfer with experience in this area. The countertop should be installed by a professional. The risk of damaging or misaligning a very costly system can prove too great for the inexperienced worker. Beginners should add 100% to the estimated time for all tasks that they have done before, and more if the job is new to them. Intermediates should add 50% to the labor-hours for the removal of the old facility, site reconditioning, placing the underlayment, and the painting. Experts should seek advice in the cabinet installation and should add 20% to the professional time for the rough work and the painting. If they tackle the quarry tile installation, they should add 50% to the estimated time to allow for a slow, methodical operation. All plumbing and electrical tasks should be done by a qualified tradesperson unless the homeowner is an expert with substantial experience in these areas.

WHAT TO WATCH OUT FOR

Cabinets in many older dwellings are not constructed in the contemporary modular design, and may be difficult to remove, often leaving holes and large voids that must be repaired before the new cabinets are installed.

Solid surface countertop materials must be installed by a professional. To find a contractor you might check with your local lumberyard, tile distributor, or a kitchen/bath designer.

Deluxe Corridor Kitchen

Description	Quantity/Unit	Labor-Hours	Material
Blocking, for mounting cabinets, 2 x 4	38 L.F.	1.2	16.42
Plywood, underlayment-grade, 1/2" thick, 4' x 8' sheets	128 S.F.	1.4	98.30
Flooring, quarry tile, 6" x 6", mud set	96 S.F.	11.0	335.23
Wall cabinets, 12" deep, 36" wide, hardwood	3 Ea.	3.0	841.68
Base cabinets, 24" deep, 36" wide, hardwood	3 Ea.	3.3	1,426.32
Countertop, solid surface, 1-1/4" thick, with 4" backsplash	18 L.F.	10.7	1,036.80
Valance board over sink	4 L.F.	0.2	37.20
Trim, crown molding, stock pine, 9/16" x 3-5/8"	42 L.F.	1.3	76.61
Paint, ceiling & walls, primer	300 S.F.	1.2	14.40
Paint, ceiling & walls, 1 coat	300 S.F.	1.9	18.00
Paint, cornice trim, simple design, incl. puttying, primer	42 L.F.	0.5	1.01
Paint, cornice trim, simple design, one coat	42 L.F.	0.5	1.51
Cooking range, built-in, 30" wide, one oven	1 Ea.	8.0	1,290.00
Hood for range, 2-speed, 30" wide	1 Ea.	3.3	540.00
Refrigerator, no frost, side-by-side with ice/water dispenser	1 Ea.	5.3	3,330.00
Sink, solid surface, integral with countertop	1 Ea.	0.7	60.60
Kitchen sink faucet	1 Ea.	0.8	60.00
Basket strainer with tail piece	1 Ea.		14.58
Dishwasher, built-in, 2 cycle	1 Ea.	5.0	338.40
Rough-in, supply, waste & vent for sink	2 Ea.	15.0	204.00
Totals		74.3	$9,741.06

Project Size	11' x 11'-6"	Contractor's Fee Including Materials	$17,432

Key to Abbreviations
C.Y.–cubic yard   Ea.–each   L.F.–linear foot   Pr.–pair   Sq.–square (100 square feet of area)
S.F.–square foot   S.Y.–square yard   V.L.F.–vertical linear foot   M.S.F.–thousand square feet

COUNTERTOP OPTIONS
Cost per Linear Foot, 24" deep, Installed

Description	
Plastic Laminate	$ 27.50
Solid Surface	$112.00
Granite	$140.00
Maple	$ 49.50
Marble	$ 50.00

SUMMARY

If your home warrants this improvement and if the deluxe additions are within your budget, you will benefit from the investment in the convenience, appearance, and durability of your kitchen.

For other options or further details regarding options shown, see

> *Electrical system: Light fixtures*
> *Electrical system: Receptacles*
> *Flooring*
> *Painting and wallpapering*
> *Standard corridor kitchen*
> *Standard window installation**

* In Exterior Home Improvement Costs

STANDARD L-SHAPED KITCHEN

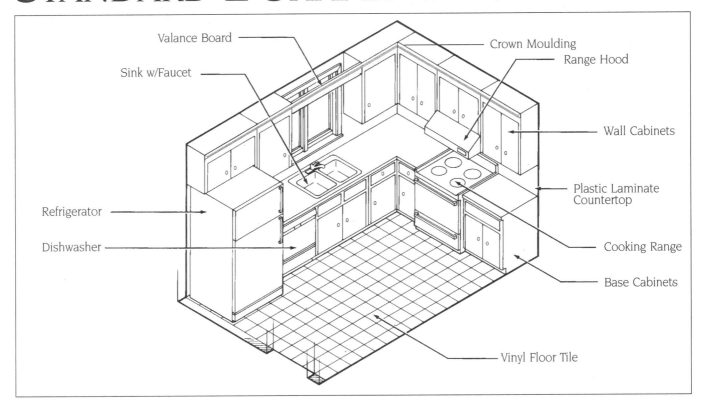

Valance Board
Crown Moulding
Range Hood
Sink w/Faucet
Wall Cabinets
Refrigerator
Plastic Laminate Countertop
Dishwasher
Cooking Range
Base Cabinets
Vinyl Floor Tile

An L-shaped layout consolidates the kitchen's work and appliance area into two walls and opens up floor space for either an eating area or an unobstructed entryway into an informal dining or living area. L-shaped facilities require an efficient layout to make the best use of a limited space. This standard L-shaped kitchen plan demonstrates how you can turn a small kitchen into a convenient and comfortable facility for a modest price.

MATERIALS

The materials included in this plan have been selected for their economy, durability, practicality, and attractive appearance. The cabinets are an economy model featuring quality hardware. Plastic-faced particleboard cabinets are available in many popular styles and colors. Their finishes are durable and easy to keep clean. They are the current popular economy choice. Laminate units are another option. They are durable, low-maintenance, and attractive. Be sure to investigate all of the options before purchasing the cabinets, as there is a wide choice in price and quality. After you have decided on the price range, design,

and preferred material, get the facts about the construction of the specific models under consideration. Quality cabinets will last longer and will also bring some immediate rewards in terms of their prominent effect on the appearance of the kitchen.

The coverings for the countertop, backsplash, and floor in the kitchen are conventional and popular materials. Plastic laminates make ideal kitchen countertops because they are durable, heat- and water-resistant, and easy to clean and maintain. They are also available in a wide variety of textures and colors. The flooring material suggested in the plan consists of 12" x 12" vinyl tiles, but other resilient tile and sheet flooring materials can be installed at varying costs. If you plan to do the flooring for the project, you will find the tiles easier to lay than sheet flooring. Be sure that the subfloor is in good condition and that the underlayment is adequately prepared before you begin the tile placement. Follow carefully the adhesive manufacturer's guidelines and recommendations during the installation process. Also, remember that the finished floor should be laid under all appliances, but not under the cabinets.

If the floor of the L-shaped kitchen terminates at the opening of the entryway, make provisions, whenever possible, to avoid a height discrepancy between the edges of the abutting floors.

The appliances in this kitchen include a standard, mid-priced cooking range, a 19-cubic-foot refrigerator, and a standard 2-cycle dishwasher. The range and refrigerator are freestanding models that can be purchased at the homeowner's convenience and put into place after the remodeling work is finished. However, if the range is a gas model, it should be professionally installed. Because the dishwasher is built in and needs professional electrical and plumbing hook-ups, it should be installed during the construction process, after the countertop has been put in place. Additional expense will have to be figured in for the electrical and/or plumbing rough-in if the range hood or sink have to be relocated.

LEVEL OF DIFFICULTY

This kitchen plan is a manageable project for expert do-it-yourselfers, but it is too advanced for most beginners and intermediates. Like any other kitchen renovation, the critical task of installing the cabinets is the most difficult undertaking. Since this kitchen involves two relatively short runs of cabinet units, installation is simple; however, the alignment of the inside corner cabinets requires precise and skillful workmanship. The plumbing and electrical tasks, especially rough-in relocations or new hook-ups, should be performed by a professional unless you have a good deal of experience in these areas. Experts should add 20%-30% to the professional time for all aspects of this project that they are prepared to undertake.

Intermediates with limited remodeling experience should plan on an additional 50% for any rough work, like the wall and ceiling reconditioning and painting and the placing of underlayment. If they tackle the tile floor installation, they should double the professional time estimate. Beginners should not attempt the cabinets, countertop, plumbing, and electrical installations, and they should add 100% to the estimated labor-hours for all other tasks, except the finished flooring. They should triple the estimated time for this job and seek professional instruction and advice before starting.

WHAT TO WATCH OUT FOR

One of the most neglected components of kitchen installations is the range hood and its accompanying exhaust fan and filtering system. If at all possible, the unit should vent to the outside of the house to rid the kitchen area of cooking odors, smoke, and moisture. If the hood and fan are located on an outside wall, the venting is easily accomplished directly through the wall. Most range hoods come with standard fittings for this type of installation, including a weathertight external port. If the hood is located on an interior wall, duct work should be run to the closest or most accessible outside wall. You may have to use some ingenuity to hide the duct in the wall cabinets or in the back of a closet on the opposite side of the partition, but it can be done. Remember that the labor and materials costs will be higher, though, for this extra work. If external venting cannot be accomplished, then make sure that the hood is equipped with the correct filters for a non-vented model.

SUMMARY

A more efficient use of space is the primary goal of this standard L-shaped kitchen renovation. At the same time, it incorporates various components that improve the appearance and convenience of your kitchen and add value to your home. Because some of the work can be done by the homeowner, this project can also be economical.

For other options or further details regarding options shown, see

> *Deluxe L-shaped kitchen*
>
> *Flooring*
>
> *Electrical system: Light fixtures*
>
> *Electrical system: Receptacles*
>
> *Island unit*
>
> *Painting and wallpapering*
>
> *Range hood*
>
> *Standard window installation**

* In Exterior Home Improvement Costs

Standard L-Shaped Kitchen

Description	Quantity/ Unit	Labor-Hours	Material
Blocking, for mounting cabinets, 2 x 4	32 L.F.	1.0	13.82
Plywood, underlayment-grade, 1/2" thick, 4' x 8' sheets	96 S.F.	1.1	73.73
Floor tile, 12" x 12" x 1/8" thick, vinyl composition, solid color	72 S.F.	1.2	387.07
Wall cabinets, 12" deep, 36" wide, 224" high, economy	5 Ea.	3.5	750.00
Base cabinets, 24" deep, 36" wide, economy	4 Ea.	3.2	1,017.60
Countertop, 1-1/4" thick, plastic laminate, w/backsplash	13 L.F.	3.5	119.34
Valance board over sink	4 L.F.	0.2	37.20
Trim, cornice molding, stock pine, 9/16" x 2-1/4"	20 L.F.	0.5	20.64
Paint, ceiling & walls, primer	250 S.F.	1.0	12.00
Paint, ceiling & walls, 1 coat	250 S.F.	1.5	15.00
Paint, cornice trim, simple design, incl. puttying, primer	20 L.F.	0.3	0.48
Paint, cornice trim, simple design, one coat	20 L.F.	0.3	0.72
Cooking range, freestanding, 30" wide, one oven	1 Ea.	1.6	304.80
Hood for range, 2-speed, 30" wide	1 Ea.	2.0	42.60
Refrigerator, no frost, 19 C.F.	1 Ea.	2.0	600.00
Dishwasher, 2 cycle	1 Ea.	2.5	304.80
Sink w/ faucets & drain, SS, self-rimming, 43" x 22", dbl. bowl	1 Ea.	3.3	618.00
Rough-in, supply, waste & vent for sink	1 Ea.	7.5	102.00
Ceiling fixture, pendant, 150 watt	1 Ea.	0.4	140.40
Totals		36.6	$4,560.20

Project Size	8' x 12'	Contractor's Fee Including Materials	$8,263

Key to Abbreviations
C.Y.–cubic yard   Ea.–each   L.F.–linear foot   Pr.–pair   Sq.–square (100 square feet of area)
S.F.–square foot   S.Y.–square yard   V.L.F.–vertical linear foot   M.S.F.–thousand square feet

DELUXE L-SHAPED KITCHEN

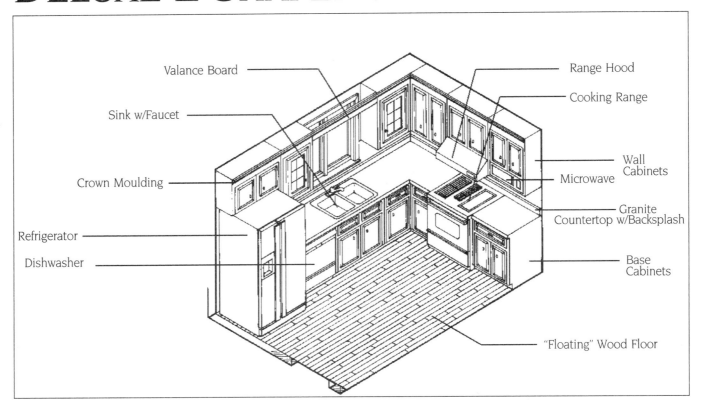

Valance Board — Range Hood

Sink w/Faucet — Cooking Range

Crown Moulding — Wall Cabinets

Microwave

Refrigerator — Granite Countertop w/Backsplash

Dishwasher — Base Cabinets

"Floating" Wood Floor

Corner, or L-shaped, kitchen arrangements are often used in situations where the food preparation area is adjacent to or clearly visible from the eating area. For example, in some cases, the open floor area of the kitchen itself is occupied by a table and chairs; in other situations, one or both of the vacant walls may be open to a remote eating or living area. Because of the visibility of the cabinets and appliances, attractive, deluxe products may be chosen to enhance the kitchen's appearance. This kitchen plan demonstrates how a conventional corner or L-shaped facility can be upgraded to deluxe standards.

MATERIALS

Like other deluxe kitchen plans included in this section, the L-shaped model utilizes top-of-the-line products in place of economical, standard-grade components. The homeowner's choice of these top-quality cabinets, appliances, and finish materials should be based on several factors, including budget, the decor and furnishings of other rooms in the house, and personal preference. There is room for compromise, which is to

say that a kitchen facility does not have to be exclusively standard or deluxe.

This L-shaped kitchen plan includes top-quality prefabricated hardwood wall and base cabinets. Paneled doors and drawer facings are common in deluxe cabinet systems, but other options are also available. Plastic laminates, for example, provide durable, easily-maintained surface coverings for quality cabinets. Remember that kitchen cabinets are permanent installations that you will use daily for many years, so do some research and see what is available before you buy. After you have purchased the cabinets, it is best to have them installed by a professional. Only expert do-it-yourselfers who have done extensive cabinet work should consider tackling this part of the project.

The flooring and countertop surface coverings also involve some important decisions. Like the cabinets, these surfaces require materials that are functional, attractive, and complementary to the color and design of the rest of the kitchen. We have selected a "floating" floor for this project because it combines the look and feel of natural wood plank with a highly durable and easily maintained surface. Floating floor systems

utilize wood composite tongue-and-groove boards, faced with a wood-patterned plastic laminate 30 times harder than the product used for countertops. These boards are glued together and installed over a thin sponge-foam base sheet. Placing the floor requires patience and a reasonable level of skill, but should be no problem for most intermediates. Like any other flooring job, one of your first and most important concerns should be the thorough inspection and proper reconditioning and preparation of the subfloor.

Floating floors can be installed over existing flooring if the surface is smooth and solid. Fill any holes or cracks with a liquid floor leveler. Follow the manufacturer's installation instructions carefully, and use all specified materials to preserve the manufacturer's warranty.

Ceramic tile provides a deluxe alternative to the commonly used plastic laminate countertop and backsplash material. Like plastic laminate, it is durable, low-maintenance, and attractive. With the right color grout, ceramic tile gives the countertop a neat, up-to-date appearance. Remember that because tiled countertops must be fabricated on the site, the charge for their installation is significantly more than for shop-prepared plastic laminate units. Other countertop materials to consider include marble, granite, wood, and solid surface composites.

The appliances included in the deluxe, L-shaped kitchen include a top-of-the-line refrigerator, dishwasher, and built-in range. Many options are available in the selection of these items, depending on the size of the kitchen area and personal preference. Be sure to allow for the additional expense of appliance installations that require relocation of rough-in or new hookups. Call in a plumber or electrician for all major plumbing or electrical tasks.

LEVEL OF DIFFICULTY

Because of the high cost of the deluxe appliances and other expensive materials used in this kitchen, most beginners and intermediates should hire a remodeling specialist or general contractor to do the better part of the work. Removing the old cabinets, stripping the floor, and preparing the walls and ceiling are jobs that the homeowner can accomplish to save on labor costs, but even these tasks require knowledge of tools and some remodeling experience. Get some advice from a professional or an expert do-it-yourselfer before you attempt any operations that are new to you. Also, removal of the old cabinets and appliances usually requires two workers, so plan accordingly and line up some help well in advance of the start of the project. Although the new floor should be a comfortable task for intermediate and expert do-it-yourselfers, the cabinet and countertop installations for this renovation should be left to an experienced tradesperson. Even accomplished do-it-yourselfers should consider restricting their efforts to tasks other than these. As a general rule, beginners should add at least 100% to the labor-hour estimates for the jobs that they tackle; intermediates, 50%-70%; and experts, 20%.

WHAT TO WATCH OUT FOR

Corner cabinets can waste considerable space if built and installed in the conventional manner. There are several ways to reclaim some or all of this often wasted cabinet area. One is to install a "lazy Susan" or rotating cabinet fixture to provide shelf and storage space that is immediately accessible. These are commonly installed in base corner units, but they are a special-order item and will add to the cost of the cabinets.

SUMMARY

If the value of your home is in keeping with this project and the renovation suits your budget, this plan is a good way to increase the efficiency and to improve the appearance of your L-shaped kitchen. The cost of top-quality components and professional installation is an investment in a highly visible and heavily utilized area of your home.

For other options or further details regarding options shown, see

> *Deluxe corridor kitchen*
>
> *Deluxe U-shaped kitchen*
>
> *Electrical system: Light fixtures*
>
> *Electrical system: Receptacles*
>
> *Flooring*
>
> *Island unit*
>
> *Painting and wallpapering*
>
> *Standard L-shaped kitchen*
>
> *Standard window installation**

> * *In Exterior Home Improvement Costs*

Deluxe L-Shaped Kitchen

Description	Quantity/Unit	Labor-Hours	Material
Blocking, for mounting cabinets, 2 x 4	32 L.F.	1.0	13.82
Plywood, underlayment-grade, 1/2" thick, 4' x 8' sheets	96 S.F.	1.1	73.73
Flooring, floating wood strip	72 S.F.	4.3	290.30
Wall cabinets, 12" deep, 36" wide, hardwood	5 Ea.	4.9	1,402.80
Base cabinets, 24" deep, 36" wide, hardwood	4 Ea.	4.4	1,901.76
Granite countertop, average, 1-1/4" thick	13 L.F.	8.0	1,443.00
Valance board over sink	4 L.F.	0.2	37.20
Trim, cornice molding, stock pine, 9/16" x 2-1/4"	20 L.F.	0.5	20.64
Paint, ceiling, primer	96 S.F.	0.4	4.61
Paint, ceiling, 1 coat	96 S.F.	0.6	5.76
Paint, cornice trim, simple design, incl. puttying, primer	20 L.F.	0.3	0.48
Paint, cornice trim, simple design, one coat	20 L.F.	0.3	0.72
Cooking range, 30" wide, downdraft with grille	1 Ea.	2.5	1,092.00
Hood for range, 2-speed, 30" wide	1 Ea.	3.3	540.00
Refrigerator, no frost, side-by-side with ice/water dispenser	1 Ea.	5.3	3,330.00
Sink w/faucets & drain, SS, self-rimming, 43" x 22", dbl. bowl	1 Ea.	3.3	618.00
Dishwasher, built-in, 7 cycle	1 Ea.	1.5	338.40
Undercabinet lights, 24" fluorescent strip	3 Ea.	1.0	185.40
Pendant lights	2 Ea.	0.8	280.80
Rough-in, supply, waste & vent for sink & dishwasher	2 Ea.	15.0	204.00
Totals		58.7	$11,783.42

Project Size	8' x 12'	Contractor's Fee Including Materials	$19,593

Key to Abbreviations
C.Y.–cubic yard   Ea.–each   L.F.–linear foot   Pr.–pair   Sq.–square (100 square feet of area)
S.F.–square foot   S.Y.–square yard   V.L.F.–vertical linear foot   M.S.F.–thousand square feet

STANDARD U-SHAPED KITCHEN

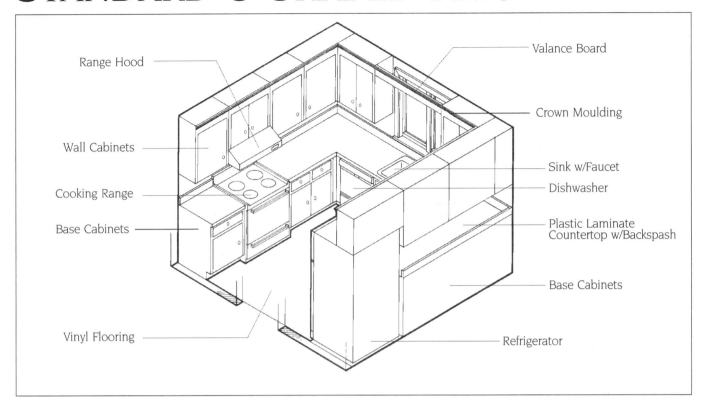

Range Hood

Wall Cabinets

Cooking Range

Base Cabinets

Vinyl Flooring

Valance Board

Crown Moulding

Sink w/Faucet

Dishwasher

Plastic Laminate
Countertop w/Backsplash

Base Cabinets

Refrigerator

One of the most efficient layouts for kitchen design is the U-shaped kitchen. All three walls of the facility are equipped with upper and lower cabinet units, with the range, sink, and refrigerator each on separate walls. This triangular setup of primary components is often recommended by experts as one of the best operational layouts for the kitchen, providing that there are not more than 7 or 8 feet between major appliances. If your kitchen area has the necessary space and shape requirements, this U-shaped layout can be a most efficient arrangement.

MATERIALS

The cabinet system in this kitchen plan is comprised of standard, economy wall and base units, complete with appropriate trim. Because many cabinet options are available, it pays to do some research and to shop for a system that meets decorative and durability standards while remaining within budget limitations. Look into the fine points of the cabinet units, including the quality of their construction, type of hardware, quality of finish, and the cost of installing them. If the retailer does not provide an

installation service, a qualified carpenter or kitchen remodeler can do the work. If you plan to install them on your own, consider the expense of the materials and the skillful placement required for this job, and work slowly and patiently. Do-it-yourselfers who are not experienced in cabinet work should get professional advice before they start.

The plastic laminate countertop, vinyl flooring, and painted ceiling and wall surfaces are conventional kitchen finishes that will provide years of service. Be sure to coordinate the design and color of these surface materials with the cabinets and appliances before the project starts. Following a master plan, which includes the specific complementary colors and designs for all components, ensures uniformity of appearance and predictable materials costs. Selecting and buying as you go can cause piecemeal results as well as unanticipated fluctuations in the expense of materials. Whenever possible, do the painting of the ceiling and walls before the cabinets, countertop, and floor are in place. Painting at the end of the project will take longer and require more brushwork, covering, and taping around the finished surfaces.

Be sure that the subfloor has been thoroughly reconditioned before the underlayment and vinyl flooring are placed. If the existing underlayment is in good condition after the old surface flooring has been removed, leave it in place and prepare it for the new vinyl. If the existing underlayment is not in good condition, you will have to tear up some or all of it and replace it. Double-check the thickness of the old underlayment before you purchase the new material.

The appliances in this standard U-shaped kitchen include a free-standing range, standard-sized refrigerator, and built-in dishwasher. The estimated prices for these items are based on white-finish models in the middle of the price spectrum. If you want the range and refrigerator in color, plan on an additional 10% over the standard models. Before selecting colored models, check to see that the dishwasher panel is available in the same color. If the panel has to be special ordered in the appropriate color, the cost will be higher. Expenses for the basic plumbing rough-in and sink trim are included in the estimate, but relocation costs, whether plumbing or electrical, have not been figured in. Any major electrical and plumbing work should be done by qualified professionals.

LEVEL OF DIFFICULTY

Remodeling a kitchen is a big undertaking for the most experienced do-it-yourselfer. Planning the new facility, removing the old one, reconditioning the site, and installing the new materials require many different remodeling skills. The various steps in the renovation process should be carefully coordinated in a sequence that will minimize delays and inconvenience. The most challenging task of the project is the cabinet planning and installation. Accurately measuring the linear footage of your space and determining the correct number of standard-sized cabinets are critical tasks at the outset. The subsequent installation of the system requires thorough carpentry know-how, and thus only experts should tackle this job. Even they should consider hiring a professional if they are inexperienced in cabinet placement. Experts should add 50% to the estimated labor-hours for the countertop, cabinets, and flooring operations and tack on 20% for all other jobs. Intermediates and beginners should use discretion in attempting any unfamiliar tasks and add 50% and 100%, respectively, to basic work like reconditioning the site and painting.

WHAT TO WATCH OUT FOR

The ceiling height for this kitchen plan is the standard 7'-6" found in most modern homes. Older homes, and even some newer ones, often have higher ceilings that require the placement of a soffit to finish the space between the top of the cabinets and the ceiling. This element is often overlooked when planning the cabinet system and may become an expensive extra if it has not been included in the plan and installation agreement with the retailer. Be sure to clarify this issue before you finalize the cabinet and installation contracts. Another option is to leave the soffit area above the cabinets open, leaving a space for storage or decorative options. Most cabinet manufacturers sell crown moulding or other trims that can be added to the tops of cabinets for a finished look.

SUMMARY

The standard U-shaped kitchen is a functional, attractive, and economical food preparation facility. Remodeling or rearranging an old three-wall kitchen with new products can provide years of dependable and enjoyable use while adding value to your home.

For other options or further details regarding options shown, see

> *Deluxe U-shaped kitchen*
> *Electrical system: Light fixtures*
> *Electrical system: Receptacles*
> *Painting and wallpapering*
> *Standard window installation**
> *Vinyl sheet flooring*
>
> ** In Exterior Home Improvement Costs*

Standard U-Shaped Kitchen

Description	Quantity/ Unit	Labor- Hours	Material
Blocking, for mounting cabinets, 2 x 4	50 L.F.	1.6	21.60
Plywood, underlayment-grade, 1/2" thick, 4' x 8' sheets	128 S.F.	1.4	98.30
Flooring, vinyl sheet goods, 0.093" thick	85 S.F.	3.0	204.00
Wall cabinets, 12" deep, 36" wide, economy	7 Ea.	4.9	1,050.00
Base cabinets, 24" deep, 36" wide, economy	5 Ea.	3.9	1,272.00
Countertop, 1-1/4" thick Formica w/backsplash	21 L.F.	6.0	579.60
Valance board over sink	4 L.F.	0.2	37.20
Trim, cornice molding, stock pine, 9/16" x 2-1/4"	30 L.F.	0.8	30.96
Paint, ceiling & walls, primer	320 S.F.	1.3	15.36
Paint, ceiling & walls, 1 coat	320 S.F.	2.0	19.20
Paint, cornice trim, simple design, incl. puttying, primer	30 L.F.	0.4	0.72
Paint, cornice trim, simple design, one coat	30 L.F.	0.4	1.08
Cooking range, freestanding, 30" wide, one oven	1 Ea.	1.6	304.80
Hood for range, 2-speed, 30" wide	1 Ea.	2.0	42.60
Refrigerator, no frost, 19 C.F.	1 Ea.	2.0	600.00
Sink w/faucets & drain, SS, self-rimming, 43" x 22", dbl. bowl	1 Ea.	3.3	618.00
Dishwasher, built-in, 2 cycle	1 Ea.	2.5	304.80
Rough-in, supply, waste & vent for sink & dishwasher	2 Ea.	15.0	204.00
Undercabinet light, 24" fluorescent strip	4 Ea.	1.3	247.20
Pendant light, 150 watt	1 Ea.	0.4	140.40
Totals		54.0	$5,791.82

Project Size	9'-6" x 10'-6"	Contractor's Fee Including Materials	$10,888

Key to Abbreviations
C.Y.–cubic yard   Ea.–each   L.F.–linear foot   Pr.–pair   Sq.–square (100 square feet of area)
S.F.–square foot   S.Y.–square yard   V.L.F.–vertical linear foot   M.S.F.–thousand square feet

DELUXE U-SHAPED KITCHEN

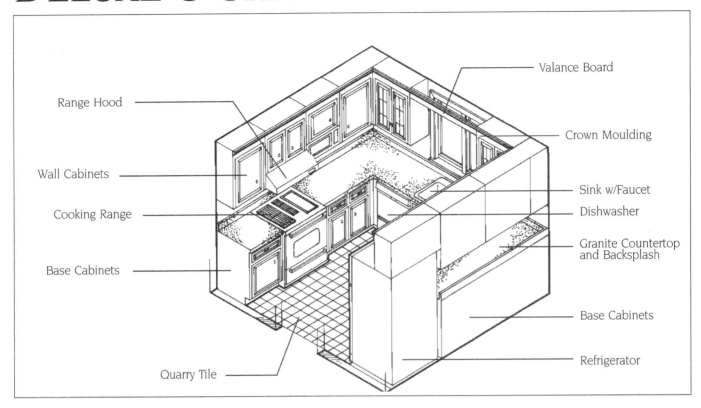

Range Hood

Wall Cabinets

Cooking Range

Base Cabinets

Quarry Tile

Valance Board

Crown Moulding

Sink w/Faucet

Dishwasher

Granite Countertop and Backsplash

Base Cabinets

Refrigerator

The deluxe facility presented in this plan demonstrates how a basic three-wall kitchen layout can take on an elegant appearance while meeting the practical demands of the kitchen. Amenities like a premium quality range and granite countertop add to the cost of the renovation, but they return to the homeowner benefits of appearance and efficiency, while increasing the value of the home. Although this kitchen is more expensive to install than a standard facility of comparable size, homeowners can reduce the costs by doing some of the work on their own.

MATERIALS

The materials and fixtures in this deluxe facility are the highest quality of ready-made components. The cabinet system, surface coverings, and appliances are available in designs and colors that will complement any decor. Seek the advice of a kitchen expert who can suggest the best combinations of materials and save you unnecessary expense and frustration.

The selection of materials for the kitchen should begin with the cabinets, since they serve as the focal point of the new room.

Deluxe kitchens usually include paneled or decorative hardwood cabinets that are durably constructed and carefully finished, but laminates and softwoods are also commonly installed at varying costs. When shopping for the cabinet systems, be sure to get several estimates on identical or similar units. By reducing the number of variables, you will be able to compare prices more accurately. Be sure to get an itemized written agreement that includes the basic units and all of the auxiliary cabinet materials. If installation of the cabinets is part of the contract, be sure to include in the agreement the labor cost for installing any trim items. Unless you are an expert do-it-yourselfer, the cabinets for the U-shaped kitchen should be placed by a professional.

The kitchen floor and countertop should match the deluxe quality of the cabinet system. The countertop material suggested in this plan is granite, but plastic laminate is an attractive, less expensive substitute. Other deluxe materials include marble, ceramic tile, and solid surface composites.

Selecting the color and design of the floor tile is an important decision, but you will also want to make sure that the

color of the grouting coordinates well with the tile and the room. The preparation and painting of the ceiling and walls should be completed before any other finish materials are placed.

If you are upgrading the present facility from standard to deluxe, additional electrical and, possibly, plumbing work will be required at extra cost. Also, electric-to-gas and gas-to-electric range conversions involve considerable work by qualified tradespeople, so plan accordingly. The refrigerator and dishwasher should not require additional installation costs, providing that they are not relocated and the rough-in and hookups are already in place. The ice maker or cold water option for a deluxe refrigerator requires extra plumbing to connect the water supply tubing. The conventional range hood has been replaced by an under-counter exhaust and vent system. These units are an expensive option on some built-in ranges, but they work more efficiently than conventional hoods and open up more wall cabinet space above the range. Because all of these appliances demand professional quality installations, do-it-yourselfers who are not experienced in electrical and plumbing work should hire capable people to perform the necessary tasks.

LEVEL OF DIFFICULTY

The basic tasks of removing the old facility, reconditioning and preparing the site for the new products, and applying paint and stain can be accomplished by beginners if they are given ample guidance as they proceed with these operations.

They should add 100% to the professional time estimates for these jobs. Intermediates should add 50% to the labor-hours for the same tasks and plan on doubling the professional time for the flooring. The installation of tile flooring and cabinets should be left to experts and tradespeople. Experts should add 50% to the estimated labor-hours for the flooring work, at least 75% for the cabinet installation, and 20% for the more basic operations. Except in the case of experts

who have h
experienc
work sh
profes
be in

Ou

Granite provi
and easily maintai
countertops, but it does
One is that the material itse
expensive than ceramic tile and p
laminates. Also, because the material is
hard and so heavy, the labor to prepare
and install it are higher than for the other
materials. Granite countertops, especially
light colors, require resealing every few
years to keep them stain resistant.

Deluxe U-Shaped Kitchen

Description	Quantity/Unit	Labor-Hours	Material
Blocking, for mounting cabinets, 2 x 4	50 L.F.	1.6	21.60
Plywood, underlayment-grade, 1/2″ thick, 4′ x 8′ sheets	128 S.F.	1.4	98.30
Flooring, quarry tile, 6″ x 6″, mud set	85 S.F.	9.7	296.82
Wall cabinets, 12″ deep, 36″ wide, hardwood	7 Ea.	6.9	1,963.92
Base cabinets, 24″ deep, 36″ wide, hardwood	5 Ea.	5.5	2,377.20
Granite countertop, average, 1-1/4″ thick	21 L.F.	12.9	2,331.00
Valance board over sink	4 L.F.	0.2	37.20
Trim, cornice molding, stock pine, 9/16″ x 2-1/4″	30 L.F.	0.8	30.96
Paint, ceiling, primer	100 S.F.	0.4	4.80
Paint, ceiling, 1 coat	100 S.F.	0.6	6.00
Paint, cornice trim, simple design, incl. puttying, primer	30 L.F.	0.4	0.72
Paint, cornice trim, simple design, one coat	30 L.F.	0.4	1.08
Cooking range, built-in, 30″ wide, downdraft with grille	1 Ea.	1.6	840.00
Hood for range, 2-speed, 30″ wide	1 Ea.	3.3	540.00
Microwave	1 Ea.	1.0	492.00
Refrigerator, 2 door, side-by-side with ice/water dispenser	1 Ea.	5.3	3,330.00
Sink w/faucets & drain, SS, self-rimming, 43″ x 22″, dbl. bowl	1 Ea.	3.3	618.00
Dishwasher, built-in, 7 cycle	1 Ea.	1.5	338.40
Rough-in, supply, waste & vent for sink & dishwasher	2 Ea.	15.0	204.00
Undercabinet lights, 24″ fluorescent strip	4 Ea.	1.3	247.20
Ceiling light, recessed, 100 watt	2 Ea.	2.0	133.20
Totals		75.1	$13,912.40

Project Size	9′-6″ x 10′-6″		Contractor's Fee Including Materials	$23,396

Key to Abbreviations
C.Y.–cubic yard Ea.–each L.F.–linear foot Pr.–pair Sq.–square (100 square feet of area)
S.F.–square foot S.Y.–square yard V.L.F.–vertical linear foot M.S.F.–thousand square feet

CABINET OPTIONS
Cost per Linear Foot, Installed

Description	
Plastic Laminate	$168.00
Thermo-Foil	$189.00
Oak	$205.00
Maple	$221.00

SUMMARY

Upgrading a U-shaped kitchen to deluxe standards is a challenging undertaking that requires a considerable commitment of both time and money by the homeowner. With some do-it-yourself cost cutting, however, the renovation can be completed with professional results and at somewhat reduced expense.

For other options or further details regarding options shown, see

Electrical system: Light fixtures

Electrical system: Receptacles

Flooring

Standard U-shaped kitchen

Standard window installation*

* In Exterior Home Improvement Costs

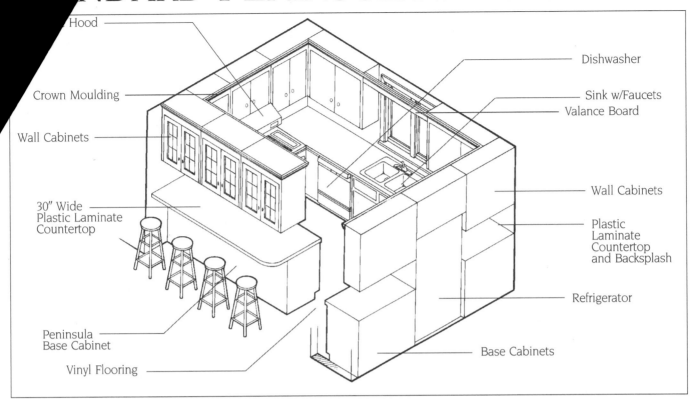

Hood — Dishwasher

Crown Moulding — Sink w/Faucets — Valance Board

Wall Cabinets — Wall Cabinets

30" Wide Plastic Laminate Countertop — Plastic Laminate Countertop and Backsplash

Refrigerator

Peninsula Base Cabinet

Vinyl Flooring — Base Cabinets

A peninsula kitchen design offers an attractive and functional alternative to conventional layouts. Although it requires more floor space, cabinets, and countertop area, this configuration pays off in terms of its efficiency and the bonus of a built-in eating or serving area.

MATERIALS

Even though this kitchen plan features standard-grade components, it is still an expensive project because all four sides of the kitchen are being equipped. Whether you are redoing an old peninsula area or creating a new one from a U-shaped kitchen, the practicality and convenience of this layout can make the extra investment worthwhile.

The design of the cabinets on the three walls of the "U" follows the basic layout of 12"-deep wall units and 24"-deep base cabinets, with standard cutouts for the various appliances and windows. The peninsula area provides the opportunity for some extra options in cabinet planning. While these modifications might boost your investment at the start, they are far easier to install at this stage than they would be later. The overhead

peninsula cabinets, for example, can be designed with doors opening to both sides of the divider. The base peninsula cabinet can also be cleverly designed to provide for the greatest degree of utility. The side facing away from the kitchen is generally left solid, but the side facing the kitchen can be imaginatively designed with many possible combinations of drawers and storage cabinets. These peninsula cabinet extras do increase the cost of the unit significantly, but the convenience they add to the kitchen area may be worth it. Before you determine the design, materials, and specific layout of the cabinets for this kitchen plan, compile a list of the features that you want in the cabinet system and make a rough drawing. Then, sit down with a kitchen specialist and plan the system. Because the cabinet work for this kitchen is so complex, it's best not to attempt the plan entirely on your own.

The choice of countertop and floor coverings should also be given careful consideration. Plastic laminate is still one of the best and most economical kitchen countertop materials because of its water- and heat-resistant properties, easy maintenance, and durability. The color and design of the floor should complement that of the countertop and

cabinets. Vinyl tiles or sheet goods are a good choice, as they come in a wide range of designs and colors and are reasonably priced. Before the finished floor is laid, make sure that the subfloor and its supports are in good condition. The area under the peninsula should be carefully checked for strength, and the subflooring around the sink should be examined for signs of wetness and rot. If either of these areas needs attention, repairs should be made before any new installations are performed; the extra cost will have to be figured in for materials and/or contractor fees.

The appliances for the standard peninsula kitchen consist of a mid-priced, freestanding electric cooking range and a 19-cubic-foot refrigerator. Both of these units can be installed by the homeowner or delivery person, as long as the electrical hookups are in place. The range hood should be installed by an electrician because the unit has to be direct-wired to the supply cable. The double-bowl sink should also be tied in by a qualified tradesperson, particularly if a new rough-in is required. If additional electrical or plumbing rough-in is needed, it should be completed before the finish materials are placed. These jobs take less time when the walls, ceiling, and floor are unfinished and accessible to the worker. As a result, they will cost you less if they are done beforehand.

LEVEL OF DIFFICULTY

The peninsula kitchen is a very demanding remodeling project that requires building skills and expertise in the use of tools. If you are a beginner, you should limit your efforts to the basic tasks of preparing and painting the walls and ceiling and installing the underlayment for the floor.

Intermediates should restrict their involvement to these basic tasks, but might also try the removal of the old cabinets if they have had some remodeling experience. Experts might want to stop short of the cabinet installation, but they should be able to handle most of the other tasks, providing that they have had some experience in those areas. Beginners ought to double the professional time estimate for the basic tasks, and intermediates should add 50%-70%, depending on their level of ability. Experts should add about 20% to the labor-hours for basic jobs like painting and installing the underlayment, and 50%-75% for specialized tasks like flooring, plumbing, and cabinetwork.

WHAT TO WATCH OUT FOR

Major remodeling projects like kitchen renovations often provide the opportunity to check the existing electrical system in a house. If you hire a professional to do the electrical work on your new kitchen, it might be worth the extra expense to have him inspect the existing kitchen wiring and the electrical service throughout your house. This may give you the chance to have any needed repairs made at a convenient time.

SUMMARY

The standard peninsula kitchen plan can be an economical source of new convenience and a more pleasant atmosphere, providing it fits into your available kitchen space and suits your lifestyle. It does require some professional installations, but with good planning, the do-it-yourselfer can save money by performing some of the other tasks.

For other options or further details regarding options shown, see

> *Deluxe peninsula kitchen*
> *Electrical system: Light fixtures*
> *Electrical system: Receptacles*
> *Flooring*
> *Painting and wallpapering*
> *Standard window installation**
>
> * *In Exterior Home Improvement Costs*

Standard Peninsula Kitchen

Description	Quantity/ Unit	Labor- Hours	Material
Blocking, for mounting cabinets, 2 x 4	72 L.F.	2.3	31.10
Plywood, underlayment-grade, 1/2" thick, 4' x 8' sheets	128 S.F.	1.4	98.30
Flooring, vinyl sheet goods, 0.125" thick	100 S.F.	3.5	270.00
Wall cabinets, economy, 12" deep, 36" wide	8 Ea.	5.6	1,200.00
Corner, 12" deep, 36" wide	3 Ea.	2.9	648.00
Base cabinets, 24" deep, 36" wide, economy	7 Ea.	5.5	1,780.80
Peninsula base cabinet, custom-built, 24" deep x 9' long	1 Ea.	1.9	656.10
Countertop, 1-1/4" thick plastic laminate, w/backsplash	23 L.F.	6.1	211.14
Countertop for peninsula, 1-1/4" thick plastic laminate, 30" wide	9 L.F.	2.6	199.80
Valance board over sink	4 L.F.	0.2	37.20
Trim, cornice molding, stock pine, 9/16" x 2-1/4"	46 L.F.	1.2	47.47
Paint, ceiling & walls, primer	600 S.F.	2.4	28.80
Paint, ceiling & walls, 1 coat	600 S.F.	3.7	36.00
Paint, cornice trim, simple design, incl. puttying, primer	46 L.F.	0.6	1.10
Paint, cornice trim, simple design, one coat	46 L.F.	0.6	1.66
Cooking range, built-in, 30" wide, one oven	1 Ea.	1.3	384.00
Microwave	1 Ea.	2.0	105.60
Hood for range, 2-speed, 30" wide, vented	1 Ea.	2.0	42.60
Refrigerator, no frost, 19 C.F.	1 Ea.	2.0	600.00
Dishwasher, built-in, 5-cycle	1 Ea.	5.0	338.40
Sink w/faucets & drain, SS, self-rimming, 33" x 22", dbl. bowl	1 Ea.	3.3	534.00
Rough-in, supply, waste & vent for sink	1 Ea.	7.5	102.00
Undercabinet lights, 24" fluorescent strip	4 Ea.	1.3	247.20
Totals		64.9	$7,601.27

Project Size   9'-6" x 13'

Contractor's Fee Including Materials **$14,058**

Key to Abbreviations
C.Y.–cubic yard   Ea.–each   L.F.–linear foot   Pr.–pair   Sq.–square (100 square feet of area)
S.F.–square foot   S.Y.–square yard   V.L.F.–vertical linear foot   M.S.F.–thousand square feet

DELUXE PENINSULA KITCHEN

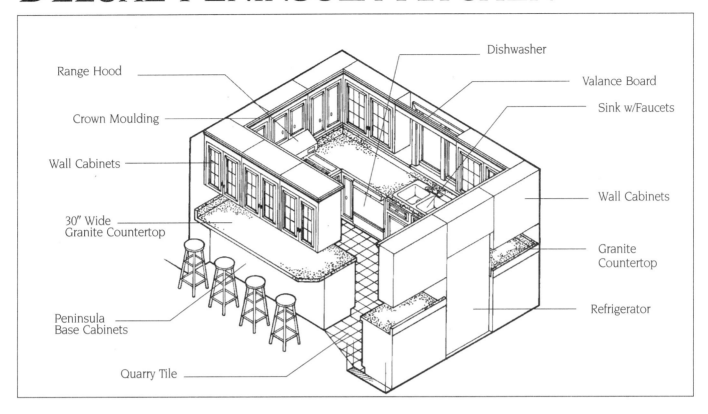

Range Hood

Crown Moulding

Wall Cabinets

30" Wide
Granite Countertop

Peninsula
Base Cabinets

Quarry Tile

Dishwasher

Valance Board

Sink w/Faucets

Wall Cabinets

Granite
Countertop

Refrigerator

The layout of the peninsula kitchen lends itself to being a social and activity center in the home as well as the primary eating and food-preparation area. Top-quality cabinets, appliances, and surface coverings can turn an old kitchen into a delightful eating and gathering place. This model can be used to renovate an existing peninsula design, or to improve a conventional U-shaped layout by adding the peninsula unit. Because of the size of the facility and the high cost of its components, most do-it-yourselfers should use discretion in tackling the specialized tasks.

MATERIALS

The materials included in the deluxe peninsula kitchen are standard products that have been upgraded. Substitutions of comparable deluxe products are readily available to meet your particular needs and to satisfy your personal taste. A qualified kitchen expert can help you select the cabinets, surface coverings, and appliances from a range of deluxe components. The large number of cabinets used in the peninsula kitchen, and the complexity of their layout, make them the most important element of

the facility. To maintain deluxe standards, the cabinets should be made of quality hard or softwood with paneled or other decorative features on the doors and drawer facings. Included in the cost of the cabinets in this plan are high-quality hardware and the custom features, including glass doors on the peninsula unit. However, the cost for any other cabinet amenities and specialized units must be tacked on to the estimate. A lazy Susan, for example, is a useful device for gaining easier access to the dead storage space found in the corners of both the wall and base units. Cabinet extras like this one are usually expensive, but they complement the other deluxe products in the kitchen and add to its operational convenience. Even if you plan to do some of the work in this kitchen renovation, you should leave the cabinet installation to a professional unless you are an expert in home remodeling. The cost and magnitude of a four-wall system demand precise installation by a skilled worker.

The surface of the countertop for this deluxe kitchen may be covered with a high quality ceramic tile, an appropriate plastic laminate veneer, or other deluxe materials such as marble or solid surfacing. All of these products are heat-

and water-resistant, durable, and attractive. Although ceramic tile adds to the deluxe appearance of the facility, it is more expensive than plastic laminate and harder to install. If you plan to do this part of the project on your own, be sure to get some advice and installation tips from a knowledgeable person. Also, before you choose tile for the peninsula countertop, consider the intended use of this area. For example, if it is to be used as a writing surface as well as for eating, ceramic tile might not be a good choice. Plastic laminate or even hardwood could be used instead for the peninsula unit surface, and you could still put ceramic tile on the other countertops in the kitchen.

Before the cabinets, countertop, flooring, and appliances have been installed, recondition the ceiling and walls and then roll them with a primer and a finish coat of paint.

Remember that additional electrical work may have to be completed before the built-in range, dishwasher, and range hood are installed. The sink and dishwasher will also require new plumbing rough-in if they are being relocated. These tasks should be accomplished by qualified professionals or by expert intermediates who have considerable electrical and plumbing experience.

LEVEL OF DIFFICULTY

Most of the tasks involved in remodeling a peninsula kitchen are highly specialized and require considerable ability in the use of tools. Experts who have extensive remodeling experience should be able to handle most of the work, but they should proceed slowly on highly demanding jobs like hanging the cabinets and installing the tile. If they plan to complete these tasks, they should allow ample time – perhaps double the professional time estimate – as the materials are expensive and quality workmanship is called for. Experts should add 20% to the labor-hour estimates for all other tasks and, depending on the amount of their experience, more for the flooring, plumbing, and electrical work. Intermediates and beginners should restrict their involvement to basic tasks like the removal of the old kitchen, reconditioning of the site, and painting. They should add 50% and 100%, respectively, to the professional time for all of these tasks.

WHAT TO WATCH OUT FOR

As noted earlier, the peninsula unit provides the homeowner with the opportunity to design an attractive and practical eating and utility area. One way to enhance both its appearance and function is to add electrical conveniences and lighting fixtures. There are many possibilities; for example, install one or several duplex outlets in convenient locations on and around the unit, or build direct or indirect lighting into the underside of the overhead unit. Both of these electrical improvements can be done for a reasonable cost, yet they provide a lot of convenience. A warming plate can also be installed in the surface of the base unit, but this electrical amenity will add significantly to the cost. Be sure to have the wiring capacity for the existing kitchen circuits checked before you add these or other new electrical fixtures. These items should be carefully planned in advance, as spur-of-the-moment modifications will cost more for installation after the finished materials are in place.

SUMMARY

The deluxe peninsula kitchen is an expensive and complex remodeling project that can enhance the quality of life in your home. The high grade of its components can add value to your dwelling, while improving its appearance and versatility.

Deluxe Peninsula Kitchen

Description	Quantity/ Unit	Labor- Hours	Material
Blocking, for mounting cabinets, 2 x 4	72 L.F.	2.3	31.10
Plywood, underlayment-grade, 1/2" thick, 4' x 8' sheets	128 S.F.	1.4	98.30
Flooring, quarry tile, 6" x 6", mud set	100 S.F.	11.4	349.20
Wall cabinets, 12" deep, 36" wide, hardwood	12 Ea.	11.8	3,366.72
Base cabinets, 24" deep, 36" wide, hardwood	7 Ea.	7.7	3,328.08
Peninsula base cabinet, custom-built, 24" deep x 9' long	9 L.F.	7.6	3,510.00
Granite countertop, average, 1-1/4" thick, 24" deep	23 L.F.	14.1	2,553.00
Granite countertop for peninsula, 30" wide	9 L.F.	6.8	1,221.00
Valance board over sink	4 L.F.	0.2	37.20
Trim, cornice molding, stock pine, 9/16" x 2-1/4"	46 L.F.	1.2	47.47
Paint, ceiling, primer	125 S.F.	0.5	6.00
Paint, ceiling, 1 coat	125 S.F.	0.8	7.50
Paint, cornice trim, simple design, incl. puttying, primer	46 L.F.	0.6	1.10
Paint, cornice trim, simple design, one coat	46 L.F.	0.6	1.66
Cooking range, built-in, 30" wide, one oven, ceramic elements	1 Ea.	1.6	840.00
Microwave	1 Ea.	1.0	492.00
Hood for range, 2-speed, 30" wide, vented	1 Ea.	3.3	540.00
Refrigerator, no frost, side-by-side with ice/water dispenser	1 Ea.	5.3	3,330.00
Sink w/faucets & drain, SS, self-rimming, 33" x 22", dbl. bowl	1 Ea.	3.3	534.00
Dishwasher, built-in, 7-cycle	1 Ea.	1.5	338.40
Rough-in, supply, waste & vent for sink & dishwasher	2 Ea.	15.0	204.00
Undercabinet lights, 24" fluorescent strip	6 Ea.	2.0	370.80
Recessed ceiling light, 100 watt	6 Ea.	6.0	399.60
Totals		106.0	$21,607.13

Project Size: 9'-6" x 13'

Contractor's Fee Including Materials: **$35,885**

For other options or further details regarding options shown, see

Electrical system: Light fixtures

Electrical system: Receptacles

Flooring

Painting and wallpapering

Standard peninsula kitchen

Standard window installation*

* In Exterior Home Improvement Costs

Key to Abbreviations
C.Y.–cubic yard Ea.–each L.F.–linear foot Pr.–pair Sq.–square (100 square feet of area)
S.F.–square foot S.Y.–square yard V.L.F.–vertical linear foot M.S.F.–thousand square feet

STANDARD ISLAND KITCHEN

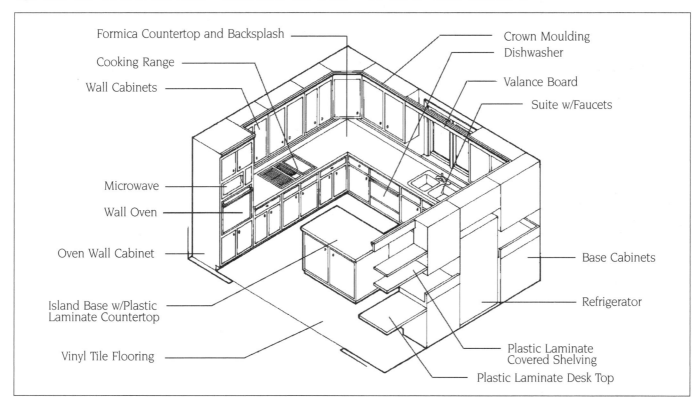

Formica Countertop and Backsplash

Cooking Range

Wall Cabinets

Microwave

Wall Oven

Oven Wall Cabinet

Island Base w/Plastic Laminate Countertop

Vinyl Tile Flooring

Crown Moulding

Dishwasher

Valance Board

Suite w/Faucets

Base Cabinets

Refrigerator

Plastic Laminate Covered Shelving

Plastic Laminate Desk Top

Based on square footage, the kitchen is one of the most expensive rooms in your home to renovate, primarily because of the cost of the specialized appliances, cabinets and flooring systems used. It is also one of the most challenging rooms for the homeowner to tackle for a variety of reasons. Because the kitchen's function is vital to the routine of home life, a long and drawn-out remodeling project can only add to the inconvenience during the renovation process. Kitchen renovations demand a wide range of construction tasks, some of them too specialized for the average homeowner to undertake.

MATERIALS

The kitchen, more than any room in your house, contains materials and appliances that vary greatly in quality and, therefore, cost. Cabinets alone constitute an area of kitchen materials that run the gamut from economical mass-produced particleboard units to custom-made soft or hardwood designs with raised panels or other decorative features.

The floor and walls of this kitchen include standard materials that can be installed

by most intermediates. Make sure that you carefully check the existing subflooring for rot before laying the new underlayment and vinyl tiling. Also keep in mind that the vinyl flooring surface should extend under appliances, but not under the cabinets in the new kitchen. The walls and ceiling should be coated with a quality primer and finish coat before the floor, cabinets, and appliances are installed.

Applying the paint first saves time because you don't have to worry about masking the cabinets and covering the appliances and floor while rolling the ceiling or doing brush work.

The plumbing and electrical work required in this kitchen renovation will vary, depending on the extent to which appliances and the sink are moved from their old locations. If your remodeling project includes relocation of the sink to another wall, then you'll have to allow for more roughing-in materials, time, and expense. The same type of cost variables will also be included in the relocation of a gas or electric range or a dishwasher. Beginners should leave all of the plumbing and electrical work to professionals; intermediates should do the same, unless they have had extensive

experience in these areas. Installing and relocating rough plumbing for a double-bowl sink for example, is a job that should be undertaken only by an expert do-it-yourselfer or a qualified plumber.

The cabinet work in this standard island kitchen involves a complete set of top and bottom units on sides of the room's perimeter surrounding the island unit. A custom plastic laminate countertop ties the units together to form a versatile, continuous work surface. Custom shelving for a bookcase and desk area is also included. The quality of the cabinet materials, as well as the design of the cabinet door and drawer facings, significantly affect their cost. Quality prefabricated units, for example, generally cost less than similar quality, custom-made units. If you are doing the work yourself and have not had experience with cabinet installation, seek some assistance. Seemingly small problems like a wall corner slightly out of plumb or out of square can raise havoc with the alignment of the cabinet facings and countertop. Skilled intermediates and experts should be able to perform the cabinet installation, but beginners and even reasonably experienced do-it-yourselfers should leave this part of the project to a professional.

The appliances for any kitchen renovation, like the cabinets, can raise or lower the cost of the job depending on the quality of the item and the number of extra features you desire. A less expensive alternative to the cooktop unit and wall ovens would be to use a standard one-oven, four-burner range. Refrigerators vary greatly in capacity and optional functions, but a standard 19-cubic-foot, frost-free unit with freezer above generally fulfills a family's needs. A dishwasher is considered a necessity today. If you're doing the electrical and plumbing work on your own, remember that the dishwasher will require both plumbing and electrical connections.

LEVEL OF DIFFICULTY

A complete kitchen renovation is a demanding remodeling job, but one that an intermediate can tackle. Tearing out the old kitchen, reconditioning the site, and installing the new components are jobs that take time, effort and an extensive knowledge of tools and building skills. Beginners should be able to accomplish some of the easier tasks like the laying of subflooring and the painting, but should get some professional assistance before starting tasks that are new to them. The more sophisticated jobs involved in the project, such as plumbing, electrical, and cabinet and countertop installations, should be left to expert do-it-yourselfers or professional contractors. Beginners should double or triple the estimated labor-hours, even for the basic tasks. Intermediates should add about 75% to the estimated times of those tasks that they feel they can accomplish. Experts should tack on about 40% to 50% to the labor-hour estimates throughout the project.

WHAT TO WATCH OUT FOR

Renovating the kitchen in an older home often presents the challenges of sagging or sloped floors, bowed or out-of-plumb walls, and other problems that can test even professional builders. The problems become magnified for the intermediate who has not had experience in dealing with them. Don't hesitate to seek advice from a qualified person who knows, for example, how to scribe a countertop into a skewed wall or how to make adjustments to allow for a dip or bulge in the floor.

SUMMARY

Modernizing and rearranging an old kitchen is an exciting way to improve your home. A project like this adds considerable value to your house and improves the overall quality of living for your family.

For other options or further details regarding options shown, see

> Deluxe island kitchen
>
> Flooring
>
> Electrical system: Light fixtures
>
> Electrical system: Receptacles
>
> Painting and wallpapering
>
> Standard window installation*
>
> * In Exterior Home Improvement Costs

Standard Island Kitchen

Description	Quantity/ Unit	Labor- Hours	Material
Blocking, for mounting cabinets, 2 x 4	68 L.F.	2.2	29.38
Plywood, underlayment-grade, 1/2" thick, 4' x 8' sheets	192 S.F.	2.1	147.46
Floor tile, 12" x 12" x 1/16" thick, vinyl composition	125 S.F.	2.0	133.50
Wall cabinets, 12" deep, 36" wide, economy	9 Ea.	6.3	1,350.00
Base cabinets, 24" deep, 36" wide, economy	10 Ea.	7.9	2,544.00
Island cabinet, 30" x 48" base	1 Ea.	0.9	291.60
Countertop, 1-1/4" thick, plastic laminate, w/backsplash	24 L.F.	6.4	220.32
Island top, 1-1/4" thick, 32" x 50"	4 L.F.	1.1	36.72
Valance boards over sink & book shelf	8 L.F.	0.3	74.40
Desktop, 1-1/4" thick, plastic laminate covered	3 L.F.	1.0	18.36
Shelving, 3/4" thick, plastic laminate covered	8 L.F.	0.9	35.04
Trim, cornice molding, stock pine, 9/16" x 2-1/4"	36 L.F.	1.0	37.15
Paint, ceiling & walls, primer	450 S.F.	1.8	21.60
Paint, ceiling & walls, 1 coat	450 S.F.	2.8	27.00
Paint, cornice trim, simple design, incl. puttying, primer	36 L.F.	0.4	0.86
Paint, cornice trim, simple design, one coat	36 L.F.	0.4	1.30
Microwave	1 Ea.	2.0	105.60
Countertop cooktop, 4 burner	1 Ea.	1.3	229.20
Hood for range, 2-speed, 30" wide	1 Ea.	3.3	906.00
Wall oven	1 Ea.	2.0	444.00
Refrigerator, no frost, 19 C.F.	1 Ea.	2.0	600.00
Sink w/ faucets & drain, SS, self-rimming, 43" x 22", dbl. bowl	1 Ea.	3.3	618.00
Dishwasher, built-in, 2 cycle	1 Ea.	2.5	304.80
Rough-in, supply, waste & vent for sink and dishwasher	2 Ea.	15.0	204.00
Totals		68.9	$8,380.29

Project Size	11'-6" x 14'-6"	Contractor's Fee Including Materials	$15,361

Key to Abbreviations
C.Y.–cubic yard Ea.–each L.F.–linear foot Pr.–pair Sq.–square (100 square feet of area)
S.F.–square foot S.Y.–square yard V.L.F.–vertical linear foot M.S.F.–thousand square feet

DELUXE ISLAND KITCHEN

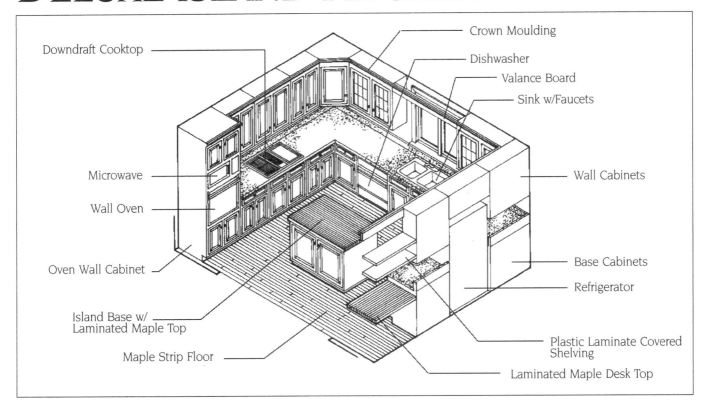

Crown Moulding
Dishwasher
Valance Board
Sink w/Faucets
Downdraft Cooktop
Microwave
Wall Oven
Oven Wall Cabinet
Island Base w/ Laminated Maple Top
Maple Strip Floor
Wall Cabinets
Base Cabinets
Refrigerator
Plastic Laminate Covered Shelving
Laminated Maple Desk Top

If it is within your budget and fits in with your home and lifestyle, a deluxe island kitchen may be the appropriate remodeling design to follow. This plan outlines a luxurious alternative to simpler, more standard arrangements. Good planning and professional workmanship will be rewarded with a beautiful, durable, and functional kitchen area.

MATERIALS

Because this kitchen plan involves deluxe appliances, cabinets, and other materials, its cost is substantially higher than that of a standard facility of the same size. All of the basic components of the kitchen can be replaced by less expensive items, and substitutions can be made to reduce the cost to a compromise level somewhere between that of a standard facility and this one. Remember, though, that any replacements should be made with discretion, as the aesthetic balance of the kitchen's appearance is important. Your personal taste and the decor of your home should serve as your guides when determining the materials for the new facility.

One of the most pleasing features of this kitchen plan is the U-shaped cabinet layout, with a convenient island located in the center of the room. This layout requires a fairly large kitchen area to ensure enough floor and counter space without cramming its design and creating a closed-in feeling. The island cabinet puts normally wasted floor space to practical use as a food preparation or utility work surface. Depending on the intended function of the island, a variety of surface materials can be used, from laminated hardwoods in butcher-block design to stone, ceramic tile, or plastic laminate coverings. The wall and base cabinets that border the island cabinet can be constructed from a variety of hard or soft woods with paneled or decorative drawer and door facings. The desk top and shelving should be included in the design and manufacture of the cabinet system.

The materials used for the floor and countertop of the new kitchen are products that complement the elegance of the cabinets. This plan includes hardwood maple strips for the floor surface. The granite countertop and backsplash provide a durable and attractive alternative to plastic laminate coverings. Other deluxe options include marble, hardwood, or Corian.

The design and capacity of the appliances to be included in the deluxe island kitchen should be determined by such variables as the size of your family, extent of anticipated use of the kitchen, and food preparation methods. Such considerations as preference for gas or electric range, refrigerator and freezer capacity, and general use of appliances should be carefully weighed before you select the fixtures and decide on their location in the new kitchen. Conversion from an electric to gas range, for example, requires significant rough-in preparation that should be built into both the design and the cost of the overall kitchen plan before the work begins. Other appliance, plumbing, and electrical relocations should be similarly planned and incorporated into the cost of the facility before you begin the renovation. Changing your mind on such matters after the work has begun will usually cost you more money and aggravation.

LEVEL OF DIFFICULTY

The kitchen is a challenging room for the intermediate to remodel because carpentry, plumbing, electrical, flooring, and other specialized building skills are required. Also, most of the materials used in kitchen construction are expensive, and many are difficult to install. For example, the planning, measuring, and placing of the cabinet and countertop systems in this plan involve three walls totaling 30 linear feet, with appropriate allowances for a window, sink, under-counter dishwasher, refrigerator, built-in range, oven and hood, as well as the island unit. Overseeing and accomplishing this cabinet installation requires the skill of a professional. Experts should have prior experience with cabinet installation if they plan to tackle the project. Beginners and intermediates can save on labor costs by removing the old facility, placing the flooring underlayment, and reconditioning and finishing the walls and ceiling, but they should leave the cabinet and appliance installation, the tile work and the countertop installation to professionals. They should add 100% and 50%, respectively, to the professional time for laying the plywood and painting. Experts should add 20% to the time estimates for these tasks and 50% to 100% to the time for the more specialized jobs within the kitchen.

WHAT TO WATCH OUT FOR

If you plan to do some or all of the work on this remodeling project, take stock of your building skills and the amount of free time available in your personal schedule. Kitchens, like baths, are difficult facilities to remodel because their operation is vital to the daily routine of home life. Disturbing their function can cause considerable inconvenience, especially if the interruption is prolonged and the work done piecemeal on weekends, evenings, and holidays. In most cases, a concentrated and limited period of remodeling is desirable for kitchen projects. Make plans ahead of time to arrange for alternative meal preparation and to allow for the inconvenience. If you are doing the work on your own, build in some extra time for those hidden problems that always surface. Keep in mind the long lead times for delivery of special order materials, such as granite countertops.

Deluxe Island Kitchen

Description	Quantity/Unit	Labor-Hours	Material
Blocking, for mounting cabinets, 2 x 4	68 L.F.	2.2	29.38
Plywood, underlayment-grade, 1/2" thick, 4' x 8' sheets	192 S.F.	2.1	147.46
Flooring, maple strip, 25/32" x 2-1/4"	125 S.F.	5.9	571.50
Wall cabinets, 24" deep, 36" wide, hardwood	9 Ea.	8.9	2,525.04
Base cabinets, 24" deep, 36" wide, hardwood	9 Ea.	9.9	4,278.96
Island cabinet, 30" x 48" base	1 Ea.	0.9	390.00
Granite countertop, average, 1-1/4" thick	24 L.F.	14.8	2,664.00
Island top, maple, solid laminated, 1-1/2" thick	4 L.F.	1.4	186.00
Valance boards over sink & book shelf	8 L.F.	0.3	74.40
Desktop, maple, solid laminated, 1-1/2" thick	3 L.F.	0.9	111.60
Shelving, 3/4" thick, plastic laminate covered	8 L.F.	0.9	35.04
Trim, cornice molding, stock pine, 9/16" x 2-1/4"	36 L.F.	1.0	37.15
Paint, ceiling & walls, primer	170 S.F.	0.7	8.16
Paint, ceiling & walls, 1 coat	170 S.F.	1.1	10.20
Paint, cornice trim, simple design, incl. puttying, primer	36 L.F.	0.4	0.86
Paint, cornice trim, simple design, one coat	36 L.F.	0.4	1.30
Cooktop with griddle, 30" wide	1 Ea.	2.7	642.00
Hood for range, 2-speed, 42" wide	1 Ea.	3.3	906.00
Wall oven	1 Ea.	4.0	1,026.00
Microwave	1 Ea.	1.0	492.00
Refrigerator, no frost, side-by-side with ice/water dispenser	1 Ea.	5.3	3,330.00
Sink w/faucets & drain, SS, self-rimming, 43" x 22", dbl. bowl	1 Ea.	3.3	534.00
Dishwasher, built-in, 7 cycle	1 Ea.	1.5	338.40
Undercabinet lights, 24" fluorescent strip	4 Ea.	1.3	247.20
Recessed ceiling light, 100 watt	6 Ea.	6.0	399.60
Rough-in, supply, waste & vent for sink and dishwasher	2 Ea.	15.0	204.00
Floor finish, 2 coats polyurethane	125 S.F.	3.4	97.50
Totals		98.6	$19,287.75

Project Size	11'-6" x 14'-6"	Contractor's Fee Including Materials	$32,324

Key to Abbreviations
C.Y.–cubic yard   Ea.–each   L.F.–linear foot   Pr.–pair   Sq.–square (100 square feet of area)
S.F.–square foot   S.Y.–square yard   V.L.F.–vertical linear foot   M.S.F.–thousand square feet

SUMMARY

If this kitchen plan is in keeping with your lifestyle and fits into your available space, the project can increase the value of your house. It will also make this central element in your living space more functional, attractive, and convenient.

For other options or further details regarding options shown, see

> *Flooring*
>
> *Electrical system: Light fixtures*
>
> *Electrical system: Receptacles*
>
> *Island unit*
>
> *Painting and wallpapering*
>
> *Standard island kitchen*
>
> *Standard window installation**

* In Exterior Home Improvement Costs

ISLAND UNIT

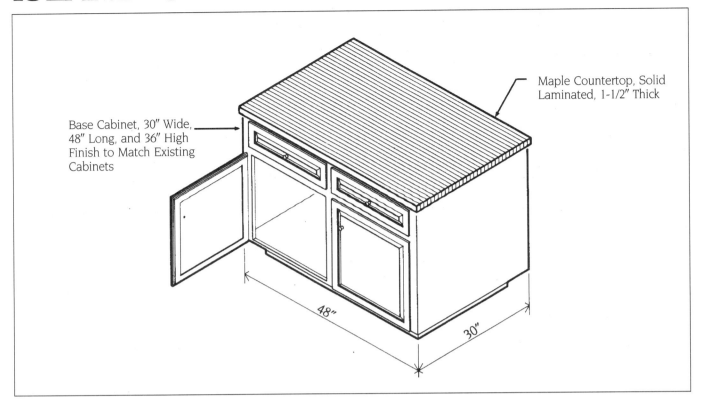

Base Cabinet, 30″ Wide, 48″ Long, and 36″ High Finish to Match Existing Cabinets

Maple Countertop, Solid Laminated, 1-1/2″ Thick

48″

30″

Most people agree that you can never have too much kitchen counter space. Food preparation calls for plenty of flat working surfaces, in addition to the thousand-and-one other jobs that a typical family undertakes in that all-purpose utility room, the kitchen. An island cabinet unit provides a useful and flexible countertop area as well as convenient storage space below. Being custom designed, it can be built to accommodate the size and shape of your kitchen – and your budget.

MATERIALS

A project like this calls for a good deal of analysis and planning. The size and layout of your kitchen largely determine the form the unit will take. The primary consideration is, of course, floor area. A permanently installed cabinet requires that adequate floor space be left around it both for ease of movement and to prevent the kitchen from looking cramped and closed in. (The industry standard is 3′ clear between counter and island.) Smaller kitchens that preclude built-in islands may be improved by the addition of a smaller, moveable unit.

The built-in island unit offers the advantage of being anchored to the floor, thereby providing a solid base for the more vigorous aspects of cooking – the chopping, pounding, beating, and mashing that T.V. chefs seem to take such delight in. More importantly, it allows you to run wiring and/or plumbing to service a number of kitchen fixtures. A cooktop or a warming plate can be built in, along with outlets to power mixers, toasters, blenders, food processors and the like. A second sink could be installed with specialty features such as a tall faucet, superheated water, or a drinking fountain. Except for the outlets, these and similar amenities definitely fall into the category of luxury items and would significantly increase material and installation costs over those shown in this project.

Kitchens lacking the floor space for a large, permanent unit might accommodate a "push around" cabinet as small as a chopping block or as large as a breakfast counter. Mounted on locking wheels or piano casters, these auxiliary counters provide flexible work centers, enabling you to move food and dishes around the kitchen, from the refrigerator to the range to the sink. If the unit is intended to be moved very

frequently, as opposed to a semi-permanent arrangement, don't design it so large that, when filled with stores and cooking gear, it is too heavy to maneuver.

Depending on its primary intended function, the island's surface material can be selected from a variety, including stainless steel, marble, butcher-block-style laminated hardwood, ceramic or quarry tile, and plastic laminate coverings. Your choice will depend on your planned use for the unit. If it will provide a surface for baking, consider marble; for slicing, chopping and carving, consider butcher block. If the island will be used for dining and as a place for the children's homework, consider creating a 9″ overhang and using a plastic laminate surface. A kitchen design center can provide you with advice and information to help you in selecting the most appropriate countertop material.

The entire island unit, base and top, should be designed to match or blend with the existing cabinets. This island will be a highly visible component of your kitchen, subject to regular, heavy use, and exposed to the rigors of everyday family life. You must assess your situation to determine the importance of durability versus appearance when choosing the type and quality of the materials for this unit.

Installation difficulties will depend on the design of the unit and the kitchen flooring material. It may be necessary to attach to the floor a rectangular mounting base of 2 x 4 lumber, over which the island is placed and to which it is secured with screws. It may also be possible to remove the bottom shelf and anchor the unit to the floor with angle brackets. On a tile floor, you will have to drill pilot holes for the screws with a masonry bit. When setting the unit permanently, care should be taken to align the height of the new unit with the top of existing cabinetry. Feeds for wiring and/or plumbing should be relatively easy to run from the basement or crawlspace below.

would be well advised to order such a unit from a supply house or custom shop. Installation could be handled by a beginner with a little guidance from an experienced hand. Wiring and plumbing are jobs for the expert or the professional contractor. A cooktop requires its own circuit ("home run" in electrician's jargon), and outlets located near a sink must be GFI rated. A sink needs to be tied in to the existing drain and might possibly require its own vent pipe. A garbage disposal calls for both plumbing and electrical skills. None of these tasks is suited to the beginner. Installing a prefab island unit will take a beginner at least twice as long as the professional. The intermediate should add 50% to the time. The expert should add 10% to the installation and 30% to 40% to electrical or plumbing tasks.

it also offers significant storage potential. Realizing and maximizing this potential calls for a little imagination and creativity, going beyond simple drawers and shelves. Visiting design centers or glancing through catalogs and magazines can turn up a number of conveniences and amenities worth considering for your island cabinet: pull-out cutting boards and hot plates; sliding shelves for anything from small appliances and utensils to spices and trash cans; bins for staples like flour and sugar; specialty drawers for bread, knives, or dish towels; lazy Susans for pots and pans or canned goods; space-saving rack systems for bags, trays, or baking pans; and on and on.

Before installing the unit, check the floor for level in both directions. If necessary, use shim shingles to bring the island level. Any spaces that are created at the base of the cabinet by shimming may be concealed by adding a wood or vinyl base after the unit has been leveled and secured. You may find it convenient to mount cabinet-leveler feet on the bottom of the unit. These are adjustable with a screwdriver. Access holes for adjustment can be drilled in the bottom shelf and later covered with removable plugs.

LEVEL OF DIFFICULTY

An intermediate with the proper woodworking tools and cabinetry skills might want to attempt building an island unit from scratch. Most homeowners

WHAT TO WATCH OUT FOR

Aside from the obvious additional countertop area an island unit provides,

SUMMARY

Not every kitchen layout can accommodate this freestanding island cabinet unit, but few would not be enhanced by the additional working surface and storage spaces it provides. The kitchen is the heart of the home, and any improvement made to it translates directly to a more useful and enjoyable environment for the entire family.

For other options or further details regarding options shown, see

Cabinet/countertop upgrade

Island Unit

Description	Quantity/ Unit	Labor- Hours	Material
Base cabinet, 30" x 48", two doors, two drawers	1 Ea.	1.2	546.00
Island top, maple, solid, lam., 1-1/2" thick x 32" wide x 50" long	4.20 L.F.	1.2	156.24
Totals		2.4	$702.24

Project Size	30" x 48"	Contractor's Fee Including Materials	$1,112

Key to Abbreviations
C.Y.–cubic yard   Ea.–each   L.F.–linear foot   Pr.–pair   Sq.–square (100 square feet of area)
S.F.–square foot   S.Y.–square yard   V.L.F.–vertical linear foot   M.S.F.–thousand square feet

PANTRY CABINET

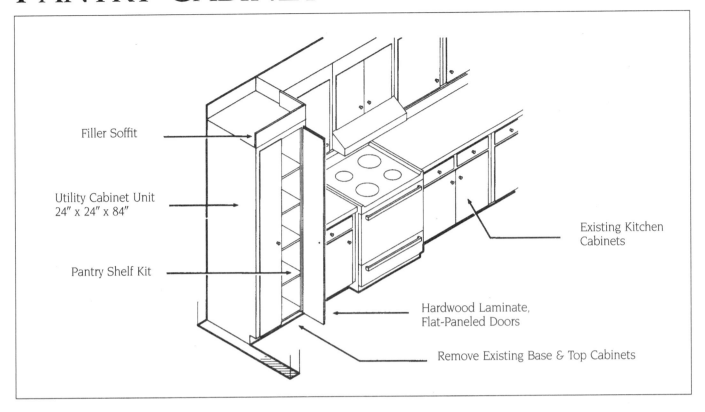

Filler Soffit

Utility Cabinet Unit
24″ x 24″ x 84″

Pantry Shelf Kit

Existing Kitchen
Cabinets

Hardwood Laminate,
Flat-Paneled Doors

Remove Existing Base & Top Cabinets

The term "pantry" conjures up images of butlers and housemaids bustling about to serve the needs and whims of the homeowner. Few of us have the luxury of such a room off the kitchen, but there is no denying how convenient and useful one would be. In today's typical kitchen, storage is provided almost exclusively by wall and base cabinets, and when they become filled, the extra foodstuffs and household supplies overflow into whatever space will have them – broom closets, basements, back hallways, and even the basement stairs. A pantry cabinet offers a welcome compromise. It may not be the elegant, walk-in butler's pantry of a Victorian baronial estate but, in the best Yankee tradition, it gets the job done.

MATERIALS

The first question to be answered is "Where?" The factory-made unit shown here calls for enough floor and wall space to accommodate its dimensions: 24″ deep, 24″ wide, and 84″ tall. If you don't have the room to simply place it in your kitchen as is, you could remove a section of the countertop, base and wall cabinets, and, in effect, incorporate it into your cabinet layout. Or you might have a closet

in another room that backs up to a kitchen wall. You could wall off a section of this closet, break through from the kitchen, and frame out an appropriate alcove. The first scenario will cost you counter and cabinet space; the second, closet space. Both will mean a good bit of additional work.

Many kitchens have broom closets whose space is often poorly utilized. The contents of most broom closets – brooms, mops, rags, buckets, etc. – are used neither daily nor exclusively in the kitchen, and could conveniently be stored elsewhere. The closet could then be converted to a pantry stocked with canned goods and other food items, thus better utilizing its proximity to the kitchen work area.

It may be that your kitchen will not lend itself to a stock, prefabricated unit, and will instead require a custom-made cabinet. You could build it yourself, if you have the woodworking tools and skills, or you could hire a local cabinetmaker to do the work. A custom cabinet accommodates itself to your kitchen, rather than the other way around, as is the case with a standard, factory unit, but custom work generally costs more.

You may be able to order a pantry to duplicate the style, color, and finish of your existing kitchen cabinets. If not, the advantage of wood is that it pretty much goes with anything, its warm look complementing a wide range of decors from traditional to contemporary.

A pantry cabinet will get constant daily use and be exposed to the usual hazards of normal kitchen activity. It should, therefore, be strong and sturdy, and equipped with high-quality hardware. Roll-out trays should be supported by full-length runners on each side. The best kind are full-suspension slides that allow you to pull out the tray its full length and that support it well for the whole distance. Ballbearing tracks are solid, smooth, quiet – and expensive. But remember that these trays will be loaded with cans, bottles, and boxes, and their combined weight can be considerable. The access sliding trays provide is almost totally negated if they are too weak to work properly, or are so flimsy that they soon sag and break down. You can, of course, eliminate these moving parts by installing simple, adjustable, open shelves, opting for economy and low maintenance instead of expense and convenience. The same thing applies to lazy Susans: if you want them, get the best quality; cheap ones don't last and are almost impossible to repair.

Many cabinets today come equipped with complicated "European" hinges, which allow the doors to be adjusted in three directions and make it possible for them to open 180°. Again, you will save in the long run by buying the best quality. For a more traditional look, simple pin hinges might be best. They show in the room, and work best with overlay type doors, but they will last and are relatively inexpensive.

LEVEL OF DIFFICULTY

Installing a freestanding pantry cabinet is not, in itself, a difficult job. The cabinet must be plumbed and leveled, and may possibly need to be scribed to one or two walls, depending on its location. The doors are prehung at the factory, although they might be demounted and shipped separately for ease of handling; likewise with drawer runners and trays. The project gets more difficult and demanding if space has to be created by removing cabinets, walls, or other existing fixtures and features. How much more difficult depends, of course, on the nature and scope of the work. If this kind of remodeling is necessary, a beginner should get experienced advice and assistance.

For a simple installation in an existing space, a beginner should need help only if scribing is necessary, and it should take no more than 50% more than the time estimate given. An intermediate should take no more than 10% more time; and an expert should be able to accomplish it within the time given below.

WHAT TO WATCH OUT FOR

Most cabinets come fitted with scribing strips on the back edges. The scribing strips allow you to make the adjustments necessary to ensure that the doors, shelves and drawers work properly and line up with each other or have equal clearance around the face frame.

If the cabinet does not fit the contour of the wall or the wall is out of plumb, you run the risk of "wracking" the cabinet and making the parts dysfunctional.

Some manufacturers offer adjustable, screw-type leveler feet on freestanding cabinets like this. These feet can be adjusted with a screwdriver through access holes in the cabinet's bottom, and the holes covered later with plastic plugs. Adjustable feet are very convenient accessories, and are much easier to work with than shims. A length of base shoe molding can be installed to hide any gap between the cabinet and the floor.

SUMMARY

A pantry cabinet is such an obvious and logical component of or adjunct to a kitchen that it seems strange for so many homes to be without one. A pantry should be located conveniently close to the kitchen work area. If such an appropriate space does not exist, you can possibly create one, with some imagination and a willingness to undertake the extra work. A freestanding, factory-made unit should be strong and sturdy enough to stand up to the use and abuse it will receive as an important fixture of the kitchen. Any hardware (drawer runners, hinges, doorknobs, and drawer pulls) should be of high quality, and solid enough to provide years of trouble-free use. Additional space in the kitchen for storing foodstuffs and general provisions is, if not a true necessity, a valuable and useful amenity.

Pantry Cabinet

Description	Quantity/ Unit	Labor-Hours	Material
Custom cabinet, 24" deep x 24" wide x 84" high	1 Ea.	2.9	799.20
Custom shelving units	20 S.F.	5.0	708.00
Totals		7.9	$1,507.20

Project Size	24" x 24" x 84"	Contractor's Fee Including Materials	$2,535

Key to Abbreviations
C.Y.–cubic yard   Ea.–each   L.F.–linear foot   Pr.–pair   Sq.–square (100 square feet of area)
S.F.–square foot   S.Y.–square yard   V.L.F.–vertical linear foot   M.S.F.–thousand square feet

SINK REPLACEMENT

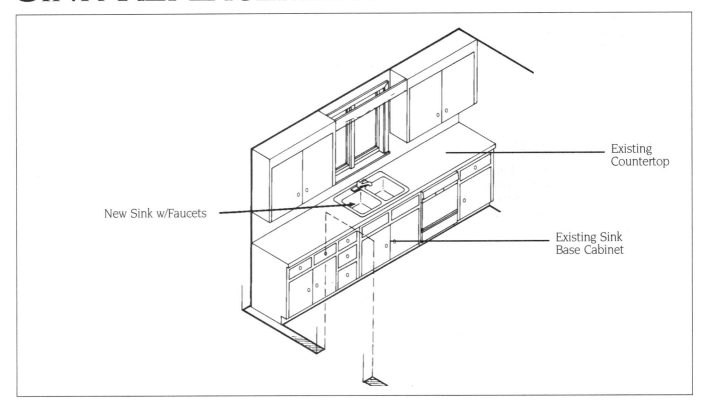

New Sink w/Faucets

Existing Countertop

Existing Sink Base Cabinet

The most frequently used plumbing fixture in the average house is the kitchen sink. As the main source of drinking, cooking, and cleaning water, it is subject to constant daily use, years of which often result in dripping faucets, leaking drains, and a generally worn and dingy appearance. Drains and faucets can, of course, be repaired or replaced, but there may come a time when you want a new sink and faucet assembly to upgrade or improve your kitchen.

In choosing a new sink, you are somewhat limited in size and shape to the existing hole in the countertop. To cover the hole, your new sink cannot be smaller than the old one, and although the hole can be made bigger, the new sink cannot be larger than can be accommodated by the available space in the base cabinet. Given these limitations, you will no doubt find a wide variety of sink and faucet styles offered by kitchen centers and plumbing supply houses.

MATERIALS

When shopping for a new sink and faucet, you won't find any shortage of products from which to choose. The first decision involves material – stainless or enameled steel, porcelain-covered cast iron, acrylic, solid surfacing material, or vitreous china – and whether to buy a rim type (the rim is separate from the sink) or self-rimming model (sink and rim are one piece). Stainless steel, self-rimming models are generally preferred for ease of both installation and maintenance, but they do have a rather plain appearance. Design-conscious homeowners often prefer enameled or china sinks because of their pleasing looks and the availability of a wide range of colors. Others prefer sinks that match or are integral with new countertop materials. Once you have decided on material, you must choose from among the various styles, which include single, double, or triple compartments; dual level; and extra-deep or deep tubs.

Faucets come in an even more bewildering array of styles and finishes. Chrome-plated is the most common finish, but if your home requires gold-plated with a crystal handle, it's available. Your choice of faucets and other materials, such as a spray nozzle, soap dispenser, and superhot or filtered water faucets, is somewhat limited by the number and configuration of the holes in the sink. Stainless steel offers the most flexibility here because holes can be drilled in it to accommodate extra fittings. Faucet types include stem, tipping-valve, disk, rotary-ball, and sleeve-cartridge. The average homeowner would do well to seek advice and information from a reliable kitchen design center or plumbing supply house before making any final decisions. Sometimes you can save a significant amount of money if you do a little comparison shopping. But do so with care; often seemingly identical fixtures resemble each other in outward appearance only.

To remove the old sink you must first close the shut-off valves. These are usually located in the base cabinet directly below the sink, but are sometimes found in the basement. It is not uncommon in older houses, and in some poorly-plumbed newer ones as well, to find the shut-offs inoperable because of corrosion, or to discover that the sink has no separate shut-offs at all. They must be replaced or installed new, and the additional expense chalked up to the ever-surprising home renovation experience.

Disconnect the hot and cold supply lines (called "risers") and loosen the drain fitting. If there is a garbage disposal, it is wise to trip its circuit breaker or remove its fuse while doing this work. There may be clips holding the sink to the countertop from underneath. Removing these often seems to require three hands and the body of a contortionist, along with the proper tools. You will also find a flashlight useful, as well as safety glasses and soft padding to protect your back. If the clips offer the choice, it is generally easier to remove them with a nut driver rather than a screwdriver. It would be the rarest of good luck if the new sink and faucet were to line up exactly with the old risers and waste line. Even if they do, the old pressure fittings will probably leak. Replacing the old risers and adding a flexible trap connector can be done at minimal cost. Flexible risers come in several lengths in chrome-plated soft copper, corrugated copper, polybutylene, and soft PVC plastic with a nylon or stainless steel braided exterior covering. A corrugated polypropylene trap connector can be bent slightly and compressed or extended to make a fit. If the sink's drain and the existing waste line are too far off-center for this to work, a more extensive plumbing adjustment must be made. When in doubt, get the advice or assistance of a professional contractor.

LEVEL OF DIFFICULTY

Removing the old sink can be a wet and neck-wrenching adventure if the clips and fittings are heavily corroded and unusually hard to reach. With a bit of luck, this task can be performed quickly and easily, even by an intermediate, provided he or she is armed with good advice and the right tools. So too with the installation of the new fixture. It could take as little as a couple of hours or so with a few screwdrivers, pliers, and wrenches but, as with so many projects of this sort, Murphy's Law is usually in effect and will call for a lot of patience, creativity, most of the tools you own, and much more time than you planned to take. All that being said, a beginner is advised to hire a professional. Domestic plumbing disasters can make for hilarious stories years later, but living through them is most unpleasant, not to mention costly. The intermediate is advised to solicit experienced help, and should add 100% to the time for both removal and installation. The expert should add 50%, to cover the influence of Mr. Murphy.

WHAT TO WATCH OUT FOR

If the countertop is plastic laminate (e.g., Formica) and the opening must be enlarged, be sure to score it with a parrot-beak linoleum knife prior to cutting it with a saber saw.

Before the sink is set into the opening, plumbers make the faucet and drain assembly hookups. Kitchen sinks have basket strainers that tie the sink to the drain. It is a good idea to salvage the old one if possible. You may be able to use it if fitting problems occur with the new unit, or if a garbage disposal is involved. Lay a bead of plumber's putty around the sink outlet. Next, place the basket, with a washer, into the outlet. Slip the other washer and the locknut onto the strainer's threaded shank and tighten. A trap connector (also called a "tailpiece") and trap will join the sink to the waste line.

To set a self-rimming sink, apply a bead of silicone adhesive caulk around the underside of the flange, about 1/4" from the edge. Lower the fixture into the opening, taking care to align it correctly. Press down firmly and evenly all around the edges of the fixture to force the caulk to ooze out between it and the countertop. Smooth this excess with a wet finger.

If you are working with an enameled or china sink, handle it very carefully. Sinks like these are quite heavy and can crack or chip. Repairs are often impossible, always difficult, and rarely invisible.

SUMMARY

Kitchen sinks and faucets must, at some time, be replaced, whether for aesthetic or purely practical reasons. Existing conditions may place certain limits on your choice of fixtures, and the removal and installation may not always be as simple as it looks. As a result, a good deal more planning and shopping around may be required than the relatively small scope of the job would seem to warrant. However, the end result of kitchen improvements like this almost always justifies the time and expense they take to complete properly.

For other options or further details regarding options shown, see

Cabinet/countertop upgrade

Sink Replacement

Description	Quantity/ Unit	Labor- Hours	Material
Remove old sink	1 Job	1.0	
Kitchen sink w/faucets & drain, SS, self-rim, 43" x 22" dbl. bowl	1 Ea.	3.3	618.00
Totals		4.3	$618.00

Contractor's Fee Including Materials	**$1,095**

Key to Abbreviations
C.Y.–cubic yard   Ea.–each   L.F.–linear foot   Pr.–pair   Sq.–square (100 square feet of area)
S.F.–square foot   S.Y.–square yard   V.L.F.–vertical linear foot   M.S.F.–thousand square feet

RANGE HOOD

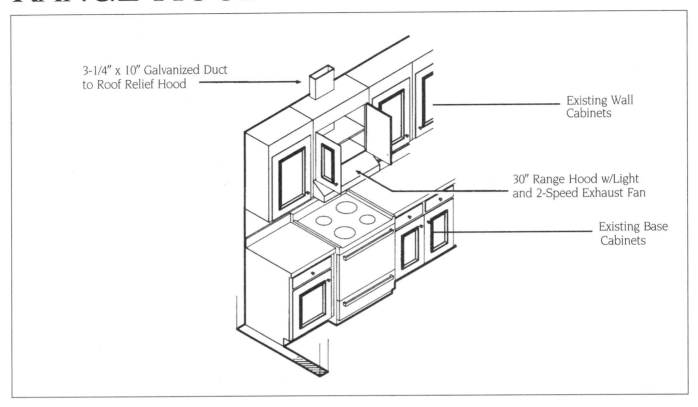

3-1/4" x 10" Galvanized Duct
to Roof Relief Hood

Existing Wall
Cabinets

30" Range Hood w/Light
and 2-Speed Exhaust Fan

Existing Base
Cabinets

It has often been noted that kitchen appliances are perhaps the hardest-working electrical components in your home. Generally overlooked, however, is the range hood – the exhaust fan and filtration unit that eliminates the heat, moisture, grease, and odors produced by cooking. This fixture is switched on almost every time the stove is used, and is usually left running after the stove is shut off, to remove the last traces of smoke and odor. Although modern range hoods are built for this kind of regular and extended use, it is not surprising that fan motors and bearings do, eventually, wear out and require replacement.

MATERIALS

If your old range hood has performed well over the years, it makes sense to replace it with a unit of the same, or similar, make and design. A standard hood measures 30" wide, fits under a 12" deep wall cabinet, and vents out the top or the back through a 3-1/4" x 10" rectangular duct. If, for any reason, you are unsure of the specifications of your existing unit, switch off the power to its circuit and remove it for analysis and measurement. The most critical distance is that from

the wall to the center of the exhaust duct. Even though range hood widths are the standard 30" mentioned earlier, the distance from the back edge of the unit to the center of its exhaust duct can vary among manufacturers. The closer you can come in duplicating these two measurements in the new and old fixtures, the easier the installation will be, obviating the need to bend the ductwork or to use ductwork elbows.

Hoods are available in different colors, but it may not be possible to exactly match the color of your range and other appliances. It will more likely be a matter of trying to find a neutral color to blend with the general color scheme of your kitchen. Other options are a one- or two-speed fan and a light. The latter does not add much to the cost, and is very useful.

Check the manufacturer's specifications for the fan's sound and air-flow ratings. The sound level of the fan is rated in units called "sones." One sone of sound is about as loud as a very quiet refrigerator. The fan's air-flow rating is given in terms of the number of cubic feet per minute (cfm) it can move. A more powerful and efficient fan will have a higher air-flow rating than a smaller, weaker one, but not

necessarily a higher sone rating. Generally speaking, a less expensive fan will have a correspondingly low air-flow rating and a higher sone rating than will a better designed, sturdier built and, hence, more expensive unit. In most cases, the latter is the better choice because it will run quieter, move more air, and have a longer trouble-free working life.

If you are planning a new range hood installation rather than a simple replacement, you will have to contend with the added expense and labor of running ductwork to the nearest outside wall. Duct runs, which are made of galvanized metal, should be short, with as few angles as possible, and go down and out instead of up and out, if there is a choice. This last will help prevent both heat loss from the kitchen and moisture build-up in the ducts. You might be tempted to install a ductless hood, which is supposed to recirculate the air while trapping grease and odors, but really does neither of these things very well. A ductless hood may be better than none at all, but if you can, install a ducted hood – it is worth the extra expense.

Be sure the power is off to the range hood's circuit before removing the old unit and installing the new one. You will need someone, or something, to support each hood in turn so you have both hands free to work on the electrical connections. Follow the wiring directions carefully, joining like-to-like with wire nuts and electrical tape, and making sure the ground connection is secure. Range hoods are usually attached to the wall cabinet with four screws, one in each corner.

LEVEL OF DIFFICULTY

Simple replacement of a range hood with one that matches the existing ductwork is a very easy task. A hood is made up of a small fan mounted in a thin metal housing, and is quite light. It is helpful to have an assistant to support the units – old and new – while you are doing the electrical work. Be sure to follow the wiring directions carefully, and if you are the least bit uncertain of what you are supposed to do, get professional or experienced help.

A new installation is a much more difficult undertaking. Cutting holes in cabinets and walls and running ductwork to the outside calls for careful planning and precise measurements. Sheet metal ductwork, whether rectangular, round, or a combination of both, can be quite difficult to deal with. This is not a job for a beginner, and an intermediate would be wise to get some expert advice before attempting it.

For a range hood replacement, the beginner should add 100% to the time estimate; the intermediate, 30%. Running ductwork for a new installation is not advisable for a beginner. The intermediate should plan the job with expert advice and add 100% to the time listed. The expert should add 20% to the time.

WHAT TO WATCH OUT FOR

Before declaring your old range hood defunct, do a little trouble-shooting to be sure that its problems are not just minor, repairable breakdowns or malfunctions. The fan's failure to work may simply be the result of a broken switch, bad contact, loose connection, or other short circuit. If it is a wiring problem, find out what caused it and be certain that any repair you do actually corrects it, and does not merely disguise it. Again, any question in this regard should be addressed to a licensed electrician.

The best-engineered exhaust fan cannot do its job if its filter is clogged or the ductwork is blocked. These conditions may have contributed to the demise of your old range hood and, if left uncorrected, they can do the same to your new unit. Check to see that the ductwork is free of obstruction and that the flap on the outside duct cap is operating freely. It should close to seal the duct, and pop open easily when the fan is turned on. The filter is designed to trap grease and airborne debris. It should be inspected periodically and, when seen to be clogged, replaced. If you are unsure of whether it is too clogged to work, replace it anyway – it is cheap enough maintenance.

SUMMARY

Range hoods provide kitchens with exhaust ventilation to remove air laden with the moisture, smoke, grease, and odors produced by cooking. If your hood fan has broken down, or is inadequately clearing the air, a replacement and/or upgrading is called for. Ideally, the dimension of the new unit should match the old. The manufacturer's specifications will give the fan's air-flow rating (cfm) as well as its sound rating (sone). Generally speaking, the higher the cfm number and the lower the sone number, the better. Be prepared to pay more to get top quality and high performance. A simple range hood replacement is an easy job, and having a new unit with a powerful but quiet fan will do much to maintain a pleasant atmosphere in your kitchen.

Range Hood

Description	Quantity/ Unit	Labor- Hours	Material
Remove existing fan hood	1 Job	4.0	
Hood for range, 2-speed, vented, 30" wide	1 Ea.	3.3	540.00
Totals		7.3	$540.00

Contractor's Fee Including Materials	$1,161

Key to Abbreviations
C.Y.–cubic yard   Ea.–each   L.F.–linear foot   Pr.–pair   Sq.–square (100 square feet of area)
S.F.–square foot   S.Y.–square yard   V.L.F.–vertical linear foot   M.S.F.–thousand square feet

Section Seven

INTERIOR ADAPTATIONS FOR SPECIAL NEEDS

Extensive remodeling to accommodate special needs should not be undertaken without extensive research and consultation involving the disabled person, concerned family members, physicians, therapists, caregivers, and support organizations – anyone whose insights and advice can help lead to the best choices among all the options. Also keep in mind the following considerations.

- Look for the best possible ways to secure new equipment (such as grab bars) to the walls. You may want to consider opening up the wall to install blocking.

- Make sure you carefully review every option in your plan. Walk through – or have the occupant with special needs go through – the project several times to ensure that you have not left anything out.

- Pay attention to color schemes and low maintenance finishes. For example, consider corner guards and bumpers to protect both a wheelchair user and the walls.

- Do not use soft woods on the floor or trim in areas where a wheelchair will be used.

- L-shaped or U-shaped kitchens work best for handicapped residents, since they allow objects to slide on a continuous countertop without having to be picked up.

INTERIOR ADAPTATIONS FOR SPECIAL NEEDS

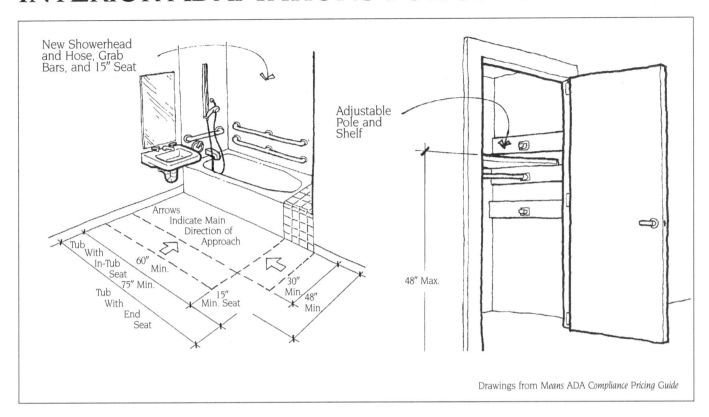

New Showerhead and Hose, Grab Bars, and 15" Seat

Adjustable Pole and Shelf

Arrows Indicate Main Direction of Approach

Tub With In-Tub Seat — 60" Min.

75" Min.

Tub With End Seat

15" Min. Seat

30" Min.

48" Min.

48" Max.

Drawings from *Means ADA Compliance Pricing Guide*

Within the past 20 years, advances in technology and architectural design have created new independence for people with physical limitations – both at home and in the workplace.

MATERIALS

Adapting a home for special needs can be relatively easy or fraught with difficulties, depending on the size and layout of the house and the extent and nature of the disability of its occupant. Extreme conditions in either case could necessitate drastic changes – a complete remodeling, the building of an addition, or even the purchase of a more suitable property.

In many homes, a number of more or less minor adaptations can be made to allow a person with physical limitations to live as independently as possible, for as long as possible.

Throughout the house, interior doors can be eliminated and, if necessary, doorways widened to accommodate a wheelchair. Doors needed for privacy can have lever handles installed to allow them to be opened with minimal effort. These doors can also be mounted on

offset hinges that swing them out and away from the frame to allow wheelchair access. For widening existing doors, see the project "Frame and Finish Partition Wall."

Electrical receptacles and light switches can be installed wherever necessary, at whatever height makes them easily reachable. Electrical outlets should be at least 15" above the floor. Controls (e.g., light switches) should be no more than 54" above the floor – specialists in design for disabled people recommend a maximum of 48". Also, both outlets and controls should be at least 18" from an inside corner. There should be adequate clear floor space in front of controls for approach in a wheelchair.

Heat-sensitive switches can be installed to automatically turn on lights when a person enters a room, and adaptors can be added to standard metal lamps to provide touch control. Overhead lighting fixtures can be placed in dark areas, or anywhere additional illumination would provide an extra measure of safety, such as in a stairwell.

A stairway lift or elevator will facilitate getting to the second floor. There are a number of lifts available for straight, curved, or winding staircases. Some are

simple seats mounted on a track; others are wheelchair platforms capable of lifting up to 400 pounds. Although they are more expensive and require a suitable space, standard shaft-type elevators are available for residential installation. Closets can be made accessible to wheelchair users by lowering shelves and poles to 48" above the floor.

An intercom system can be installed to allow quick and easy audio monitoring and communication throughout the house. For the hearing impaired, telephones can be wired to lights that flash with each ring, and amplifiers can be added to the handsets. Roll-out wire or wooden bins can be placed in closets to provide easier access to clothing.

Many accommodations can be made in the kitchen. Refer to the next project for some suggestions.

In the bathroom, a standard shower head can be removed and replaced by one that can be adjusted to various heights or by a hand-held model. The countertop and lavatory can be raised to reduce bending, or lowered, with the vanity cabinet removed, for wheelchair access. Sinks should be equipped with accessible faucets. Exposed pipes must be wrapped with insulation or otherwise protected from contact. A small bathroom with an in-swing door can become a trap in the event of a fall; change the door to out-swing. Grab bars should be installed near the bathtub, shower stall, toilet, and anywhere such a stabilizer is needed. Grab bars should be 33" to 36" above the floor. Be certain that grab bars are attached to solid bearing. Open the wall, if necessary, and install blocking. Nonslip mats should be placed in the tub or shower, and on a tile floor. An anti-scald adaptor that automatically mixes hot and cold water to a safe temperature can be added to the shower or tub. Toilet seats should be 17" to 19" from the floor. Removable and adjustable seats can be attached to the toilet or placed in the tub or shower to facilitate the use of these fixtures. The bathroom should also have 60" diameter clear turning space to accommodate a person in a wheelchair.

For estimates of modification projects to meet the needs of physically challenged people, see Means ADA Compliance Pricing Guide, *also published by R.S. Means Company. The book* contains 75 *of the most commonly needed alterations, and many variations on each, illustrated and with detailed cost estimates.*

LEVEL OF DIFFICULTY

Many of the devices mentioned here – and hundreds of others – can be installed easily by the average homeowner with standard household tools and the ability to read and follow instructions. A number of electrical and plumbing adaptors simply plug in or snap on to existing fixtures, but work that requires running new wires and pipes, or installing new fixtures, should be done only by an expert or a professional contractor. Structural adaptations, such as removing doors, widening passageways, or repositioning counters, can be accomplished by the average intermediate. Installing specialized machinery such as stair lifts and elevators should be done only by a professional, preferably one who is licensed. Be sure that any changes you make in the house conform to all applicable codes; do not compromise safety for convenience or thrift.

In general, the beginner should increase the time estimates by 100%; the intermediate, by 50%; and the expert, by 15%.

WHAT TO WATCH OUT FOR

Many of the new devices and procedures that allow people to cope with physical limitations are the indirect results of the motion and time study research done years ago by the industrial engineers Frank and Lillian Gilbreth. They pioneered the development of a scientific approach to the study of ways in which industrial workers might save time and energy. They used a six-question approach to the study of work simplification, which can help in solving problems related to a disabled person performing daily tasks around the house.

- What, precisely, is the job to be done?
- Why should the job be done; is it necessary?
- Where should it be done? Could energy or time be saved if it were done elsewhere?
- When should it be done?
- Who should do the job? Could it be done more efficiently and economically by someone else – a family member or hired help?
- How should it be done? Must there be some adaptation of the house or its mechanical appliances, or is there a need for new tools or appliances?

SUMMARY

Consultation with experts in the field (for example, members of the National Council on Independent Living) can provide information and support to the disabled person and his or her family. It is possible that a major home remodeling job might be called for, but in many cases, outfitting the house with a number of small, relatively inexpensive and easy-to-install electrical and mechanical devices can provide enough access and convenience to enable a physically challenged person to enjoy a full and satisfying independent life.

For other options or further details regarding options shown, see

> *Wheel-chair accessible kitchen*
>
> *Wheelchair ramp**
>
> * *In* Exterior Home Improvement Costs

Interior Adaptations for Special Needs

Description	Quantity/ Unit	Labor- Hours	Material
4 duplex receptacles, including 20 ft. of wire & conduit, for each	4 Ea.	6.0	78.48
Non-keyed passage set	2 Ea.	1.3	108.00
Handicapped lever	2 Ea.		316.80
Swing clear hinges	1 Pr.		230.40
Surface mounted 60 W economy light	1 Ea.	0.3	61.80
Wiring and switch for light	1 Ea.	0.3	8.88
Residential Stair Climber (chair lift), single seat	1 Ea.	16.0	4,710.00
Telephone flashing adapter with bell	1 Ea.		36.00
Telephone handset amplifier	1 Ea.		24.00
Grab bar, straight 1-1/4" diameter, stainless steel, 30" long	2 Ea.	0.7	60.00
Shower seat adjustable, with back	1 Ea.	0.3	158.40
Clamp-on raised toilet seat	1 Ea.	0.3	111.60
Hand-held shower head	1 Ea.	0.4	76.20
Totals		25.6	$5,980.56

Contractor's Fee Including Materials	**$9,926**

Key to Abbreviations
C.Y.–cubic yard   Ea.–each   L.F.–linear foot   Pr.–pair   Sq.–square (100 square feet of area)
S.F.–square foot   S.Y.–square yard   V.L.F.–vertical linear foot   M.S.F.–thousand square feet

WHEELCHAIR-ACCESSIBLE KITCHEN

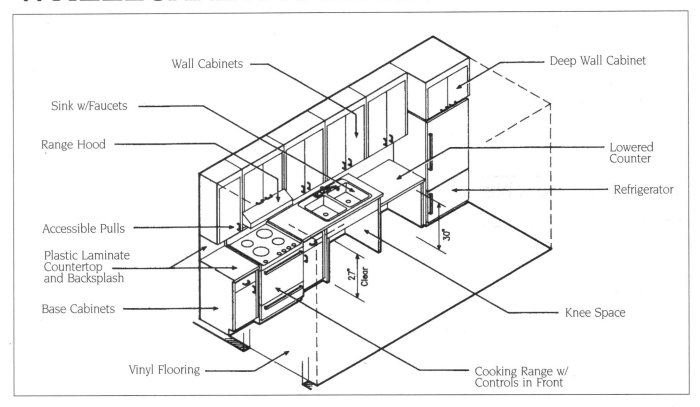

Wall Cabinets

Sink w/Faucets

Range Hood

Accessible Pulls

Plastic Laminate Countertop and Backsplash

Base Cabinets

Vinyl Flooring

Deep Wall Cabinet

Lowered Counter

Refrigerator

Knee Space

Cooking Range w/ Controls in Front

30"

27" Clear

Adapting a straight-wall kitchen for special needs can be relatively easy or quite difficult, depending on the level of accessibility that is needed. A good first step is to familiarize yourself with standard requirements (such as those established in the Americans with Disabilities Act)[1] regarding accessible kitchen spaces. The primary requirement is that the kitchen must be on an accessible route, with an accessible entrance.

It is unlikely that you will adapt only one area of your home for special needs. Modifying a kitchen will probably be done in conjunction with other projects. The level or extent of your work can be directly proportionate to the special needs of the user.

MATERIALS

To be accessible to people in wheelchairs, there must be at least 30" x 48" of clear floor space for a frontal or parallel approach to all features (appliances, counters, etc.). Countertops and sinks must be no more than 34" above the finished floor. A counter lowered to 30" above the floor will provide usable work space for wheelchair occupants. At least

50 percent of the shelf space in the cabinets and in the refrigerator/freezer must be within reach of the disabled person. Controls and handles should be accessible (ideally, operable with a closed fist). Finally, the flooring material should be slip-resistant.

Some modifications are straightforward and can be done by a beginner or intermediate do-it-yourselfer. For example, if the cabinet floor under the sink is not installed and the finish floor continues for the full depth, removing the cabinet doors can provide knee space. Base cabinets shouldn't structurally support the sink, so they can be removed for knee space if necessary.

Lever-handled faucets and a hose at the sink are recommended, even at sinks with a faucet located near the front. Loop-type handles on drawers and cabinets allow for ease of use (cabinets with routed holds are difficult for people with diminished fine-motor control to use). Pull-out drawers and lazy Susans prevent the need for reaching to the back of storage spaces.

Some standard appliances are easier to use by a wide range of people. Side-by-side refrigerator/freezers allow a range of storage space on both sides. Stove and

range controls should be located in front, and staggered burners prevent having to reach over a hot surface from a seated position.

Depending on your needs, it may also be important to make the kitchen usable for people with low vision. Lighting levels should be high at work stations, and placed so as not to cast shadows on the work space. Light-colored finishes, matte surfaces, and sharp color contrasts also help people with low vision.

LEVEL OF DIFFICULTY

The level of difficulty of this project depends on your accessibility needs. Most beginners can undertake the cabinet and storage modifications. The sink and stove alterations can be done by intermediates. Only expert remodelers should attempt a total kitchen rehab.

For basic changes that do not require a large amount of cutting and patching to match existing materials, a beginner should add 50% to the time estimated here. For a major renovation or the addition of new cabinets, add 100% to the time, and plan to work with a skilled carpenter or do-it-yourselfer. Intermediate-to-expert remodelers should add 50% to the time required for a full renovation.

WHAT TO WATCH OUT FOR

When developing your plan, check the dimensions of existing cabinets to ensure that you can remove and replace them without a lot of additional work. If your cabinetry is built-in or custom, you may want to forego cutting and patching and simply replace the modified unit with new cabinetry.

It is recommended that demolition be done carefully and as much undamaged material as possible saved for future use. At some point, you may be unable to match an existing piece and may need to retrofit an old piece.

Even basic retrofitting will require cutting and removing of existing cabinet components. Some manufacturers use systems that include hidden hardware or fasteners. Check for hidden fasteners and ways to brace the cabinetry pieces that will remain after the cut has been made.

Painted cabinets tend to chip when cut. Use tape to protect the surface before cutting and be sure to select the proper blades for cutting, using information on the packaging, or by asking your hardware dealer.

SUMMARY

Creating a work area for special needs is a very worthwhile and rewarding project. The adaptations really do not need to cause the overall look of your house to change. Careful planning and selection of materials and equipment can enhance the project. Some alterations can be temporary and accomplished easily; others require professional help, especially in the plumbing and electrical areas.

For other options or further details regarding options shown, see

> *In-law apartment*
>
> *Interior adaptations for special needs*

[1] Available from the U.S. Department of Justice in Washington.

Wheelchair-Accessible Kitchen

Description	Quantity/Unit	Labor-Hours	Material
Blocking, for mounting cabinets, 2 x 4	18 L.F.	0.6	7.78
Plywood, underlayment-grade, 3/8" thick, 4' x 8' sheets	128 S.F.	1.4	73.73
Flooring, vinyl composition tile, 12" x 12" x 1/16", plain	65 S.F.	1.0	78.00
Wall cabinets, w/ wire pull handles, 12" deep, 30" wide, hardwood	4 Ea.	3.7	1,061.28
Wall cabinets, w/ wire pull handles, 15" deep, 15" wide, hardwood	1 Ea.	0.8	178.20
Base cabinets, w/ wire pull handles, 24" deep, 15" wide, hardwood	2 Ea.	1.5	456.72
Countertop, 1-1/4" thick, plastic laminate, w/ backsplash	8.75 L.F.	2.3	80.33
Trim, crown molding, stock pine, 9/16" x 3-5/8"	28 L.F.	0.9	51.07
Paint, ceiling & walls, primer	285 S.F.	1.1	13.68
Paint, ceiling & walls, 1 coat	285 S.F.	1.8	17.10
Paint, cornice trim, simple design, incl. puttying, primer	28 L.F.	0.3	0.67
Paint, cornice trim, simple design, 1 coat	28 L.F.	0.3	1.01
Cooking range, freestanding, 30" wide, one oven	1 Ea.	1.6	304.80
Hood for range, 2-speed, 30" wide	1 Ea.	2.0	42.60
Refrigerator, no frost, side-by-side	1 Ea.	5.3	3,330.00
Sink w/gooseneck faucet & blade levers, 33" x 22", dbl. bowl	1 Ea.	4.0	640.80
Rough-in, supply, waste & vent for sink	1 Ea.	7.5	102.00
Totals		36.1	$6,439.77

Project Size	6'-6" X 14'	Contractor's Fee Including Materials	$10,877

Key to Abbreviations
C.Y.–cubic yard   Ea.–each   L.F.–linear foot   Pr.–pair   Sq.–square (100 square feet of area)
S.F.–square foot   S.Y.–square yard   V.L.F.–vertical linear foot   M.S.F.–thousand square feet

Section Eight
INTERIOR FINISHES

Changing your home's interior finishes can alter the atmosphere dramatically. The following tips can help you to make the right choices regarding materials, color, and installation.

- When you choose a flooring material, check the manufacturer's instructions for acceptable substrates and proper adhesive.

- Underlayment options for flooring include the following:
  - Ceramic tile: old ceramic tiles in good condition; concrete slab; cement board; underlayment-grade plywood.
  - Vinyl flooring: unembossed vinyl or linoleum floor in good condition; underlayment-grade plywood; lauan plywood.
  - Wood flooring: old vinyl or linoleum floor in good condition; underlayment-grade plywood; lauan plywood; hardboard.

- Wood flooring gives a room warmth and character. When properly sealed, wood floors can be fairly resistant to moisture.

- Vinyl flooring used to be considered inferior to ceramic tile. Today you can spend as much on some seamless vinyl flooring as on tile. There are subtle differences in vinyl grades.

- Old vinyl flooring may contain asbestos, which can cause cancer. Have a certified professional test the flooring material and eventually remove and dispose of it, if necessary.

- Consider linoleum for its all-natural ingredients.

- Wood-look "floating" floors are another option.

- When choosing paint and wallcoverings, remember that lighter colors help to make a room appear larger and brighter. Darker shades tend to shrink rooms and can make them appear more cozy. If you want a long narrow room to look wider, paint one or both of the short walls a bright or dark color and the other walls a pale color.

- Homes built before 1978 often contain lead-based paint, which should be handled only by properly trained, certified, or licensed abatement contractors. Lead exposure to children under age 6 poses a serious health threat, so be sure to keep young children away from a work site where lead paint is being removed.

- Bargain paints sometimes require two or three coats to provide the same coverage as one coat of a quality paint.

- Because walls are much larger than a paint swatch, you will find that the color you select will seem to darken and intensify as you spread it over the wall.

- Many paint materials and solvents need to be used with extreme caution and disposed of with care. Dispose of empty paint cans according to your local disposal regulations.

- Install interior gypsum wallboard only after all rough inspections by the building department have been completed.

- Because colors appear different depending on the light source, look at ceramic tiles, vinyl floor tiles, and paint swatches in daylight as well as under the electrical lighting that exists in the room.

INTERIOR DOORS

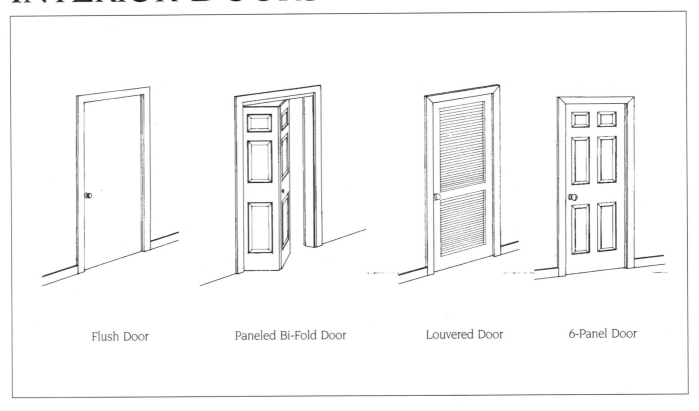

Flush Door   Paneled Bi-Fold Door   Louvered Door   6-Panel Door

Interior doors are in constant use in our homes. They provide privacy and security for ourselves and our possessions, and they reflect the character and decor of the spaces they separate.

Many types and styles of doors are available, with a wide range of prices and quality. But appearances can be misleading. A quality door is not cheap, and hanging one properly takes time. Before you purchase, comparison shop and talk to a professional or two.

Try to avoid buying hollow core, precased units known as "split-jambs." Because the casings are in the way, it is all but impossible to shim the jambs into the trimmer studs. The manufacturer's instructions call for shims, but everyone, including the manufacturer, knows that they are rarely used. Instead, these precased doors are nailed to the walls through the casings, and that's that. Before long, the nails work loose, or the casings split; the door drops out of alignment and binds, rubs, rattles, or won't stay shut. This worst-case scenario may not apply to your home, but similar problems may exist, or you may simply not like the looks of your interior doors.

New doors can be a considerable expense, and installing them takes time. But, even if you decide to replace every door in your house, this type of remodeling is minimally disruptive, and you can do it one or two doors at a time.

MATERIALS

The kind of replacement door you choose is determined by the overall style and design of your house. If you are at all unsure, get some advice. Any sort of remodeling project should enhance the appearance and value of your home, and that can occur only when the materials match the house and blend into the overall look.

Unless you intend to increase or decrease the door opening (a much more involved undertaking), the size of the replacement door is, obviously, that of the existing door. Interior doors are generally 1-3/8" thick. (Bi-fold doors, a very different type of installation, are usually 1-1/8" thick.) Most doors come in either 6'6" or 6'8" heights, and in widths ranging from 1' to 3', in increments of 2". A door can be trimmed a little, within the limits imposed by design and construction, to fit an odd-sized opening. Stock sizes depend

on the type and style of the door. Consult manufacturers' catalogs with a knowledgeable salesperson or contractor to determine what is available.

Interior swing doors may be classified as flush, louvered, sash, or paneled. The simplest and most common door style is flush, which means flat and plain. Flush doors are made with hollow or solid cores. Hollow-core doors contain a gridwork of thin wood or cardboard, surrounded by a light frame of wood and covered by a thin layer of species plywood ranging from lauan mahogany to birch, oak, ash, and others. Solid-core construction uses composition board to support the outer plywood layer. This results in a heavier, stronger unit with a much more substantial feel.

Louvers are horizontal slats installed in the door at an angle, in such a way as to obstruct vision but permit the flow of air. These are generally used on clothes closets, or where air circulation is desirable.

Sash doors, also called French doors, come in pairs, each door containing one large or a number of small glass panes. They are commonly used in entrances to porches and sun rooms, or between rooms, to lend a touch of elegance and formality.

Paneled doors consist of a frame (made up of stiles, rails, and mullions) containing a varying number of flat or raised panels. One form of paneled door is the pressed hardboard unit. This is basically a hollow-core construction with the surface molded to look like panels. Wood paneled doors are made of small pieces of solid wood glued together and covered with a wood veneer. If you plan to stain the door, be sure it is classified as stain-grade, to ensure that the veneer is high quality and well-joined.

This project assumes that your existing door jamb is properly installed and still reasonably square, level, and plumb. It also assumes that the existing casing will match the looks and quality of the new door. If your house is afflicted with the split-jamb doors mentioned above, or if the jambs are badly damaged or askew, the whole unit – door, jambs, and casing – should be removed and replaced. A premium-grade prehung door is recommended for easiest installation. If new casing is desired or necessary, be sure it is compatible with both the new door and any existing trim in the room.

If the old door works well and the hardware is in good condition, you can use the door as a template for locating the hinges and the lockset on the new door. It is critical that these locations be transferred to the new door exactly. Tolerances are very close, and a seemingly small discrepancy can cause operational problems when the door is hung.

LEVEL OF DIFFICULTY

Hanging a door is one of the more difficult jobs of finish carpentry. It is made somewhat easier by the use of specialized tools such as butt gauges and hinge markers, but a well-functioning door is more the result of knowledge skillfully applied. A door is a very expensive piece of material to practice on, and a fairly small mistake can be ruinous. A beginner should not attempt this job without over-the-shoulder guidance from an expert. Even a competent intermediate is advised to proceed cautiously, observing the old adage, "measure twice; cut once." In fact, the truly prudent carpenter will recheck measurements more than twice before setting tool to wood.

Hanging a door is also notoriously slow, labor-intensive work. A beginner, working under an experienced, watchful eye, should not be surprised if the job takes three times as long as the estimate given here. The intermediate should add 100% to the time; the expert, 50%. If your project calls for many doors to be replaced, it is reasonable to expect that the time required will gradually decrease as experience is gained.

WHAT TO WATCH OUT FOR

Most hollow-core doors are light enough to operate well on two hinges. If your replacement door is of the heavier solid-core or wood paneled type, adding a third hinge is a good idea. This may mean that you'll not be able to use the existing hinges if you can't find a new hinge to match. If the door is very heavy, you would do well to not only add a third hinge, but also increase the size from the standard 3-1/2" to 4" hinges. If you do need to replace the hinges, be sure the color and surface finish match others in the room. This also applies to the style and finish of the lockset. Be sure that hinged doors swing in to dead walls, away from electrical outlets, switches, and traffic flow through the room.

If you are installing a prehung unit, it must be shimmed into the rough opening to be made plumb, square, and level. Use the narrow shim shingles that are made for this purpose, and put three or four sets on the hinge side, and two or three on the lock side. To provide a nailing surface parallel to both the frame and the jamb, insert one shingle from either side of the jamb (the door having been removed), and overlap them equally until the desired thickness is achieved. Be sure to nail through the shims. Cut the shingle ends back far enough so they won't interfere with the casings.

Using solid masonite doors instead of six-panel pine doors can save approximately $75 to $100 per door. Hollow masonite doors cost even less. You can also buy a pine veneer door, with pine only on the door surface, for a price in between the costs of the hollow masonite and solid masonite doors.

INTERIOR DOOR OPTIONS
Cost Each, Installed

Description	
Hollow Core, Flush	$180.00
6-Panel	$289.00
French	$270.00
Louver	$226.50
Bi-fold, 4' wide	$183.25

Interior Door Replacement

Description	Quantity/Unit	Labor-Hours	Material
Door, six panel, solid, 1-3/8" thick, pine, 2'-8" x 6'-8"	1 Ea.	1.1	147.60
Paint, panel door and frame, 3' x 7' per side, oil, primer	1 Ea.	1.3	2.36
Paint, panel door and frame, 3' x 7' per side, oil, 1 coat, brush	1 Ea.	1.3	2.11
Totals		3.7	$152.07

Project Size	2'-8" x 6'-8"	Contractor's Fee Including Materials	$394

Key to Abbreviations
C.Y.–cubic yard   Ea.–each   L.F.–linear foot   Pr.–pair   Sq.–square (100 square feet of area)
S.F.–square foot   S.Y.–square yard   V.L.F.–vertical linear foot   M.S.F.–thousand square feet

SUMMARY

Doors are operational components of your home's decor; that is, they must both look good and function well. Cheap doors rarely do either. The replacement door you choose should be appropriate to its location, compatible in design to the look and style of your house, and of sufficient quality to rate as a true improvement over the existing unit. The materials and labor involved justify, and may even demand, the services of a professional carpenter. The upgrade new doors can provide to your home can certainly balance the expense incurred.

WAINSCOTING, CHAIR RAIL, & MOULDINGS

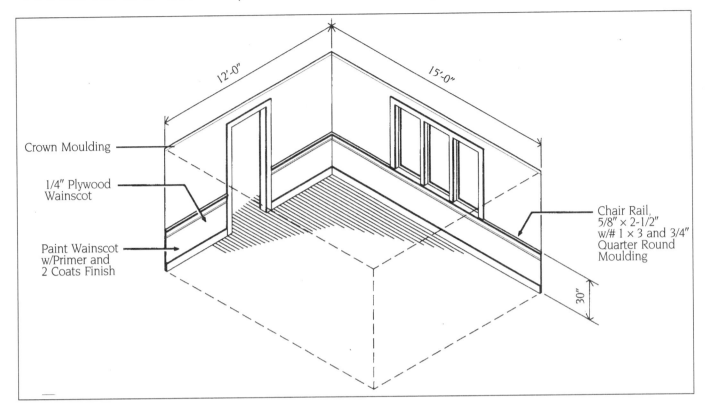

Crown Moulding

1/4" Plywood Wainscot

Paint Wainscot w/Primer and 2 Coats Finish

12'-0"

15'-0"

Chair Rail, 5/8" × 2-1/2" w/# 1 × 3 and 3/4" Quarter Round Moulding

30"

Many features of traditional house design seem at first glance to be merely decorative, but upon closer inspection reveal a practical aspect. Painted wood wainscoting and chair rails have added charm and visual interest to the first-floor (heavy-use) rooms in houses from the Colonial period through the Victorian Age to the present. They were intended to absorb the inevitable bumps and scrapes of chairs and tables (and, more romantically, spurs and scabbards), preserving the less durable and more expensive plaster and wallpaper on the upper portions of the walls.

As a result of changes in taste and architecture, and in the economics of construction, these features have almost completely disappeared. First-floor walls may appear somewhat bland and empty, and are, in fact, unprotected. Lining the lower sections of these walls with appropriate-looking wood can dress up a room with a variety of textures, lines, and colors, while shielding the walls from wear and tear. This is an idea whose time has come, again.

MATERIALS

Wainscoting, capped by a chair rail, usually covers the bottom third of a wall. This is only an approximation; the actual height of the rail can be anywhere from 30" to 36" above the floor. Where you establish this line largely depends on the location of the room and its overall dimensions, the heights of the ceiling and window sills, the size and style of other mouldings, etc. You should be guided by your eye and your sense of balance and proportion. These same considerations apply to your choice of materials. If you feel you lack the judgment to make these decisions, or if you have doubts about the appropriateness of your choices, ask the opinions of others whose experience, taste, and decorative sense you trust. Because the area in question is in the "public" part of your house (living room, dining room, hall, parlor), it may be worth the expense of hiring a design expert to help you plan the project.

There are basically two types of wainscoting materials available: individual, solid-wood boards, or laminated-wood sheet product. Individual boards – for example, 1 x 4 tongue-and-groove, beaded boards – provide an

informal, country look, common in late 19th- and early 20th-century farmhouses. This type of lumber is readily available, reasonably priced, and can be stained or painted. The best method of installing it involves a good deal of work, but the reward is the built-in, substantial look of the finished product. Essentially, what is required is to cut and remove the wallboard in the area to be covered, let in 1 x 4 nailers across the studs, install the vertical wainscoting, and cap it with the chair rail. Too bad it can't be done as easily as it can be described! Door (and possibly window) casings, baseboards, and window aprons must be removed, as well as outlet plates and anything else that may interfere with the removal of the wallboard.

Establishing a level line at the chair rail height around the room is most important. If you intend to install wainscoting in more than one room, and the ceilings are a consistent height, it makes sense to measure down from the ceilings to establish the chair rail height so as to avoid problems that could possibly arise because of discrepancies in floor levels from room to room. Snap a chalk line, cut along it with a keyhole or drywall saw, and pry off the wallboard.

Expect a mess; plaster dust has a way of drifting everywhere. Snap lines on the exposed studs and cut notches 3/4" deep to accommodate two 4" nailer strips, top and middle. The sole plate, 2 x 4 pieces attached to it, will serve as the bottom nailer. If you cut through the vapor barrier in an outside, insulated wall, be sure to staple a layer of 6 mil polyethylene over the area before nailing up the wainscoting.

This same installation can be used for sheet products, such as 3/4" AC plywood or hardboard. When painted, the result is a more formal, antique look, such as can be seen in elegant houses of the late Colonial and Federalist periods. Thinner sheet materials (1/4" or 3/8") of either smooth or textured surface can be installed in a quicker, more economical, and less messy way by simply gluing and tacking it directly over the wallboard, using construction adhesive and 1-1/4" brads. Be sure to locate all the studs for nailing and, of course, all wires and water pipes for where you want to avoid nailing.

The chair rail itself can be a simple one-piece moulding or a more elaborate design made up of several different mouldings. Again, you must let the overall decor and style of the room and its existing finish detailing dictate what is going to look proportional and appropriate. Stained and varnished chair rails will add an elegant look to a room. Paint, which is used in this project, can also be elegant, especially if the moulding is painted a contrasting color from the walls. Real antique chair rails were often

quite substantial in depth, if not also in width, extending 1-1/2" or more beyond the surface of the wall. You might well decide, however, that a less costly and more basic treatment, such as a plain wainscot moulding, will look perfectly acceptable.

LEVEL OF DIFFICULTY

This project involves the tools and skills of basic finish carpentry. Outside corners will require 90° miter joints, and should be glued as well as nailed. (Clean off all excess glue if you intend to stain the wood.) Inside corners should be coped and scribed. This detail prevents separation of the joint by laying one piece of trim over the other and cutting it to fit exactly. If you plan to paint, you gain a little working leeway because you can fill any small gaps with caulk, but if the wood is to be stained, your joinery skills will be on display for all to see. You will need a miter box and back saw or, if you prefer to rent one, a power miter saw. Once you get used to it, you'll be able to work faster and more accurately. Removing the existing mouldings and baseboards, and tearing out the lower wall, calls for more care than skill. Cutting the notches in the studs for the nailers requires operating a circular saw horizontally, and it can kick back if not handled properly.

A beginner should get informed advice on the correct layout and the selection of appropriate materials, and should probably be guided through all phases

of preparation and installation. Time estimates should be increased by 100%, 50%, and 20%, respectively, for beginner, intermediate, and expert.

WHAT TO WATCH OUT FOR

If you choose to install any type of solid wood wainscoting, the boards should be stacked inside the house for a week or so prior to installation. Separate the boards with spacers so air can circulate around them and acclimate the wood to the humidity level of your house. This will help minimize any shrinkage that might open up unsightly gaps between the boards after they are nailed up. It is also a good idea to apply a coat of primer or sealer to the backs of the boards before installing them. Once they are up and the faces finished, the humidity absorption will be more even, front and back, and the likelihood of cupping and warping will be reduced.

The thickness of the wainscoting may require that you move electrical boxes out far enough to be flush to the new surface of the wood. If for any reason this cannot be done, an extended collar may be used. This is a metal device that slips around the receptacle and is secured with longer versions of the same screws that fasten the receptacle to the box. Just be sure the collar's "ears" catch the wood, top and bottom. This will be no problem if you have been careful to cut the hole in the wainscoting accurately, both for size and location.

It is important to realize that adding moulding is not always a wise idea. For instance, it could make a small room look smaller instead of more elegant.

SUMMARY

Wainscoting, capped with a chair rail, is both attractive and functional. It adds interest to a room and helps protect the walls from the nicks and scars of traffic and normal household activity. A wall-removal type of installation is more labor-intensive and will produce more mess than the tack-and-glue method, but in either case the room can still be used by your family during the course of the work.

For other options or further details regarding options shown, see

Painting and wallpapering

Wainscoting & Chair Rail

Description	Quantity/ Unit	Labor-Hours	Material
Plywood panel, A2 grade, 1/4" x 4' x 8' pine	224 S.F.	8.0	365.57
Crown molding, 11/16" x 4-5/8"	60 L.F.	2.2	138.24
Chair rail, 5/8" x 2-1/2"	60 L.F.	1.8	61.20
Molding, pine, 1 x 3	60 L.F.	2.0	43.92
Molding, quarter round, 3/4" x 3/4"	60 L.F.	1.9	30.24
Paint, primer, wall and trim	224 S.F.	2.2	24.19
Paint, 2 coats, wall and trim	224 S.F.	2.8	37.63
Totals		20.9	$700.99

Project Size	12' x 15'	Contractor's Fee Including Materials	$2,024

Key to Abbreviations
C.Y.–cubic yard   Ea.–each   L.F.–linear foot   Pr.–pair   Sq.–square (100 square feet of area)
S.F.–square foot   S.Y.–square yard   V.L.F.–vertical linear foot   M.S.F.–thousand square feet

OAK STRIP FLOOR

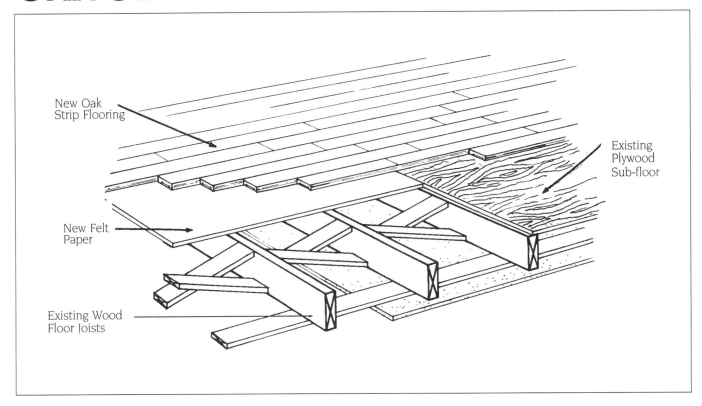

New Oak Strip Flooring

New Felt Paper

Existing Wood Floor Joists

Existing Plywood Sub-floor

Floors are, by their nature, a very durable fixture, but there are times when a complete replacement is called for. You may want to change from a nonwood flooring material such as tile, or change from one type of wood to another. Water damage resulting from flooding may have warped and buckled a floor beyond repair, or many years of foot traffic and refinishing may have finally taken their toll.

There is a wide variety of wood flooring materials available, and choosing the one best suited to your needs and the style and design of your house may take some research. A reputable lumberyard is staffed by people with the knowledge and resources to help you come to the right decision.

Most hardwood finish flooring is made from white or red oak, although beech, birch, maple, and pecan are also common. Softwood flooring includes fir, hemlock, yellow pine, and, for an authentic New England Colonial look, white pine. These softwoods are generally less expensive. Hardwood flooring comes in strip, plank, and parquet form. Strip is the most common, and usually comes tongue-and-groove style with ends and edges matched. Uniform grading

rules have been established for hardwood strip flooring based on its length and the number of defects. Some flooring varieties are available prefinished.

MATERIALS

Stack the flooring in the room where it will be laid and let it adjust to humidity conditions for about a week before installation. If you don't, it may shrink after it is laid, producing wide cracks along the joints.

The existing floor must be torn up to expose the subfloor. Carefully remove any baseboards and moulding and save them for reinstallation. Door casings may or may not have to be removed. The new floor should go under them and it may be possible to square-cut the bottoms with a backsaw in preparation. The difficulty of removing the old floor depends on its material and location. A kitchen tile floor will take a good deal more time and will be more inconvenient than a wide-pine bedroom floor. Some ingenuity and a lot of elbow grease may be required.

Make sure the subfloor is securely nailed and swept clean. If the subfloor is not plywood, you may want to install an

underlayment over it to provide a firm and smooth surface for the finish floor and to ensure against the possibility of warping or shrink-and-swell of the subfloor boards.

Apply a layer of building paper, lapped 3″ to 4″ at the seams. Flooring laid in the direction of the longest dimension of the room gives the best appearance. In any case, it must be laid perpendicular to a board subfloor that runs at right angles to the joints.

Lay out the floor so that if the two outside strips have to be "ripped," or cut to fit, they will be about the same width. The location and straight alignment of the first course is very important. Find the exact middle of each end wall and snap a chalk line between those two points. Lay the first strip exactly parallel to this line, leaving an expansion gap of about 1/2″ from the starter wall. Keep the tongue side toward the center of the room. The first two or three courses must be blind-nailed using 7d hardened, cut, or spiral-shank nails. A 20-ounce hammer will give you the necessary driving power. Take care not to miss or glance off the nail and damage the edge of the flooring. Also wear safety glasses. Set the nails using another nail as a nail set. A regular nail set will be ruined because the nails are harder than the set.

Once clear of the wall, you can use a power nailer to fasten all but the last couple of courses, which will have to be blind- and face-nailed with a regular hammer. Joints should, of course, be staggered, and any obstructions, such as a hearth, should be framed using the flooring wood with mitered corners. Tight joints are essential to a professional-looking job. Use cleats and pry bars with pieces of scrap flooring to urge bowed strips into line. The power nailer will tend to draw the strips tightly together. Never hit the tongue edge directly with a hammer; always use a scrap piece to absorb the damaging blow.

With some exceptions, flooring is not ready for finishing until it has been sanded. Rent a drum sander and a disk edger with enough paper to complete the job. Start with coarse grit, followed by medium and fine, and remove just enough wood to make the surface smooth. Run the drum sander with the grain of the wood, never across it. The edger will take care of those areas close to the walls that the drum sander can't reach. Once the floor has been finished with stain and polyurethane, replace the baseboards and moulding, spackling and touching up the paint where necessary.

LEVEL OF DIFFICULTY

Laying a new floor means a total disruption of the normal use of that room from tear-out to finish. If the room is a high-use area such as a kitchen or hallway, the daily routine of the entire household is affected. Scheduling is important. Arrange for the job to be completed as quickly as possible to minimize its negative impact. Life provides enough trials of a family's patience and goodwill without the added burden of a desultory and dragged-out improvement project.

Because of the high visibility of the finished product, a beginner should think twice about undertaking this project alone. When working with hardwood flooring, mistakes are always costly. Even a poor sanding job is difficult and costly to correct. Major errors like poor alignment and loose joints are not correctable short of tearing out and starting over.

The intermediate and expert should have no trouble managing a successful completion of this floor. The only tools they might lack experience with are the floor sanders, but if they proceed with caution and remember to keep the sanders moving, the results should be, if not professional, at least acceptable.

A beginner could apply the stain and polyurethane, and should add about 50% to the time required. Intermediates should add 60% and experts 30% to all tasks except the application of the finish, for which they should add 10%.

WHAT TO WATCH OUT FOR

Flooring is packaged in odd lengths to make it possible to lay a floor with no end joints side by side, and to give you short pieces to fill out at the end walls with a minimum of waste. Carpenters generally save the short lengths for use in closets and other small areas. Strips should be laid out in a staggered pattern with end joints at least 6" apart. Professionals even try to avoid several end joints close together. They also try to distribute long and short pieces evenly and avoid clusters of short strips.

After completing work with the drum and disk sanders, smaller belt and orbital finish sanders can be used with fine-grit paper to remove the last traces of the edger swirls. Remember, stain will bring out sanding marks that may have been virtually invisible beforehand.

Wood floors are commonly given three coats of polyurethane varnish for long-lasting and durable protection. High-gloss varnish is the hardest and may be used for the undercoats, with a final coat of the softer satin finish. Coats must be applied within a specific time or they will not bond together, so be sure to read and follow the directions on the can. Also, rub down the floor between coats with fine steel wool. Use tack rags to pick up loose filaments, brush hairs, and assorted specks of debris.

SUMMARY

Floors receive a lot of wear and tear, and their condition and overall look greatly influence the appearance of a room. Replacing a worn or inappropriate flooring material with well-chosen, properly installed, and finely finished wood adds permanently to the value of your home and will enhance your enjoyment of it for years.

For other options or further details regarding options shown, see

> *Ceramic tile flooring*
> *Parquet floor*
> *Vinyl sheet flooring*

Oak Strip Floor

Description	Quantity/ Unit	Labor-Hours	Material
Oak strip flooring, white or red, 25/32" x 2-1/4", including sanding	210 S.F.	9.9	698.04
Asphalt felt sheathing paper, 15 lb.; one roll minimum	175 S.F.	0.4	4.20
Power nailer, rent/day	1 Day		34.20
Portable electric compressor, rent/day	1 Day		33.00
Nails, 200 lbs. per box	200 lbs.		292.80
Disk sander, rent/day	1 Day		25.80
Drum sander, rent/day	1 Day		39.00
Sanding sheets	6 Ea.		7.78
Sanding disks	6 Ea.		4.82
Varnish, wood floor, brushwork	210 S.F.	0.7	15.12
Totals		11.0	$1,154.76

Project Size	12' x 16'		Contractor's Fee Including Materials	$2,209

Key to Abbreviations
C.Y.–cubic yard   Ea.–each   L.F.–linear foot   Pr.–pair   Sq.–square (100 square feet of area)
S.F.–square foot   S.Y.–square yard   V.L.F.–vertical linear foot   M.S.F.–thousand square feet

Parquet Floor

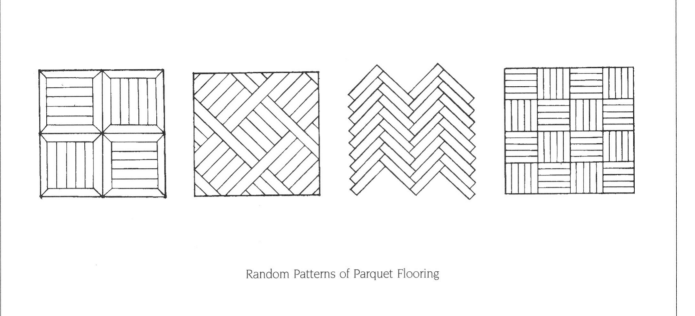

Random Patterns of Parquet Flooring

Because the floor of a room is a large, highly visible surface, subject to a great deal of abuse, much thought should be given to the selection of the material that covers it. Both appearance and durability must be considered when weighing alternatives. Through technical advances in the industry, man-made flooring and protective coatings generally rate very high in durability, allowing you to base your choice almost entirely on looks.

Flooring materials such as wood parquet, laminate floating floors, and resilient vinyl tile are very popular, especially for upgrading projects. Contemporary products are generally thin and flexible enough so that minimal surface preparation is needed. This means they can often be installed directly over an existing floor, eliminating the expense of tearing up the old flooring.

The location of the floor must be considered when choosing between vinyl, laminate, and wood. If the floor is on concrete, or below grade, check with a flooring specialist to see whether preparation is needed to seal out moisture and control the chemical action of the concrete. On suspended floors – that is, floors over open and properly vented

space – the selection should be based on the function of the room. Kitchens, baths, and laundries generally call for the moisture- and scuff-resistant properties of vinyl or laminates. First-floor areas such as hallways, living rooms, dining rooms, and dens are more fittingly enhanced by the beauty and warmth of wood.

MATERIALS

Parquet flooring is available in two forms: "pattern" and "block." The pattern type comes in short strips, with each piece cut to exact length to match each other and to match multiples of its width. Each piece is also end- and edge-matched in the tongue-and-groove manner. This type may be laid in a number of intricate patterns, such as basket weave, stone, herringbone, and French herringbone.

Parquet in block form is more common, being less expensive and much easier to install. This type is manufactured as squares or rectangles, in single pieces or composed of strips, which interlock with tongue-and-groove edges and ends. The 9" x 9" size chosen here is among the most frequently used. Blocks less than 1/2" thick are usually set in a bed of mastic applied with a notched trowel.

Thicker blocks are blind-nailed in place. Both pattern and block style parquet are generally available in a variety of colors and tones.

It is essential that the surface to be covered is clean, smooth, level, and firm. Very minor surface flaws can be bridged by blocks, but serious defects that go uncorrected will cause problems later. Use a straightedge to check for dips and bumps. Sand down high spots and fill any depressions with a recommended floor-leveling compound, sanding it smooth after it sets. Be sure the existing floor is securely nailed off and all the nails set. The floor should also have sufficient expansion gaps around the perimeter. If these are lacking, you can cut kerfs (grooves cut down to about two-thirds of the thickness of the subfloor) with a circular saw, set to the proper depth, to gain the necessary expansion space.

If the existing floor is very uneven or unstable, a plywood or hardboard underlayment will be required. Underlayment thickness may range from 1/4" to 3/4" depending on the material and the job specifications. Be sure to lay it out with proper attention to the direction of underlying seams and joints, and leave an expansion gap of about

1/8" between the panels. Seek advice if you are unsure about any aspect of this preparation.

Remove baseboards and shoe mouldings carefully so they can be reused. Trim the bottom of door jambs and casings with a handsaw so that the new flooring can slip under them. It may also be necessary to trim the bottom of the doors to permit clearance.

When spreading the adhesive, follow the directions provided, both for the correct depth of spread and the proper solvent for cleaning. Be sure not to spread too far ahead of yourself, or the adhesive may set before you get to it. If your work is interrupted, use a wide putty knife to remove the uncovered adhesive so that it won't harden. Press the parquet units into the mastic with hand pressure, interlocking the tongues and grooves and mating the pieces squarely. Do not use a mallet or roller to set the flooring.

The most common installation, and the one used in this project, is to lay the blocks square to the walls. Another method is to lay them in a diagonal pattern, with their edges at a 45° angle to the walls. This pattern is a bit harder to lay out, but it can be an attractive and even useful option – for example, helping to conceal the fact that a room is badly out of square. In either case, lay out the blocks so that the width of the borders is about equal. Snap chalk lines to mark the inside edges of the border blocks and, working from them, establish the center of the room. In a rectangular room this means the geometric center – the focal point of the room where the flooring

is most exposed and its appearance most noticeable. In a room like this, you may be forced to use unequal borders. Work with the layout until you find the one that is most satisfactory. From this central point, mark lines to the walls in all four directions. Make sure the lines intersect at exactly 90°. Start with a block at the intersection of the lines, and continue laying the blocks in a pyramid form, working outward from the center toward the walls. Wood blocks are milled to be the same size, but small variations do occur. Make adjustments as you go to prevent misalignment.

Trim the border blocks with a saw to maintain a 1/2" expansion space. This gap will be covered by the baseboard or shoe moulding. It is best to wait at least 24 hours before exposing the floor to normal traffic. If you must use the room before this, lay down wide boards or sheets of plywood to walk on, but do so carefully so the blocks are not shifted out of position.

LEVEL OF DIFFICULTY

Because of the wide variety of products available, getting professional, informed advice is an important part of making the right selection regarding flooring material and finish. Careful layout is necessary to make sure enough flooring is ordered, while allowing for cuts and waste. Correct preparation of the subsurface is critical. If the existing floor is basically sound and level, roughening it with a disk sander may be all that is needed. If an underlayment is required,

be sure you know what material to use and how to install it. Laying the blocks in a square pattern calls for a few accurate measurements to ensure 90° layout lines, and for enough care in following those guidelines to produce a good-looking floor. None of this is beyond the realm of a beginner, but those who do tackle the project are advised to seek experienced guidance. A beginner should add at least 100% to all times given. For the intermediate and expert, all tasks are fairly routine, and they can plan on increasing their time estimates by 20% and 10%, respectively.

WHAT TO WATCH OUT FOR

Determining the number of blocks or tiles you need for a job like this calls for figuring the area of the room, then dividing by the area of a block. When the area of the floor is irregular, the simplest method is to divide it into rectangles, figure the area of each rectangle, then add the results together. It is a good idea to buy a few extra blocks. Save these, and any other leftovers, to use for repairs should they become necessary later on.

If it is necessary to trim a door bottom, take down the door and lay it across a pair of padded sawhorses. Mark the amount to be trimmed and, using a metal straightedge, score the line about 1/8" deep with a sharp utility knife. Use a try-square to guide the knife when scoring the lines across the thickness of the door. When making the cut, keep the saw blade just to the right of the scored line. This will prevent the blade from tearing out the surface of the wood and leaving a ragged edge above the cut line.

SUMMARY

A full floor replacement can be a major undertaking that can disrupt the use of the room for a number of days. Upgrading by installing wood parquet blocks over an existing floor simplifies the work and can be accomplished in much less time than a total tear-out and replacement, thereby reducing both cost and inconvenience. The improvement in the appearance of the room can be quite dramatic.

For other options or further details regarding options shown, see

Ceramic tile floor

Oak strip floor

Vinyl sheet flooring

Parquet Floor

Description	Quantity/ Unit	Labor-Hours	Material
Remove baseboard	52 L.F.	0.7	
Plywood, underlayment-grade, 1/2" thick, 4' x 8' thick	192 S.F.	2.1	147.46
Latex underlayment	1 Gal.		36.00
Parquet, 5/16" thick, oak, min. T & G, 9" x 9" block, prefinished	168 S.F.	8.4	665.28
Baseboard, stock pine, 9/16" x 4-1/2"	52 L.F.	2.1	82.37
Paint, trim, including puttying, primer	52 L.F.	0.6	1.25
Paint, trim, including puttying, 2 coats	52 L.F.	1.0	3.12
Totals		14.9	$935.48

Project Size	12' x 14'	Contractor's Fee Including Materials	$2,065

Key to Abbreviations
C.Y.–cubic yard   Ea.–each   L.F.–linear foot   Pr.–pair   Sq.–square (100 square feet of area)
S.F.–square foot   S.Y.–square yard   V.L.F.–vertical linear foot   M.S.F.–thousand square feet

VINYL SHEET FLOORING

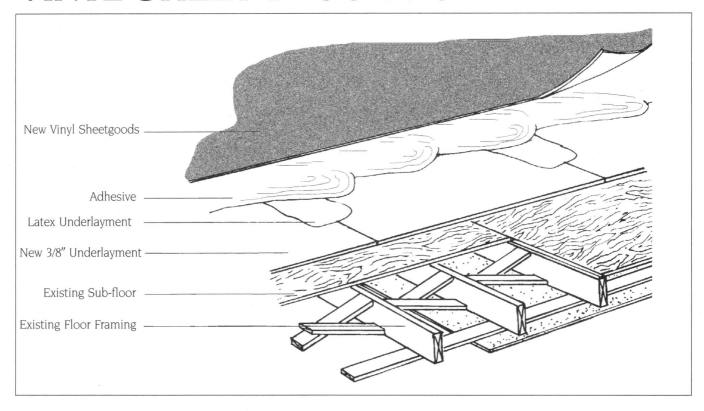

New Vinyl Sheetgoods

Adhesive

Latex Underlayment

New 3/8″ Underlayment

Existing Sub-floor

Existing Floor Framing

In any home, the two floors that receive the most wear, tear, and abuse are in the kitchen and bathroom. With that fact in mind, most builders provide these rooms with flooring materials that are high in both moisture resistance and durability. Well-laid floors of hardwood, ceramic tile, and vinyl are designed to last for many years, and usually do. But years of harsh treatment and constant scrubbing do eventually take their toll. Resilient vinyl is a good choice for a floor replacement project because it can be installed around most in-place fixtures and cabinets.

Depending on the material chosen and its installation requirements, the room involved may be unusable for a full day or more. Prepare well, and schedule the job so as to minimize the inconvenience to you and your family.

MATERIALS

Resilient sheet flooring is available in a wide variety of surface textures, colors, and patterns. It comes in rolls that are 6′, 9′, or 12′ wide, the 6′ width being the most common. You will need the advice and guidance of an experienced contractor or

salesperson to come to an informed decision on the size and style that will best meet your needs – functional, aesthetic, and economical. Take exact measurements of the floor and make a scale drawing (1″=1′ is easy to work with). Include the locations of doorways, fixtures, cabinets, and any structural irregularities in the plan. Lay out the floor on paper with three main considerations in mind: minimize the number of seams; keep seams away from heavy traffic areas; and waste as little material as possible. A single color or nonrepeating design is much easier to lay out and install than a geometric or repeating pattern or one that involves centering a design.

Sheet vinyl must be laid over a surface that is completely smooth and level. This could possibly be the existing floor. Old resilient flooring must be solid, not cushioned, and firmly bonded to the subfloor. Old wood flooring must be solidly nailed, with any gaps and depressions filled and smoothed, and any high spots sanded down. Use a disk sander to abrade the surface of either material to give the adhesive for the new floor a texture to "grab."

An underlayment of 1/4″ plywood is very often required over an old wood floor

to bridge cracks and irregularities that cannot be tolerated under a smooth flooring surface. The underlayment must be securely fastened with ring-shank nails every 6″ in both directions, driven flush with the surface. Any hammer "smiles," knotholes, "dishes," dimples, and voids should be filled and smoothed. Over a period of time any of these left unfilled could become visible as depressions in the new flooring. An underlayment of appropriate thickness can also serve to bring the new flooring flush to surrounding floors, as might be the case when old ceramic tiles are removed.

Before you put the underlayment down, remove any baseboard trim from the walls. Think economically and do this with care, pulling the finish nails through from the back side, so you can reinstall it later. Also, run the seams of the underlayment perpendicular to any joints in the floor below, and avoid seam-over-seam configurations. Never install underlayment over severely rotted wood. Cut out and replace the rotted wood and, most importantly, locate and eliminate the source of the water that caused the damage.

The condition of the old vinyl flooring may demand its removal. This can be done by cutting it into strips with a knife and prying it off with a flat bar or stiff putty knife. Heating with a propane torch or other device may soften the adhesive and make this easier to do.

The exact method of securing the new sheet vinyl to the floor depends on the type of material you choose. Some sheet flooring requires that mastic-type adhesive be spread over the entire floor. Other kinds need to be fastened only at the edges with staples, adhesive, or double-face tape. The dealer from whom you buy the flooring will give you the proper installation instructions, usually provided in leaflet form by the manufacturer. Any special tools such as spreaders and rollers can be bought, borrowed, or rented. If adhesives are involved, be sure they are the kind recommended by the flooring maker.

Any cuts should be made only after taking accurate measurements and double-checking them. Cut the flooring so that it is roughly 3″ oversize on all sides and, after the sheet is in place, use a knife and a straightedge to trim away the excess a little at a time until the sheet lies flat. Leave an expansion gap of 1/8″ on all sides. This will be covered by baseboard. Paper templates can be cut for areas with odd or irregular contours. Kits are available with do-it-yourself patterns to help with this part of the project.

If it is necessary to cut a seam, try to do so in a way that it can be hidden within the pattern, and place it in a nonprominent, low-traffic location. Overlap the two sheets by at least 2″, making sure that any patterns match.

Using a straightedge as a guide, cut through both layers without permitting either to move. After the cut has been made, lift up the flooring on either side of the seam, remove the two overlap strips, and spread adhesive under the seam. The pressure of the roller used to smooth the seam will cause some adhesive to ooze out. Clean off the excess with the recommended solvent. When the seam is dry and clean, a sealant is sometimes applied to fuse the two pieces.

LEVEL OF DIFFICULTY

Sheet flooring materials were once the exclusive province of the professional. They were stiff, making cutting and fitting difficult and placing installation beyond the capabilities of the average homeowner. The materials currently available have changed all that, and today these floors are routinely installed by homeowners.

Planning and layout should be done only with experienced advice. This also applies to decisions regarding the condition of existing flooring, its suitability as a subsurface, the need for its removal, and requirements for underlayment. The end results will directly reflect the quality of the preparation, so don't try to cut corners here.

Mistakes in a job such as this are highly visible and costly to repair, so a beginner is advised to proceed cautiously and only with some experienced guidance. Beginners should also plan on taking at least twice as long as the given time to complete the job. The intermediate and estimated times by 20% and 10% respectively.

WHAT TO WATCH OUT FOR

Contemporary vinyl flooring materials are very tough, but they can be easily cut with a pair of shears or a sharp utility or linoleum knife. Shears are best where there is an unusual shape or curve to the edge you are cutting. A knife and a straightedge are best for straight cuts, but be sure to hold the knife at exactly 90° to the surface when cutting in exposed areas. This is not necessary when trimming around the edges that will be covered by baseboard, but handle the knife with care and change blades frequently.

Manufacturers of roll goods suggest that you can achieve a better match at the seams if you switch succeeding strips end for end. This results in an excellent seam that matches the same edge from the roll. Obviously, this will not work with all patterns or designs; do it when you can.

As with wall-to-wall carpeting, save some leftover scraps of the vinyl and store it. They won't take up much space, and you'll appreciate having them in the event that serious damage requires patching the floor.

Vinyl flooring can be slippery when wet. If this is a concern, offset the hazard with throw rugs placed strategically, especially in a bathroom next to the tub and shower or in a kitchen next to the sink and dishwasher.

Keep in mind that cushioned vinyl sheet goods are not recommended for kitchen floors. Dropped knives and other sharp instruments tend to slice the finish and speed up the aging process.

Vinyl Sheet Flooring

Description	Quantity/Unit	Labor-Hours	Material
Remove baseboard	45 L.F.	0.6	
Plywood, underlayment-grade, 3/8″ thick, 4′ x 8′ sheets	128 S.F.	1.4	73.73
Latex underlayment	1 Gal.		36.00
Flooring, vinyl sheet goods, 0.08″ thick, high-quality	128 S.F.	5.1	437.76
Baseboard, stock pine, 9/16″ x 4-1/2″	45 L.F.	1.8	71.28
Paint, trim, including puttying, primer	45 L.F.	0.6	1.08
Paint, trim, including puttying, 2 coats	45 L.F.	0.9	2.70
Totals		10.4	$622.55

Project Size	11′ x 11′-6″	Contractor's Fee Including Materials	$1,373

Key to Abbreviations
C.Y.–cubic yard   Ea.–each   L.F.–linear foot   Pr.–pair   Sq.–square (100 square feet of area)
S.F.–square foot   S.Y.–square yard   V.L.F.–vertical linear foot   M.S.F.–thousand square feet

SUMMARY

Replacing a kitchen or bathroom floor with resilient vinyl sheet material can provide the room with a fairly quick and inexpensive makeover. A floor is an important and highly visible component of these rooms, and a well-chosen color and design can transform their appearance, making everything in the room look new.

For other options or further details regarding options shown, see

Ceramic tile floor

Oak strip floor

Parquet floor

CERAMIC TILE FLOOR

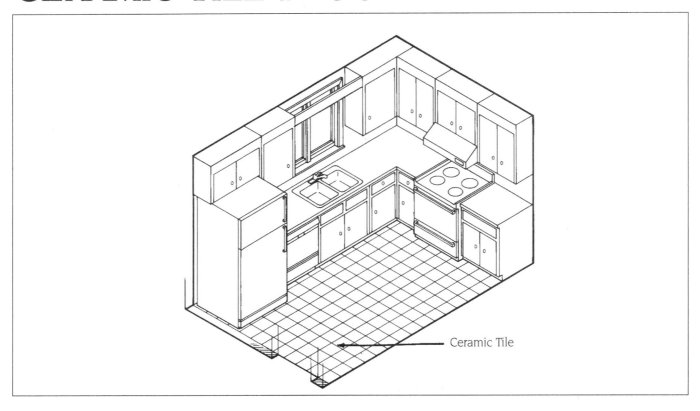

Ceramic Tile

Made of hard-fired clay, the very substance of the earth itself, ceramic tile is mar-resistant, fireproof, impervious to moisture, and easy to maintain.

There is a good deal of work involved in replacing a tile floor. This is especially true if an old floor has to be removed, which is the case more often than not. Tile floors are generally found in kitchens and baths, the two most used rooms in any house. A floor replacement in either of these rooms requires the removal of fixtures and appliances both to remove the old floor and to install the new one. Scheduling and coordinating your work with that of any tradesperson involved is very important in order to expedite the project and get the room back in service as quickly as possible.

MATERIALS

As with any job of this kind, thorough and thoughtful planning is necessary, including making the best choice from among the amazingly wide variety of products available. Factors to be considered are type, color, finish, texture, and shape. These can be somewhat determined by the location of the floor

and its surroundings, but there is still a lot of research to do, with the aid of qualified flooring experts, before you pick up a tool.

For increased weight and durability, floor tiles are generally thicker than wall tiles. Never use wall tiles on a floor because they will crack. Floor tiles come in squares, rectangles, hexagons, and octagons, as well as such shapes as Moorish and ogee.

Glazed tiles have color applied to the surface before they are fired at high heat to bake the color on. They are available in several finishes: high gloss, satin, matte, and dull. Glazed tiles are slippery when wet. Those with matte or textured surfaces provide the best traction and longest wear.

Natural colors or pigments are mixed with the clay of unglazed tiles prior to firing, so the color runs throughout the thickness of the tile. Unglazed tile, like quarry, is less slippery than glazed, and generally maintains its looks better because the color is integral to the clay and will not wear off.

Although it is possible to lay tile over an existing floor, most professionals prefer to strip the old floor to expose the subfloor and install a proper underlayment.

Tile demands a clean, solid, and level surface. Anything less, and even the most careful and painstaking tile installation will eventually crack and come apart. Removing a floor means more, and not very pleasant, work, but that may be a price you must pay to get professional results. Remember that the new floor should, if possible, be flush with the level of abutting floors to avoid trip traps in doorways and other openings. This is rarely possible without tearing up the existing floor. In a bath or kitchen, the plumbing fixtures and other appliances must be disconnected and removed. Also, carefully remove any baseboard and moulding, and set it aside to be reinstalled later.

Once the old floor has been removed, the subfloor should be checked for stability, and any loose areas repaired and nailed off. The underlayment should be laid in such a way as to most effectively stiffen the floor, and be secured with ring-shank nails every 6″ to 8″. Leave 1/8″ expansion gaps between the panels, and 1/2″ gaps along the walls. It is worth repeating that a tile floor can tolerate only the slightest unevenness, so take pains to make the underlayment a perfectly flat surface.

Lay out the tiles so that the cut border tiles are as even as possible around the perimeter. Measure from the inside edge of the border tile lines to the center of the room, and snap 90° transverse chalk lines to divide the room into quarters. On a masonry subfloor, use the chalk lines as your guides. If the subfloor is wood, you can tack batten boards along the lines to act as a rigid guide for the first rows of tile. Working from the center, lay the tile in a pyramid pattern, first in the quarter of the room that is furthest from the door, and end by tiling the quarter closest to it.

For adhesive, use thin-set mortar mixed with latex additive instead of water. This material is a bit more expensive, but it makes a very strong yet slightly flexible bond, enabling it to give without cracking when the wood floor below expands and contracts. Don't mix too large a batch of mortar at a time. It is workable for about an hour before it begins to harden and has to be thrown out. Spread the mortar with a notched trowel according to the recommendations given by the manufacturer. It is important to apply mortar in the proper thickness for the tile you are using.

The tiles may come with designed-in spacer bumps that establish the grout lines as you go. If not, you must use molded, plastic spacers to keep the tiles aligned and to maintain a grout line of constant width. Generally speaking, the narrower the grout line, the better. Grout gets dirty faster and is harder to clean than tiles.

Use hand pressure to embed the tiles in the mortar, and carefully clean off any excess that oozes out.

Also, be sure to remove the spacers once the tiles are in place, and keep the joints clear of too much mortar. If it builds up in the joints and hardens, it will prevent you from applying a thick enough layer of grout, in which case the mortar is likely to show through.

If the edges of the border tiles are going to be covered by baseboard or shoe moulding, leave a 1/4" expansion gap. Clean, straight cuts are not necessary. Anywhere the edge cuts are exposed – for example, along a threshold – careful scribing and cutting is required to give the job a professional look.

Once the mortar has set (the time required for this is indicated on the bag), the grout can be applied using a rubber-bottom float. Grout is available in a range of colors and shades, to blend with the color of the tile. Lighter tones are not recommended for use on floors because they tend to show dirt and are difficult to keep clean. As with the thin-set, mix the grout with a latex additive. Both grout and unglazed tiles should be given the protection of silicone sealer. You must allow the grout to cure completely before sealing, again following the manufacturer's instructions.

LEVEL OF DIFFICULTY

Surface preparation, including tearing out an old floor and installing the proper

underlayment, is an area where experienced advice should be sought. The actual laying of rectangular tiles square to the walls, as in this project, is a matter of working patiently and with care. Tile cutters can be borrowed or rented, and they are fairly easy to use. Installations on the diagonal or involving odd-shaped tiles are more difficult, and thus less advisable for the beginner or intermediate. Tile installations that require the use of a power-driven wet saw are generally very expensive and are best left to the expert or professional contractor.

A beginner should proceed with caution and with help, adding at least 100% to all time estimates. The intermediate should not attempt tasks beyond his or her capabilities, and should add 50% to 60% to the time. An expert should increase the time for all tasks by 20%.

WHAT TO WATCH OUT FOR

The key to laying straight rows of tile is to establish accurate working lines. Once they have been measured and marked, make a dry run by laying out a sufficient number of tiles in all directions to verify that your calculations are correct and that your transverse lines are square.

All ceramic tile can feel cold and hard underfoot. You can cushion the hardness of a tile floor with a throw rug, but make sure the rug is backed by a surface that grips the floor.

Ceramic tiles, properly installed, will last indefinitely. If and when problems occur, they will usually involve cracked grout joints or cracked and loose individual tiles.

SUMMARY

Professional advice should be sought and followed regarding the choice of the tile and its layout, with consideration given to the ambient colors and textures of the room. The permanence of this type of floor also suggests the purchase of better-grade materials, although it is good to remember that high price does not always mean high quality.

For other options or further details regarding options shown, see

> Oak strip floor
> Parquet floor
> Tile walls
> Vinyl sheet flooring

Ceramic Tile Floor

Description	Quantity/ Unit	Labor-Hours	Material
Plywood, underlayment-grade, 3/8" thick, 4' x 8' sheets	128 S.F.	1.4	73.73
Tile, natural clay, random or uniform, thin set, color group 1	96 S.F.	8.4	417.02
Spacers, 1/8"	1 Bag		5.40
One step mortar, latex, 25 lb.	1 Bag		25.80
Grout, floor grade, 25 lb.	1 Bag		22.02
Latex additive	1 Quart		14.70
Sealer	1 Quart		9.48
Totals		9.8	$568.15

Project Size	8' x 12'	Contractor's Fee Including Materials	$1,225

Key to Abbreviations
C.Y.–cubic yard   Ea.–each   L.F.–linear foot   Pr.–pair   Sq.–square (100 square feet of area)
S.F.–square foot   S.Y.–square yard   V.L.F.–vertical linear foot   M.S.F.–thousand square feet

SUSPENDED CEILING

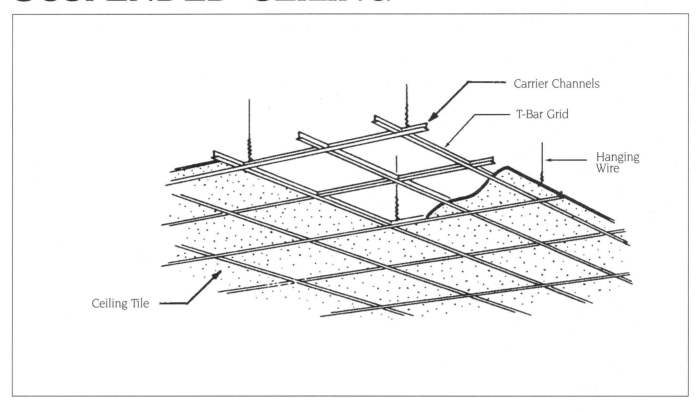

Carrier Channels

T-Bar Grid

Hanging Wire

Ceiling Tile

Suspended ceiling systems are the most popular choice for basement finishing projects and similar spaces where plumbing, wiring, ductwork, and structural features prohibit the installation of a standard sheetrock ceiling. The panels are removable, allowing convenient access for repairs or new work. A suspended ceiling offers a degree of soundproofing and insulation and, best of all, is perhaps the most economical type of ceiling available. Installing one is an ideal do-it-yourself project.

MATERIALS

A suspended ceiling consists of panels (typically 2' × 2' or 2' × 4') that are supported by a light, metal grid system made up of wall angles, main runners, and cross tees. Grid systems come prefinished in a number of colors, including wood grain. The most popular ceiling system is a white grid supporting white panels.

Wall angles are L-shaped and come in 12' lengths. They are attached to the walls to support the ends of the runners, tees, and panels. Main runners, shaped like an inverted "T", are also available in 12' lengths. They have slots at 6" intervals to

receive the cross tees. The runners have interlocking ends that allow them to be joined together. A series of punched holes along the top edge are provided for the suspension wires. Cross tees usually come in 2' or 4' lengths, and are shaped like the main runners. At each end of the tees are tabs that are inserted and locked into the slots on the main runners.

The best way to plan your ceiling system is to measure the dimensions of the room and draw them to scale on paper. Take careful measurements, especially of any odd-shaped walls or angles. Calculate the width of the border panels, indicate the location of recessed lighting fixtures, and draw the layout of the gridwork. Main runners should go at right angles to the joints, and be spaced 4' on center. If you choose to run them parallel to the joists, you must add blocking between the joists from which to hang the main runners. Lay out the cross tees 2' apart between the main runners.

To install the panels you need at least 3" of clearance – if recessed lighting is used, you must allow at least 6". The more clearance you can give yourself, the easier the installation will be. Find the lowest point of the existing construction

– joist, pipe, duct, etc. – measure down the determined number of inches from that point, and mark it on the nearest wall. Using a chalk line, snap a level line at that height on all the walls. Nail the wall angles around the room with the bottom aligned to the chalk line. Use metal snips to cut the first strip of wall angle so it starts in a corner and ends halfway on a stud. Nail the wall angles into the studs or other solid backing, locating the nails no more than 24" apart. At inside corners the wall angles may be overlapped; cut miters for outside corners.

Working from your layout drawing, mark the location of the main runners, and snap lines across the joists. Install screw eyes, not over 4' apart, along these lines for the suspension wires. Cut the ends of the main runners so that one set of slots for the cross tees lines up to accommodate the width of the border panels, parallel to the short wall. This will ensure that the grid system is square. Hang the main runners, resting the ends on the wall angles and attaching them to the suspension wires. Install all the full length cross tees between the main runners. When installed they will effectively straighten and square the grid. Cut and install the border tees and, when the grid is complete, sight down the sections and straighten them where necessary.

Ceiling panels are placed in position by tilting them slightly, lifting them above the grid, and letting them drop into place. Be very careful when handling the panels to avoid marring the finished surface, and keep your hands clean so you don't leave fingerprints. Install all the full-sized, field panels first, and then cut and install the borders. Measure each border panel individually, and cut them slightly smaller than measured so they will fit easily. Cut the panels with a sharp utility knife, and use a scrap piece of cross tee as a straightedge and cutting guide. Always cut with the finished side of the panel up.

LEVEL OF DIFFICULTY

Proper layout is important so as to accurately estimate material needs and ensure that the grid is installed with enough overhead and perimeter support to render the ceiling secure and level. It is also very important that the main runners be cut carefully so the slots for the cross tees line up. A beginner will probably need some experienced help with the layout and with getting started, but should be able to complete the remainder of the project unaided. Recessed lighting will call for an expert

or professional to determine the necessary number and layout of the fixtures, as well as to install them.

A beginner can expect to add 100% to the time required for hanging the ceiling; an intermediate, 30%. Both should hire a contractor for the electrical work. An expert should add 15% for all tasks.

WHAT TO WATCH OUT FOR

To determine the widths of the border panels, use the following calculations as applied to the size of this project (21'4" x 23'6") and adapt them to your own.

To find the width of the long wall borders, add the odd inches of the short wall's length (4") to the length of a panel (48"), and divide the result by 2. Thus: 48" + 4" = 52". 52" ÷ 2 = 26".

To find the width of the short wall borders, add the odd inches of the long wall's length (6") to the width of a panel (24"), and divide by 2. Thus: 6" + 24" = 30". 30" ÷ 2 = 15".

So, our sample ceiling calls for border panels of 26" for the long wall and 15" for the short wall.

Marking a level line for the wall angles is very important. Using a standard carpenter's level, even a long one, can produce unsatisfactory results. The quickest and most accurate way to get a level line all around a room is with a

water level. This is a long, clear, thin tube, open at both ends and nearly filled with water. One person holds the tube so the water line is on the mark for the ceiling height, while another person marks the four corners of the room at the level of the water in his or her end of the tube. After moving the tube to a new location, allow time for the water level in the tube to stabilize. It must be completely still for an accurate and truly level mark to be made. Water levels are not expensive, and using one can be the key to properly installing this ceiling system.

Instead of using regular screw eyes for the suspension wires, building suppliers usually carry special screws that have holes for the wire and are designed to be driven with an electric drill.

To make certain your ceiling will form a level plane surface, stretch lengths of mason's line from wall angle to wall angle, at the locations of the border panels. Be sure the lines are pulled very tight and that they intersect at exactly 90°. These lines can also serve as reference points when calculating the cuts for the main runners.

Some of the lighter-weight ceiling panels that are available today may tend to "lift" up and shift because of changes in air pressure when a door or window is opened or closed. This situation is easily remedied by introducing hold down clips or by placing small amounts of weight on the tiles prone to lifting.

In parts of the country where seismic codes are in effect, check with building officials for proper hangers. Remember that light fixtures must be chained or wired independent of the grid.

SUMMARY

The need for access to the various water pipes, valves, heating ducts, electrical wires, and junction boxes that are located in a basement necessitates some sort of suspended ceiling system. Fortunately, the materials for such a system are not terribly expensive, and the installation is easy enough for the average homeowner to accomplish.

Installing a suspended ceiling is a project in which a homeowner can economize while still achieving a good looking and functional improvement.

For other options or further details regarding options shown, see

    Ceiling replacement

Suspended Ceiling

Description	Quantity/ Unit	Labor-Hours	Material
2' x 2' grid, film-faced fiberglass, 500 S.F. area			
T-bar suspension system, 2' x 2' grid	501 S.F.	6.2	444.89
Ceiling board, film-faced fiberglass, 5/8" thick, 2' x 2'	501 S.F.	6.4	312.62
Carrier channels, 1-1/2" x 3/4"	501 S.F.	8.5	132.26
Hangers, #12 wire	32 Ea.	0.9	14.98
Luminous panels, prismatic, acrylic	32 S.F.	0.6	65.66
Totals		22.6	$970.41

Project Size	21'-4" x 23'-6"	Contractor's Fee Including Materials	**$2,532**

Key to Abbreviations
C.Y.–cubic yard   Ea.–each   L.F.–linear foot   Pr.–pair   Sq.–square (100 square feet of area)
S.F.–square foot   S.Y.–square yard   V.L.F.–vertical linear foot   M.S.F.–thousand square feet

CEILING REPLACEMENT

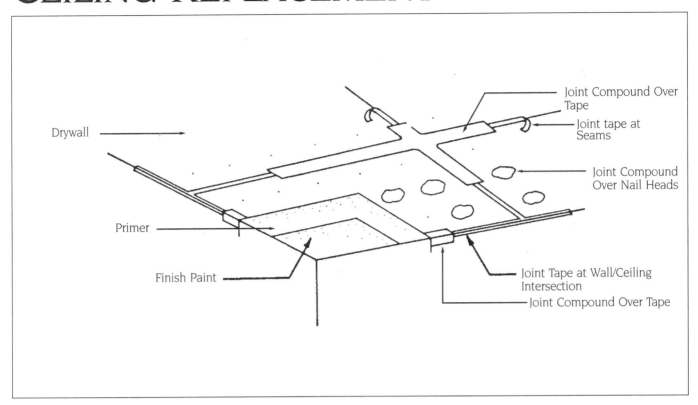

Drywall

Primer

Finish Paint

Joint Compound Over Tape

Joint tape at Seams

Joint Compound Over Nail Heads

Joint Tape at Wall/Ceiling Intersection

Joint Compound Over Tape

Deteriorating ceilings are a very common problem in older homes. Over the years the settling and shifting caused by heat, humidity, seismic activity, vibration from traffic on the floor above, and general exposure to human habitation can cause plaster to crack and sag, working loose from its lath backing. Even in newer homes, sheetrock can sag and buckle as a result of water damage or warping joists.

If the affected area is small enough, a spot repair can often be performed. Sagging plaster can be reattached to wood lath with drywall screws and ceiling buttons (washers), which are covered with a skimcoat of patching plaster or joint compound. It is also possible to remove the plaster to expose the lath for refinishing, or to cut out a section of sheetrock and patch in a new piece.

Very often, however, the damage is too extensive for this approach and a full ceiling replacement is required.

MATERIALS

Since a ceiling covers every square foot of the floor below, the entire living area of the room in question is involved. This means a disruption of normal family activities in that room for the duration of the job. If the ceiling is to be replaced in a first-floor, high-traffic room like the kitchen, the impact will be far greater than if it is an upstairs guest bedroom. In any case, be prepared for a mess. There is no way to remove or replace a ceiling neatly. This is especially true in older homes with plaster and wood or wire lath. Sheetrock removal is not as bad except in second-floor rooms where the existing ceiling supports loose insulation.

Cover the floor with cloth or plastic sheets and staple the same over any doorless openings to isolate the room and contain the mess as much as possible. Carefully remove any ceiling moulding and pull the nails through from the back side. Moulding is expensive and difficult or impossible to match, so any that you can reuse saves you money.

Check the location of any plumbing or wiring that may be affected by the demolition work. When tearing down the ceiling, be sure to wear safety glasses and a dust mask. You will be showered with accumulated dirt and dust, along with assorted spores and microbes that can raise havoc with your respiratory system.

You may find it useful to rent or borrow

a trailer or pick-up truck to haul away the demolished ceiling. Even a small ceiling, when broken up, will very quickly fill several large trash cans.

Once the ceiling is down and the dust has cleared, check the furring on the underside of the ceiling joists. Make sure it is securely nailed off. Repair or replace any damaged or missing furring strips. Then, install new 1/2" sheetrock using 1-1/4" drywall screws. Tape and finish seams, and paint the ceiling to complete the job. You may prefer to have a plaster skimcoat applied by a professional plasterer. If so, be sure to use 1/2" blueboard instead of regular sheetrock.

An alternative approach that eliminates the mess and expense of demolishing an old plaster-and-lath ceiling is to simply cover it with a layer of 3/8" sheetrock. Snap chalk lines to mark the location of the joists and/or furring and be sure to use drywall screws long enough to penetrate both layers of ceiling, old and new, and grab solidly into the wood backing. As usual, determine beforehand the location of any plumbing or wiring that may be threatened by the screws.

Installing a ceiling is a two-person job, especially if it is high enough to require stepladders. If the ceiling height is about 7'6" or lower, one person could undertake the job with the help of a pair of T-braces, known as "deadmen." But considering the difficulties of working overhead while trying to fit awkward and heavy 4' x 8' sheets squarely together, going it alone is not recommended.

LEVEL OF DIFFICULTY

As mentioned above, ceiling work is messy, awkward, and potentially, if mildly, hazardous. It should, therefore, not be attempted by a beginner except under the guidance of an experienced hand. A moderately skilled do-it-yourselfer could probably do an acceptable job with assistance, but the finish work should be done only by an expert or contracted to a professional, especially in the case of a plaster skimcoat over blueboard. Ceilings are white and very exposed, and any defect or lack of smoothness in the finish coverage of the joints and seams will only be highlighted by a coat of paint. A heavily textured finish will hide minor

flaws, but even then a ceiling is not a good place for a beginner to practice. All the mess, hard labor, and expense of the project is for naught if the final product looks anything less than professionally done. A beginner can save money at both ends by doing the demolition and painting, and hiring out the rest to skilled tradespeople.

The beginner should add 100% to the time of the tasks he or she chooses to assume. Intermediates and experts should add 30% and 10%, respectively, to the demolition and installation, and 50% and 20% to the finish work.

WHAT TO WATCH OUT FOR

Demolishing a ceiling exposes framing members that have been hidden for many years. This is a good opportunity to check for signs of rot and the presence of pests like ants, termites, and powder-post beetles, whose depredations can compromise the structural integrity of the framing. Unsoundness means

unsafeness, and any severely damaged joist should be replaced, or reinforced by "sistering" new lumber to it. Renovation work is full of surprises, few of which are pleasant, so be prepared.

If there is a squeaking floor overhead, this is your chance to locate the source (almost always a loose floorboard rubbing on the shank of a nail) and eliminate the problem with shims, cleats, and glue.

Also revealed for inspection is any wiring or plumbing running along the joists. Check soldered joints for fissures or signs of leaks, and make sure any clamps are securely fastened. Loose clamps are often the cause of rattling pipes. Old frayed or chewed (by mice) wiring should be inspected by a licensed electrician and replaced. This is also a good time to assess the general wiring and lighting in a room. With the ceiling down it is a fairly easy job to run wires for new fixtures, switches, and outlets. Remember, faulty wiring can cause a fire; leave this work to a professional.

After you have removed the failed plaster it may be necessary to fill in areas of wooden lath or add furring to which the new board will be attached. Make sure any new materials introduced to provide for proper fastening of finish materials match the existing in thickness or shim to ensure a uniform surface.

Ceiling Replacement

Description	Quantity/ Unit	Labor- Hours	Material
Remove existing plaster, lime & horsehair on wood lath, incl. lath	168 S.F.	3.8	
Drywall, 1/2" thick, standard, 4' x 8', taped and finished	192 S.F.	3.2	57.60
Corners taped and finished	52 L.F.	0.8	3.74
Paint, primer	168 S.F.	0.7	8.06
Paint, 1 coat	168 S.F.	1.0	10.08
Totals		9.5	$79.48

Project Size	12' x 14'	Contractor's Fee Including Materials	$534

Key to Abbreviations
C.Y.–cubic yard Ea.–each L.F.–linear foot Pr.–pair Sq.–square (100 square feet of area)
S.F.–square foot S.Y.–square yard V.L.F.–vertical linear foot M.S.F.–thousand square feet

SUMMARY

Deteriorating ceilings are easy to ignore and live with. After all, no one relishes the prospect of dealing with the mess and inconvenience of a full replacement. On the other hand, a more attractive interior enhances your life as well as the value of your house. All in all, this project is a good investment.

For other options or further details regarding options shown, see

Suspended ceiling

FRAME AND FINISH PARTITION WALL

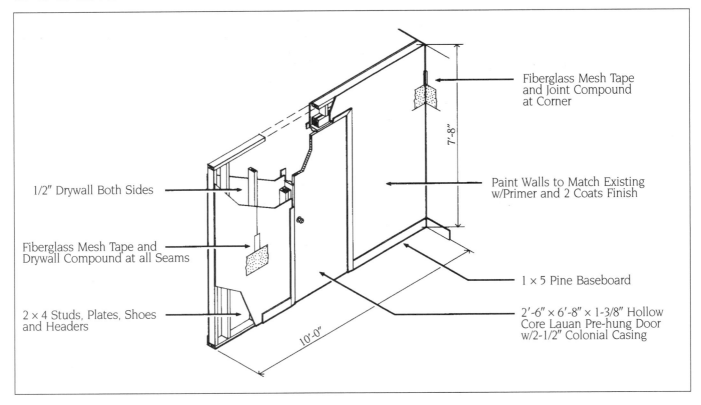

Fiberglass Mesh Tape and Joint Compound at Corner

Paint Walls to Match Existing w/Primer and 2 Coats Finish

1/2″ Drywall Both Sides

Fiberglass Mesh Tape and Drywall Compound at all Seams

2 × 4 Studs, Plates, Shoes and Headers

1 × 5 Pine Baseboard

2′-6″ × 6′-8″ × 1-3/8″ Hollow Core Lauan Pre-hung Door w/2-1/2″ Colonial Casing

7′-8″

10′-0″

The ability to frame and finish a wall is one of the most fundamental skills of rough carpentry and one that is the basis for developing many of the other skills involved in the trade. Considering the fact that no one will work on your house more cheaply than you, the more work you can do on your own, the more money you can save and the more improvements you can ultimately afford. And, of course, there is no denying the intangible rewards of the pride and satisfaction of knowing that you accomplished the project yourself.

MATERIALS

A framed wall is simply a rectangle of various 2 × 4 pieces nailed together. The top and bottom pieces are called, respectively, the top plate and the sole plate; the upright pieces are called studs. Trimmers or jack studs support the horizontal pieces making up the header, which spans the top of the door's rough opening, and which in turn supports several short studs called cripples. Carpenters generally build a wall frame flat on the floor and then raise it into position. This is common practice in new construction and it could be done

in remodeling projects such as this, but there are many variables in the structure of an existing room that can cause problems for an amateur using this method. Consequently, building the wall in place stick by stick is, in most cases, a safer approach.

Remove the baseboard, if any, and measure from wall to wall to determine the length of the sole plate. Cut and lay it in place, measuring out the proper distance from the corners of the opposite wall. Using the plate as a straight edge, mark its location on the floor. Lay out and mark the locations of all the studs 16″ O.C. (on center), as well as the location of the door opening. (Check the specs of your new door for its rough opening width.) Cut the width of the opening out of the sole plate and nail the two pieces to the marks on the floor. Plumb lines up from the ends of the plate, and snap a chalk line across the ceiling between them to mark the location of the top plate. Measure and cut this plate, and nail it up to the line. If the partition runs at right angles to the joists above, you can secure the top plate by nailing through it into the joists. If the partition runs parallel to the joists, try to place it directly below a joist for convenient nailing. (Determine this before you nail down the sole plate.)

If this is not possible, remove a section of ceiling between the two joists above the partition and install 2 × 4 blocking for nailers.

Measure and cut the single studs, toenail (nail at an angle) through the stud and into the sole plate (using the layout marks), then plumb, mark, and toenail them at the top. The two pairs of studs forming the door opening can be cut and nailed together prior to putting them in place. (Check the specs of your door for the rough opening height.) After toenailing them at the bottom, plumb the studs on the hinge side of the door, then mark and toenail at the top. Toenail the opposite pair at the bottom, but don't nail the top until the header is in place. Cut the double 2 × 4 header exactly the width of the rough opening plus 3″, and place it on the trimmer studs across the top of the opening, nail it in, and then finish toenailing the top of the last stud. Finish the frame by cutting and nailing the short cripple studs between the header and the top plate. If there is to be baseboard along the wall, cut and nail two short pieces to the bottom of the end studs to serve as nailers.

Hang the sheetrock by putting up a full sheet, starting at the corner from which you laid out the studs. If your layout was precise, the full sheet should break on the center of a stud at 96". In new construction, it is customary to cover the top portion of the wall first, so that the milled edge of the sheetrock butts against the ceiling panels, making a tight, straight corner joint. In older houses, ceilings are often anything but smooth and level, so it may be just as effective, and also somewhat easier, to start at the bottom. Use whichever approach will produce the best results. After the first full sheet is up, cut out the rough opening with a drywall saw. Hang the remaining sheets, cutting them when necessary with a sharp utility knife, and making sure to stagger the vertical joints. Also, be sure to "dimple" the nails; that is, set each nail head just below the surface of the sheetrock in a shallow depression made by the last blow of the hammer.

Fill all large gaps – more than 1/2" wide – with a smooth layer of quick-setting patching plaster, then tape all the joints, including the inside corners at the intersections of both the walls and the ceiling, unless the joint at the ceiling will be hidden by crown moulding. Work carefully to make the tape lie perfectly flat. Cover all nail dimples and joints with at least three coats of joint compound, known in the trade as "mud," overlapping each succeeding layer beyond the one beneath and feathering the edges. The final coat of mud can be mixed with a little water to a soft ice cream consistency to allow you to spread it out very thin. If you have been careful to apply each layer of mud as smooth as possible, a light sanding with fine grit paper should be enough to eliminate any light ridges left by the tools. Install all mouldings and trim work after the door is in place.

LEVEL OF DIFFICULTY

A complete novice is advised not to attempt constructing a wall without some experienced help and guidance, but anyone with the right tools and a minimal amount of skill in using them could undertake a project such as this and produce acceptable results. Of the three tasks involved, cutting and hanging the sheetrock is the easiest, mudding it the most difficult, and framing it the most critical. It is said that a good framer is worth two finish carpenters, and there is more than a little truth to it; high-quality finish work is very difficult to achieve if the framing has been done improperly and, conversely, high-quality framing can make good finish work easier to accomplish.

Beginners should add 200% to the estimated times, and should have help along the way. An intermediate should add about 50% to the framing and sheetrock work, and about 75% to the mud work. An expert should add about 20% for all tasks.

WHAT TO WATCH OUT FOR

Most intermediates own the basic carpentry tools, but few have a proper set of wall-finishing tools. They can be obtained as a boxed set from a building supply store. If you foresee future projects involving drywall work, it is worthwhile investing in a good-quality set. You will need four basic tools: a 6" taping knife for covering the nails; a 90° corner tool for applying the first two coats of mud to the inside corners; a 10" or 12" broadknife for covering the flat seams and smoothing the inside and outside corners; and a hawk, or mortarboard, for holding the mud while you work.

Hard bits of dried mud or dirt sometimes get on your knife and leave bumps or grooves in the wet mud on the wall, usually on the final smoothing stroke. This is not only frustrating, but it slows you down while you remove the offending grit and resmooth the surface. You can prevent the problem by frequently cleaning your knife as you work, rinsing off your hawk between loads of mud, and disposing of dirty mud rather than returning it to the bucket.

Frame And Finish Partition Wall

Description	Quantity/Unit	Labor-Hours	Material
Wood framing partition, 2 x 4	10 L.F.	1.6	42.24
Drywall, 1/2" thick, plasterboard, taped and finished	160 S.F.	2.7	48.00
Door, flush, hollow core lauan, finished	1 Ea.	0.8	195.60
1-3/8" x 2'-6" x 6'-8", prehung w/casing 2-1/2" wide	1 Ea.	0.5	34.20
Trim for baseboard, 1 x 5 pine	20 L.F.	0.7	27.36
Passage set, non-keyed	1 Ea.	0.7	45.60
Paint, ceiling and walls, primer	160 S.F.	0.6	7.68
Paint, ceiling and walls, 2 coats	160 S.F.	1.6	17.28
Paint trim, including putty, primer	24 L.F.	0.3	0.58
Paint trim, including putty, one coat	24 L.F.	0.3	0.86
Totals		9.8	$419.40

Project Size	10 L.F. x 7'-8" high	Contractor's Fee Including Materials	$1,075

Key to Abbreviations
C.Y.–cubic yard   Ea.–each   L.F.–linear foot   Pr.–pair   Sq.–square (100 square feet of area)
S.F.–square foot   S.Y.–square yard   V.L.F.–vertical linear foot   M.S.F.–thousand square feet

SUMMARY

By starting with something small and manageable, such as framing and finishing a stud wall, then gradually moving on to larger jobs, do-it-yourselfers can acquire considerable construction skills and the confidence to exercise them in more ambitious remodeling projects such as major conversions or full-scale additions. In doing so, they discover the satisfaction that comes from creating something useful and beautiful out of wood, and come to realize what every carpenter secretly knows – it's fun work hitting things with a hammer.

For other options or further details regarding options shown, see

> *Interior doors*
> *Painting & wallpapering*

INSULATION

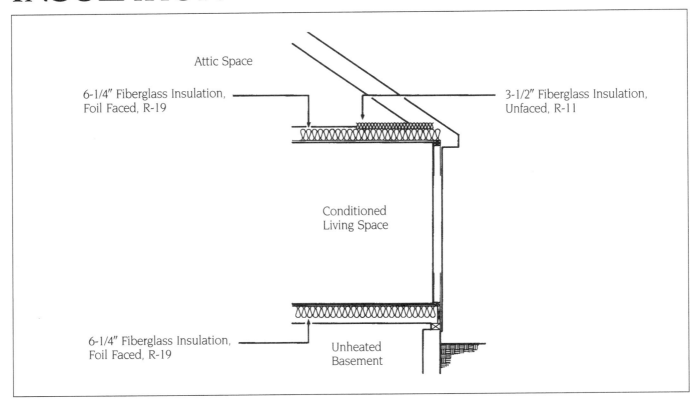

Attic Space

6-1/4″ Fiberglass Insulation, Foil Faced, R-19

3-1/2″ Fiberglass Insulation, Unfaced, R-11

Conditioned Living Space

6-1/4″ Fiberglass Insulation, Foil Faced, R-19

Unheated Basement

Now that the days of cheap energy are past, homeowners have become increasingly aware of the high cost of heating and cooling a poorly insulated house. Every time the furnace or air conditioner switches on in your home, what you hear is the sound of money burning. If your house is not properly insulated, the warm or cool air for which you pay dearly is being wasted. It's no wonder that building codes have been expanded to include minimum standards for insulation in new construction. Older homes are not required to meet these standards, but the cost- and comfort-conscious homeowner will do everything possible to ensure that his or her house is as energy-efficient as today's insulating materials can make it. Start with an energy audit of your house to determine where more insulation is needed, and then see that it gets installed there, either by you or by a professional contractor. The cost of the job will usually be recovered through energy savings within a very few years.

MATERIALS

In an existing home, do-it-yourself insulation projects are generally limited to attics, basements, and crawl spaces. Insulating finished walls is a job for a professional contractor with the knowledge and specialized equipment to blow loose insulation material into empty bays and other cavities through holes cut in the exterior or interior walls. The better contractors then use heat-scanning devices that produce images indicating heat loss to verify that all wall cavities are, in fact, filled.

Simply stated, insulation should be placed between all heated and unheated areas. How much, and what type of insulation is needed, depends on the climate and the type of heating or cooling system in your home. The insulating value of a material is referred to as its "R-value" (R = resistance to heat flow). The higher the R number, the better the insulating properties of the material.

Insulating your attic is one of the most important energy-saving home improvements you can make. (Remember, heat rises.) There are basically three types of attics to be considered: unfinished, unfinished with a floor, and finished.

If your unfinished attic has no insulation at all, lay 6-1/4″, R-19 fiberglass insulation, paper- or foil-faced, in all the bays

(spaces between the joists), with the vapor barrier (paper or foil) down. A vapor barrier should always face the heated living space. Be sure to leave a minimum 1″ air space at the eaves for ventilation. To achieve an R-30 value (recommended in the frost belt), lay 3-1/3″ R-11, unfaced fiberglass over the first layer, at right angles to the joists. If you already have insulation in the attic, you should add more if it measures less than 6″ deep. If you choose to add fiberglass over existing insulation, be sure that the fiberglass is of the unfaced variety; there should never be a vapor barrier between layers of insulation.

An unfinished attic with a floor having less than 6″ of insulation under it should be upgraded. You can remove a few boards every 6′ or 8′ (or every third piece of plywood) and push the fiberglass under the remaining boards with a stick. This can be a lot of very difficult work just getting the boards up, and some may prefer the alternative of hiring a professional to blow loose insulation under the floorboards.

A finished attic may be difficult or impossible to insulate yourself. If there is no access through the kneewalls to the space behind, nor through the ceiling to the space above, you may have no choice but to hire a contractor to do the work, and it will not be easy for him either. If you can get to those areas, the ceiling should be insulated with at least 6-1/4" fiberglass (R-19), and the kneewalls with 3-1/2" fiberglass (R-11).

Any type of attic must be properly and adequately ventilated. If yours is not, refer to the attic ventilation project earlier in this book.

If you have a heated basement that you use as a living or work space, it will pay to add a layer of insulation to the walls. To do this you must first build frame walls to which 3-1/2" R-11 paper- or foil-faced fiberglass insulation can be stapled. The insulated frame must then be covered with sheetrock, both for appearance and for fire protection. This is the first step in a finished basement project. If you choose to carry it through and completely finish the basement, see the basement projects in this book.

An unfinished, unheated basement should have an insulated ceiling. Use 6-1/4", R-19 fiberglass with the foil or paper facing up, touching the underside of the subfloor above. Hold the insulation in place with hangers, sometimes called "lightning rods," which are lengths of stiff wire jammed between the joists.

If your home, or a part of it, is over a crawlspace sealed off from the outside air in winter, the cheapest and best way to insulate it is to cover the ground with six-mil polyethylene and drape six-inch batts of unfaced fiberglass from the sill, down the foundation wall, and out onto the ground about two feet. A crawlspace vented year-round should also have its floor covered with a plastic vapor barrier, and the house floor above insulated with 6-1/4", R-19, foil-faced fiberglass vapor barrier facing up, in contact with the underside of the subfloor, as in an unheated basement.

LEVEL OF DIFFICULTY

Assuming that the uninsulated areas of the house are accessible, there is no reason why even a beginner could not perform this work, once he or she has done the necessary research and obtained sound advice on what materials to use. Fiberglass insulation is lightweight and can be cut easily using a sharp utility knife and a section of 2 x 4 for a straightedge. If paper- or foil-faced, it can be stapled to framing members. If unfaced, it is simply laid down or held in place overhead with wire insulation hangers. The most difficult aspect of the job is crawling around in confined areas such as attics and crawlspaces. The most unpleasant aspect is simply handling the insulation, especially overhead, when the tiny glass fibers begin to find their way down your collar and up your sleeves. For this reason, many carpenters will tell you that the one task they least enjoy is hanging insulation. Any completely inaccessible areas must have insulation blown in, and this should be left to a professional contractor.

A beginner should add about 150% to the estimated times – and somewhat more for a basement requiring walls, and for work in very confined spaces. An intermediate should add 50% to 75% for all tasks, and an expert, about 20%.

WHAT TO WATCH OUT FOR

Installing insulation is not the most pleasant task, but you can minimize the unpleasantness by wearing protective clothing – gloves, hat, and long-sleeved shirt tucked into long pants; goggles and a dust mask are also recommended. Dust your face, neck, and hands with talcum powder before you begin and every so often while you work; it will help keep you dry, thereby preventing the glass fibers from clinging to your skin.

Measure the square foot area to be insulated to determine the amount of material you will need. Note the distance between the framing members – usually 14-1/2" or 23-1/2" – and buy insulation sized to fit. Some older houses have such irregularly spaced framing that it may be best to buy 24"-wide unfaced insulation and cut it to fit. Be sure it fits snugly in the space, and install a vapor barrier wherever one is required.

SUMMARY

Increasing your home's energy efficiency by installing insulation, or by adding to what is already there, is a project that will begin saving you money upon completion and add to the long term resale value of the house. You can also save money by doing the work yourself in those areas that are accessible to you – unfinished attic, basement, and crawlspace.

For other options or further details regarding options shown, see

Attic ventilation*

Residing with wood clapboard*

Standard basement

Walk-out basement

* In Exterior Home Improvement Costs

Insulation

Description	Quantity/ Unit	Labor- Hours	Material
Attic, fiberglass, foil-faced batts, 6-1/4" thick R-19, 15" wide	600 S.F.	3.0	295.20
Fiberglass, unfaced batt, 3-1/2", R-11, 15" wide	600 S.F.	3.0	151.20
Basement ceiling, fiberglass, foil-faced, 6-1/4" R-19, 15" wide	500 S.F.	2.5	246.00
Crawl space, fiberglass 6" foil faced, R-19	100 S.F.	0.5	49.20
Polyethylene vapor barrier, standard 6 mil.	100 S.F.	0.2	3.60
Wire supports, 100 per package	100 Ea.	0.8	6.00
Totals		10.0	$751.20

Project Size	600 S.F.	Contractor's Fee Including Materials	$1,580

Key to Abbreviations
C.Y.–cubic yard   Ea.–each   L.F.–linear foot   Pr.–pair   Sq.–square (100 square feet of area)
S.F.–square foot   S.Y.–square yard   V.L.F.–vertical linear foot   M.S.F.–thousand square feet

PAINTING AND WALLPAPERING

Size of Room	Single Rolls of Wallpaper			Yards of Border	Rolls of Ceiling
	Wall Ceiling Height				
	8'	9'	10'		
4' × 8'	6	7	8	9	2
6' × 10'	8	9	10	12	2
8' × 12'	10	11	13	15	4
10' × 14'	12	14	15	18	5
12' × 16'	14	16	17	20	7
14' × 18'	16	18	20	23	8

Item	Coat	One Gallon Covers			Coverage in 8 Labor-Hours			Labor-Hours per 100 S.F.		
		Brush	Roller	Spray	Brush	Roller	Spray	Brush	Roller	Spray
Paint brick masonry	prime	180	135	160	750	800	1800	1.066	1.000	.444
	1st	270	225	290	815	975	2275	.981	.820	.351
	2nd	340	305	360	815	1150	2925	.981	.695	.273
Paint interior plaster or drywall	prime	400	380	495	1150	2000	3250	.695	.400	.246
	others	450	425	495	1300	2300	4000	.615	.347	.200
Paint interior doors and windows	prime	400	–	–	650	–	–	1.230	–	–
	1st	425	–	–	800	–	–	1.000	–	–
	2nd	450	–	–	975	–	–	.820	–	–

Notes:
1. A single roll of wallpaper will cover approximately 36 square feet. Allow 6 S.F. per roll for waste. Therefore, use 30 S.F. per roll in determining the number of single rolls needed.

2. For vinyls and grass cloths with no pattern to match, allow 10% waste. For patterns that require matching, allow 25% to 30% waste.

3. For large bold patterns, waste can run as high as 50% to 60%.

4. Deduct one single roll for each two doors or windows of average size.

Renovating a room with paint and/or wallpaper is the one remodeling project that, sooner or later, every homeowner is bound to undertake. You might be a first-time home buyer seeking to cover up the decorative choices of the previous owners, or perhaps you have lived in your home long enough to have grown tired of the wallpaper that looked so "special" in the dining room when both you and your paper were younger. Ideally, a home reflects the tastes and aesthetic sensibilities of its owners, and few people can live happily when surrounded by unattractive colors and patterns. A weekend or two, some elbow grease, and several gallons of paint or a few rolls of wallpaper are all that it takes to transform the dreariest room into a bright and pleasant space that is a joy to be in.

MATERIALS

The relative merits of latex vs. oil-base paint are debated, often hotly, by professional painters and decorators. Briefly stated, latex paint dries faster; it leaves a thicker, somewhat more flexible film; it can be rolled on or applied with a nylon bristle brush; and it cleans up with water. Oil-base paint flows more easily

from the brush (which must be of natural bristle); it dries more slowly and tends to leave a thinner film, and thus will not obliterate the fine, crisp details of moldings; and it cleans up with mineral spirits. When selecting paint for a bathroom or kitchen, remember that the higher the gloss, the better the finish withstands moisture. Glossy finishes, however, show imperfections more than flat finishes, so be sure to prepare the surface thoroughly.

Many professionals feel that the best results are achieved by using premium-quality paint – latex for ceilings and walls, and oil-base for wood trim, including windows and doors. In this way, the best properties of each kind of paint are used to greatest advantage.

All professionals agree that proper preparation is the key to a good-looking, long-lasting finish. The room should be cleared of furniture and all not-to-be-painted items like door knobs, window locks, and outlet plates. Cover the floor with drop cloths or newspapers. Thoroughly wash with detergent all surfaces to be painted. Scrape all loose, chipped, or peeling paint; fill dents, cracks, and holes with paintable caulk or vinyl spackle; and sand wherever

necessary to feather rough edges and slightly roughen the old surface to enable the new paint to adhere. Remove old wallpaper and all traces of adhesive using a scraper, abrasive detergent, and sandpaper, if necessary. Vacuum all surfaces, wash down the walls and woodwork, rinse, and let dry.

Paint the ceiling first with a flat latex ceiling paint. (Warning: If yours is an older house, the ceilings might be coated with calcimine – a sort of whitewash that will not hold regular paint. Special paint, formulated for calcimine, is available from some manufacturers; consult your local supplier.) Ceilings, like walls, can be rolled, but in both cases, a great deal of splattering results, which means more cleaning up. Also, even the smoothest roller leaves a slightly stippled surface that some find objectionable. Rolling, not including the extra clean up, is somewhat faster than brushing, but the latter is much neater and easier to control. You will need a brush to "cut in" on the walls and ceiling – that is, paint edge and corner areas that the roller can't reach.

If you are painting the walls, they are done next, followed by the woodwork; if you are papering, the woodwork is next. The wall paint can be rolled on or applied with a 3″ or 4″ brush. The woodwork – windows, doors, baseboard, mouldings – should be painted with a 2-1/2″ sash brush with angled bristles that fit more easily into corners. All bare wood exposed by scraping and sanding should be spot primed before the first coat of finish paint is applied. Read the labels on the paint cans to be sure that the primer is compatible with the type of finish paint you are using.

It is something of an understatement to say that wallpaper comes in a wide array of colors, prints, patterns, textures, and materials. You have only to visit a decorating center and see the stacks of sample books from dozens of manufacturers, American and European, to realize that choosing your wallpaper might well take more time than hanging it. Wallpapers come in various types, including solid vinyl, expanded vinyl, vinyl-coated, paper-backed vinyl, fabric-backed vinyl, handpainted, silk screened, textile, and metallic-look. The specific features of each – for example, typical application, ease of installation, potential problems, and cost – should be discussed with an experienced designer or knowledgeable salesperson. Be sure to provide him or her with all the pertinent details of your room (such as dimensions, durability requirements, etc.) required to assess your needs and recommend the best type of wallcovering.

Before hanging the paper, apply a coat of sizing to the walls. This will seal the surface, make it easier to adjust the paper,

and help it to adhere better. Working from a plumb line located where the first seam will fall, the wet paper is aligned, smoothed, rolled, and trimmed. Use a damp sponge to remove excess paste from the paper, the ceiling, and woodwork. Butt the edge of the second strip to the first, after you have determined the proper positioning for the pattern repeat, if any.

Hang the wallpaper from right to left if you are right-handed. When papering over and cutting around electrical outlets, turn off the power. Around doors and windows, make rough trim cuts first, then after the opening is completely surrounded, make diagonal cuts at the corner, press the paper tight against the frame, and make the final trip cuts.

LEVEL OF DIFFICULTY

Preparing a room for paint and paper is tedious work. There is a tendency for nonprofessionals to cut corners during this phase of the project, but doing so can compromise the quality of the finish, both in terms of its looks and its ability to adhere. Anyone can stick a brush in a paint pot and smear the walls, but it takes a certain degree of skill and care to do the job right, especially on the woodwork. It's worth consulting a how-to-paint manual to learn a few tricks of the trade that can make a big difference in the final product.

Hanging wallpaper calls for even more care and skill, and a beginner should read

up on the subject and work with some experienced help. The degree of difficulty depends very much on the pattern of the paper and the layout of the room. For example, a paper with a pattern repeat (the vertical distance between two points on the paper when the pattern is identical) will be more difficult to hang than a paper with a random match (no discernible pattern from top to bottom or from one strip to the next).

A beginner should have help when hanging paper, and should add 150% to the estimated times for all tasks. An intermediate should add 100% for paperhanging, and 50% for painting. An expert should add 30% and 10%, respectively.

WHAT TO WATCH OUT FOR

Be sure you order enough paper to complete the job, allowing about 10% for trimming and matching. Rolls of paper are marked with a run or dye lot number. Dye lots can vary in color, so check rolls before hanging, buy extra, and, if you have underbought, reorder quickly before your dye lot sells out.

Painting and papering a room will probably require that some work be done from a ladder. Stepladders are not generally perceived as dangerous, which is exactly what makes them so; people see no risk and get careless. A fall from a stepladder is not likely to be life threatening, but it can certainly be injurious and, if accompanied by a gallon of paint, very messy. Wooden stepladders are heavier and thus generally more stable than aluminum. A flimsy, rickety ladder with bent, broken, or missing rungs is the proverbial "accident waiting to happen," and should be replaced. Also, be aware that most falls from ladders occur because of leaning too far; if you can't reach the spot, move the ladder.

SUMMARY

Repainting or papering rooms can add a great deal to your home's charm and attractiveness, as well as to its long-term value. The special skills required to do the work are not difficult to learn, which makes this an ideal do-it-yourself project: you can put some sweat equity into your home and enjoy both the visible results of your labor and the satisfaction of having accomplished it yourself.

Painting & Wallpapering

Description	Quantity/ Unit	Labor- Hours	Material
Door & frame, 3′ x 7′ per side, oil, primer, brushwork, incl. prep.	1 Ea.	1.3	2.11
Door & frame, 3′ x 7′ per side, oil, 2 coats, brushwork, incl. prep.	1 Ea.	2.7	6.84
Window, colonial, 6/6 lites, 3′ x 5′, primer, brushwork, incl. prep.	3 Ea.	2.0	2.48
Window, colonial, 6/6 lites, 3′ x 5′, 2 coats, brushwork, incl. prep.	3 Ea.	3.4	5.44
Ceiling, prime, includes surface prep.	192 S.F.	0.8	9.22
Ceiling, 2 coats, includes surface prep.	192 S.F.	1.9	20.74
Wallpaper, medium-weight, includes surface preparation	420 S.F.	7.0	342.72
Totals		19.1	$389.55

Project Size	12′ x 16′	Contractor's Fee Including Materials	$1,425

Key to Abbreviations
C.Y.–cubic yard　　Ea.–each　　L.F.–linear foot　　Pr.–pair　　Sq.–square (100 square feet of area)
S.F.–square foot　　S.Y.–square yard　　V.L.F.–vertical linear foot　　M.S.F.–thousand square feet

Per side per coat WP 60 sqft p. hr. Door & Frame 1.3 m hous

Section Nine
STAIRWAYS

Stairway installation is not a job for a beginner, and even intermediate or expert remodelers will need assistance in these projects. Read the project descriptions thoroughly, plan your project carefully, and keep the following guidelines in mind if you decide to tackle a stairway installation.

- Stairs require a minimum headroom of 6'-8" (spiral stairs require 6'-6").

- The maximum allowable vertical distance between landings in a stairway is 12'-0". Risers can be a maximum of 7-3/4" high; treads should be a minimum of 10". Common dimensions for riser and tread are 7" rise and 10-1/2" tread.

- To avoid tripping hazards, make sure all risers and treads are equal. The first step someone takes sets their expectations for the steps that follow.

- Handrails are required for stairways with four or more risers. Handrail height must be 34" to 38" from the floor or the nosing of the tread.

- Handrail stock is generally 1-1/4" to 2" with maximum projection into the stairwell of 3-1/2".

- Doorways off of stairs should open onto a minimum 36" landing area.

- Stair edging helps to protect carpeted and hardwood tread fronts from excessive wear.

- Many stair parts, such as the handrail, can be finger-jointed, that is, made up of several pieces of lumber. When a finger-jointed part is stained, it is common to have variations in color. You can purchase solid members at greater expense.

- Stair dimensions and other requirements are specified in building codes. Before you begin a project, check with your local building department for requirements. New stairs or renovation projects involving changes in the dimensions of stairs will probably require a building permit.

CIRCULAR STAIRWAY

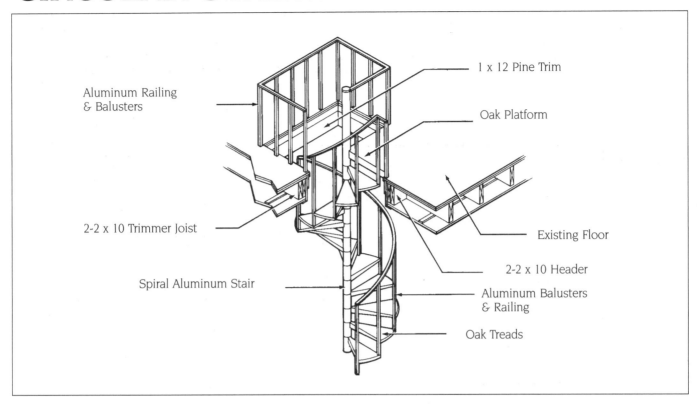

Aluminum Railing & Balusters

2-2 x 10 Trimmer Joist

Spiral Aluminum Stair

1 x 12 Pine Trim

Oak Platform

Existing Floor

2-2 x 10 Header

Aluminum Balusters & Railing

Oak Treads

While conventional staircases take up a significant area on two levels, circular stairways can be an efficient alternative, particularly in smaller dwellings where space is at a premium. A prefabricated model can simplify the project and provide a new or more efficient stairway between levels of the house.

MATERIALS

The materials used for this project include a factory-made circular stairway unit and the framing and support materials required to maintain floor and ceiling strength. The stairway itself is spiral in design, consisting of a heavy aluminum or steel center support pole and triangular oak treads that fan outward from it. Several other tread materials are available at varying costs, including steel and cast iron.

The stairway comes from the supplier in a prepackaged kit with instructions and guidelines for installation. After the ceiling opening has been cut to the dimensions specified by the manufacturer, it should be framed. The stairway is then assembled in place by sections, and plumbed, leveled, and fastened in its final location.

After the first step, which involves cutting out the required ceiling and floor spaces, the exposed joists will need to be boxed off and framed. The lumber used for the trimmer joists and headers should match the existing joists, which are usually 2 x 10s. Additional 2 x 10s may be required for blocking near the opening. These and any joist hangers that are needed will add some extra cost to the project.

Because much of the load of the stairway is concentrated on the base plate, additional support should be provided beneath the floor on which the stairway stands. Usually, doubling the existing 2 x 10s in the area directly under the unit and adding supplementary blocking is enough. Plan on spending some extra time and money for this operation if you have to cut into a finished ceiling below to install these additional floor support members. If you are inexperienced in these tasks, then seek assistance or hire a professional to do the work. Inadequate placement of these structural components can cost you more in the long run because of repairs needed to correct sagging ceilings or floors.

The materials used to finish the ceiling cut-out should match the trim work in the room below. The rough 2 x 10 box should first be faced with pine boards and then finished with one of several available types of trim. Care should be taken at the start of cutting the stairway opening to avoid the extra expense of patching the ceiling material around it later.

LEVEL OF DIFFICULTY

The installation of a circular stairway may at first appear to be a routine project, but the consequences of doing the work incorrectly make it a job for accomplished intermediates. Beginners may be able to assemble the stairway and do the finish work, but they should hire a professional contractor to cut the opening and reinforce the joists. Experts and intermediates who have experience in framing should be able to accomplish most of the work on this project. Intermediates without experience in structural carpentry should get some help and guidance before attempting those tasks. Beginners should add 150% to the estimated professional time if they plan to assemble the stairway and secure the permanent installation. They should also seek professional assistance on the task before they tackle it. Intermediates should add 90% to the labor-hour estimates for the actual stairway installation and 75% to the time for cutting and framing work if they have some experience with these kinds of tasks. They should double the time for the cutting and joist work if they have not performed these tasks before. Experts should add 20%-30% to the labor-hour estimates throughout the project.

WHAT TO WATCH OUT FOR

Spiral stairways and their minimum allowable dimensions are defined in building codes. Before attempting to build one, be sure that the stairway meets the requirements of the local building code.

There are several considerations that can save you time and money on this project. One is selecting the right location for the new stairway if the choice is open to you. A position near a corner or wall is desirable because it normally involves less reinforcing work. Remember that you have to be aware of headroom requirements at the top of the stairway, particularly if it leads to an attic or room with sloped ceilings. If the stairway leads to a second- or third-story room with a finished floor, extra materials and labor may be required to trim the floor area around the opening. Exercise care when cutting the opening; try to avoid damage to the existing ceiling and floor, as this can add needless cost and time to the finish work on the project.

Circular Stairway

Description	Quantity/ Unit	Labor- Hours	Material
Cut opening for stairway, 5' x 5'	1 Ea.	1.6	
Rough in opening for stairway, headers, double 2 x 10 stock	40 L.F.	1.9	41.76
Trimmer joists, double, 2 x 10 stock	40 L.F.	0.5	27.36
Blocking, miscellaneous to joists, 2 x 10 stock	12 L.F.	0.5	17.86
Reinforcement of first floor, double existing joists, 2 x 10 joists	40 L.F.	0.7	59.52
Hangers for header & joists, 18 ga. galv., 2 x 10 joists	4 Ea.	0.2	2.06
Stairs, spiral alum., stock unit, 5' diam., 14 oak treads & platform	1 Ea.	10.7	4,800.00
Trim, casing at opening, 11/16" x 2-1/2"	22 L.F.	0.7	20.59
Trim, pine, 1 x 12 at stair opening	22 L.F.	1.0	35.64
Totals		17.8	$5,004.79

Project Size	5' diam. x 13 risers	Contractor's Fee Including Materials	$8,007

Key to Abbreviations
C.Y.–cubic yard   Ea.–each   L.F.–linear foot   Pr.–pair   Sq.–square (100 square feet of area)
S.F.–square foot   S.Y.–square yard   V.L.F.–vertical linear foot   M.S.F.–thousand square feet

SUMMARY

If you are cramped for space and in need of a stairway to the attic or next floor in your house, a circular staircase may provide the solution to your problem. The installation requires only as much floor area as a circle 4' to 5' in diameter, and as long as care is taken in the cutting and framing, the result should be a sturdy and functional stairway.

For other options or further details regarding options shown, see

> *Disappearing stairway*
> *Standard stairway*

DISAPPEARING STAIRWAY

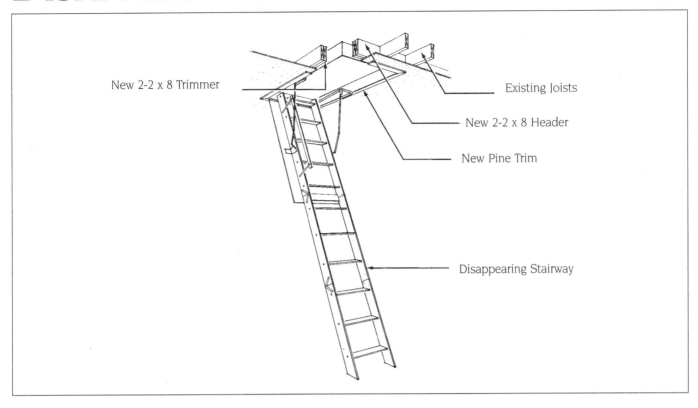

New 2-2 x 8 Trimmer

Existing Joists

New 2-2 x 8 Header

New Pine Trim

Disappearing Stairway

The disappearing or fold-down stairway can solve attic and storage area access problems without wasting valuable floor space. The prefabricated fold-down units literally disappear into the ceiling after use, leaving the area below, usually a hallway or bedroom, open for normal use. The installation of the unit is a moderately difficult operation, but one that advanced beginners could tackle and complete as a weekend project. It must be done to professional standards, however, as it can otherwise become dangerous. The various fittings, screws, nuts, and bolts also tend to loosen and require periodic tightening to keep the structure strong and safe. Because the stairway is very steep with narrow treads, it is really more like a convenient ladder than a set of stairs. Therefore, a disappearing stairway should not be installed as an economical substitute for a permanently placed staircase, and it should be used only for its intended temporary, make-do purpose.

MATERIALS

The list of materials needed for the disappearing stairway installation begins with the factory-prepared unit.

Conventional framing and support lumber will also be needed to restore the ceiling bracings that are weakened when room is made for the stairway. The unit comes from the supplier fully assembled and with complete instructions and guidelines for the installation procedure. As noted earlier, several manufacturers produce the stairway units in varying levels of quality and cost. Shop around and purchase the one that best suits your needs. Before you place and secure the stairway in its permanent position, check it over carefully by examining the condition of the wood and the tightness of the fittings and fasteners. The treads, rails, and support should be solid and free of cracks or imperfections. Some of the units come with a built-in handrail, but others require installation of an accompanying safety rail during the stairway's placement. If your stairway is the latter type, make sure that you take the time to attach the handrail carefully, as it is an essential safety item. Regardless of the quality or design of the disappearing stairway that you have selected, follow the manufacturer's installation instructions and guidelines.

The framing and support materials used to strengthen the ceiling opening for the stairway unit are basically the same

as those included for the circular stairway project. A general rule is that the joist materials should match the dimensions of the existing joists on the floor of your attic or storage room. In most newer homes, the joists or stringers are 2 x 8s or 2 x 6s depending on the structural design of the house and location of the area accessed by the stairway. If the stairway is to be placed near a supporting wall or partition, the required amount of supplies may be reduced. However, if the unit has to be placed in the middle of the ceiling, away from partition or wall support, additional joists and blocking may be needed for adequate support. Galvanized joist hangers should also be used where needed for the headers and trimmer joists. The trim pieces used to face the unit should match the design of the trim work in the room or hallway where the stairway is placed. After the unit is installed and trimmed, it can be painted to blend in with the ceiling.

LEVEL OF DIFFICULTY

This project places moderate demands on the skill level of the intermediate. With some advice and instruction from a professional, advanced beginners can handle most of the tasks. The biggest challenge involved in the installation process is the cutting of the opening and subsequent joist support placement. These are moderately difficult procedures because the ceiling location necessitates cutting over one's head or in a stooped position from the attic floor. If you don't have skill with tools, especially the use of a circular or saber saw overhead, then hire a contractor to do the cutting and joist work. Most intermediates should be able to complete this project if they have had experience with cutting joists. If you are reasonably skilled in the use of tools, but have not performed these specific tasks before, then seek some advice before you start the project. Experts should have no trouble in completing all tasks required to install the stairway and should add about 10% to the estimated professional time.

Intermediates should add about 30% to the labor-hours if they are experienced in cutting and joist support work, and about 50% if they are not. Beginners should not attempt the cutting and joist support operations unless they are advanced enough to use tools in the awkward positions required for this installation. Regardless of your level of remodeling skill, bear in mind that the installation of a disappearing stairway is a two-person job, so plan accordingly.

WHAT TO WATCH OUT FOR

The prefabricated disappearing stairway unit is ready to be placed in the prepared opening as soon as it is uncrated; but it should be handled carefully, because it is heavy and can be put out of square if dropped or mishandled. Take some time to make up temporary guides out of strapping or other scrap lumber, and tack them to the ceiling side of the unit. These guides help to keep the face of the unit flush with the ceiling and can double as temporary supports while the stairway unit is fastened from above. You can also make up a "T" to hold the assembly in place while it is being fastened. If you plan to refinish the ceiling below, you can tack two pieces of strapping across the opening and place the unit on them from above. Two workers are required for the placement of the unit, one above to do the fastening, and one below to position the stairway and to man the temporary supports. When fastening the unit, use shims to keep it square and to avoid bowing of the side pieces of the frame. Before seating the fasteners, double-check the stairway box to see that it is square. If it's not, the unit will not sit properly in the fold-up position, and the ends of the stair rails will not sit evenly on the floor when the unit is in fold-down position. Follow the manufacturer's recommendations for trimming the stair rails at the correct length and angle.

SUMMARY

A disappearing staircase is one way of solving access problems to an attic or storage area. The prefabricated unit is moderately difficult to install, but the job can be completed within a workday by most intermediates and experts.

For other options or further details regarding options shown, see

Circular stairway

Disappearing Stairway

Description	Quantity/ Unit	Labor- Hours	Material
Cut opening for stairway, 30" x 60"	1 Ea.	1.6	
Rough in opening for stairway, header, double 2 x 8 stock	12 L.F.	0.6	12.53
Trimmer, joists, fastened to existing joists, 2 x 8 stock	28 L.F.	0.4	19.15
Hangers for headers & joists, 18 ga. galv. for 2 x 8 joists	4 Ea.	0.2	2.06
Blocking, miscellaneous to joists, 2 x 8 stock	8 L.F.	0.3	8.35
Stairway, disappearing, custom grade, pine, 8'-6" ceiling	1 Ea.	2.3	118.20
Trim, casing, 11/16" x 2'-2"	16 L.F.	0.5	14.98
Totals		5.9	$175.27

Contractor's Fee Including Materials	$543

Key to Abbreviations
C.Y.–cubic yard   Ea.–each   L.F.–linear foot   Pr.–pair   Sq.–square (100 square feet of area)
S.F.–square foot   S.Y.–square yard   V.L.F.–vertical linear foot   M.S.F.–thousand square feet

STANDARD STAIRWAY

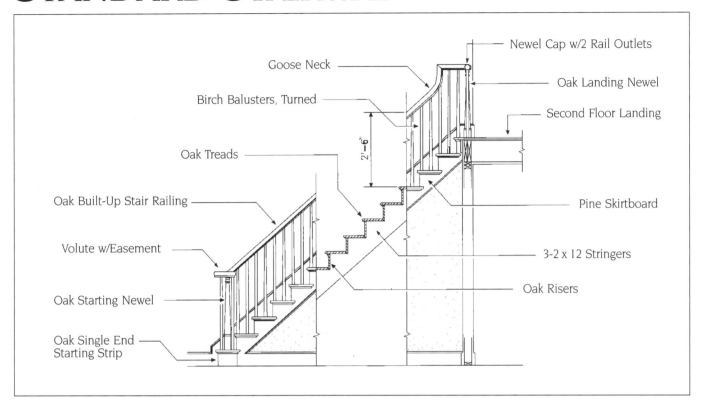

Labels, top to bottom and left to right:
- Newel Cap w/2 Rail Outlets
- Goose Neck
- Oak Landing Newel
- Birch Balusters, Turned
- Second Floor Landing
- Oak Treads
- 2'-6"
- Oak Built-Up Stair Railing
- Pine Skirtboard
- Volute w/Easement
- 3-2 x 12 Stringers
- Oak Starting Newel
- Oak Risers
- Oak Single End Starting Strip

Renovating an existing stairway requires a substantial level of carpentry skill and knowledge. For this reason, beginners and intermediates with limited experience in this area should plan to hire a carpenter for much of the work in this project. There are many reasons why stairway installations demand specialized building skills, but foremost among them is the precise cutting and fitting of hardwood treads, risers, railings, bannisters, and other milled stairway materials. Despite these very demanding aspects, there are other tasks within this project that the do-it-yourselfer can accomplish. The expense and time put into the renovation, whether the work is done by the homeowner or a professional, will result in a refreshing improvement in your home. Because a stairway is often the focal point of the entryway, hallway, or room that it enters, select your materials with both appearance and function in mind.

MATERIALS

This remodeling project has been planned around the renovation of an existing flight of stairs measuring 4' in width and rising to a height of 8'-6". The installation is typical of those found in many two-story dwellings, but variations in such factors as stairway design, height, width, and tread-to-riser proportion are as numerous as the dwellings themselves. The best course of action for the homeowner is to duplicate as closely as possible the existing stairway. Significantly altering the design, size, or direction of the staircase or attempting to match exactly a plan that differs from the original structure will add to the time and cost of the project.

The materials used in this plan consist primarily of milled wood products typical of most house stairway systems. Oak is the featured wood, but other hard or soft woods may be installed at varying costs. Under normal conditions, it is best to duplicate the material of the existing staircase as well as its design. Oak has been chosen for this plan because it is the most commonly used material for treads. It is durable, pleasing in appearance, relatively economical, and is readily available in most areas of the country. The risers in this stairway plan are also made from oak, but softwoods like fir and pine can function adequately in its place. If you are looking for a cost-cutting measure, or if the stairway is to be carpeted, the use of softwood treads and risers in place of oak will reduce your materials expense considerably. In addition to the treads and risers, a special starting step and tread nosing will be needed to complete the stairway.

The materials used for the handrail and its support are also constructed of woods milled specifically for stairways. The homeowner has a wide choice of variously priced materials, from economical softwood bannisters to expensive and ornate hardwood rails, newels, and rail starters. If you want to spruce up or change the design of your stairway, this is the place to do it, as structural alterations usually are not required to modify the stairway's railing system.

In addition to these milled woods, some rough lumber may also be required to replace damaged or warped stringers and other stairway supports. When removing the old stairs, try to avoid chipping or cracking the existing stringers. If they are straight, level, and sound, then leave them in place and save yourself the time and cost of replacement. If you are not sure about their condition, then have a knowledgeable person check them.

LEVEL OF DIFFICULTY

As noted in the description of this project, the renovation of a stairway demands a considerable level of carpentry skill and know-how. The finish tasks, particularly the cutting and fitting of the risers and treads and the setting of the railings, newel, and bannisters, require precise planning and mastery of woodworking skills. Near-perfect accuracy of measurement in all aspects of the project is also necessary to assure quality of appearance and safety in the use of the new stairway. Beginners, therefore, should not attempt any of these advanced carpentry tasks, and should restrict their efforts to the removal of the old stairway and finish work such as the staining or sealing of the stairs. Intermediates might be able to accomplish some of the finish carpentry on the stairs, but only if they have worked with hardwoods and have had some experience in finish carpentry. Intermediates should stop short of the newel, railing, and bannister installations. Experts should be able to install the treads, risers, and various trim pieces; but they, too, should leave the railing work to a carpenter unless they have the know-how, skill, and correct tools to tackle this part of the installation. Generally, beginners should double the professional time for the removal of the existing stairway and any finish work such as painting and staining. Intermediates who want to install the treads, risers, and trim on the stairs should plan to work slowly and deliberately. They should triple the professional time for these tasks and should add 50% to the labor-hours estimate for the removal of the old stairway. Experts, unless they are accomplished carpenters or woodworkers, should double the professional time for the installation of the treads and risers and leave the placement of the newels, railings, and bannisters to a professional.

WHAT TO WATCH OUT FOR

Every so often the stairways in older homes bow, twist, or sag out of alignment because of settling. It is important, therefore, to check the condition and alignment of the stringers and other supporting members under the stairway after you have removed the old treads and risers. Before you install any of the new materials, you must correct any problems that will cause the new stairway to tilt, sag, or become poorly aligned. Failure to correct these kinds of problems makes the installation more difficult and harms the appearance of the new stairs.

SUMMARY

A stairway often occupies a central place in the house, and thus its improvement can enhance the overall appearance and comfort of the home. The services of a contractor may be required for this job, as the types of skills involved are not often in the do-it-yourselfer's range of experience. Nevertheless, this project has valuable benefits in terms of added safety and aesthetic appeal.

For other options or further details regarding options shown, see

Understairs closet

Standard Stairway

Description	Quantity/ Unit	Labor- Hours	Material
Removal of existing stairway	1 Ea.	8.0	
Rough stringers, three 2 x 12, 14' long	42 L.F.	0.8	85.68
Treads, oak, 1-1/16" x 9-1/2" wide, 4' long	12 Ea.	6.1	452.40
Risers, oak, 3/4" x 7-1/2" wide, 4' long	13 Ea.	6.0	288.00
Single end starting step, oak	1 Ea.	1.0	48.00
Balusters, turned, 42" high, birch	27 Ea.	8.0	247.86
Newels, 3-1/4" wide, starting, including volute, minimum	1 Ea.	1.1	40.20
Newels, 3-1/4" wide, landing, minimum	1 Ea.	1.6	88.20
Balusters, turned, 30" high, birch, at landing	6 Ea.	1.7	45.36
Railings, oak, built-up, stair & landing	20 L.F.	2.7	156.00
Skirt board, pine, 1 x 12	14 L.F.	2.2	64.68
Totals		39.2	$1,516.38

Project Size	4' wide x 13 risers	Contractor's Fee Including Materials	$4,033

Key to Abbreviations
C.Y.–cubic yard   Ea.–each   L.F.–linear foot   Pr.–pair   Sq.–square (100 square feet of area)
S.F.–square foot   S.Y.–square yard   V.L.F.–vertical linear foot   M.S.F.–thousand square feet

Part Three
DETAILS

This part supplements the preceding project plans and descriptions. It deals with the cost of unit items rather than that of whole, predetermined projects. You can use this section to add or substitute components in one of the example projects, or to put together your own model that will suit your specific space requirements, the style of your house, your budget, and your personal taste. This information will also help you to arrive at the total cost of a smaller project, such as adding a suspended ceiling system or an open stairway.

"Details" includes a variety of available materials and their comparative costs, as well as the labor-hours involved in their installation and the corresponding contractors' fees. Everything from kitchen cabinets to insulation is listed in this part of the book, with a range of quality and price in each category. By knowing the prices of the available options, as well as their installation times and contractors' charges, you can make informed choices about the fixtures and materials you need and want for a given home improvement project. You can also determine how much of the work you are willing to take on yourself and how much should be done by a contractor. This section also has illustrations identifying the various components. This feature makes planning and purchasing that much easier and helps you to visualize the renovation in progress.

Using the information in this part can be a bit more challenging than the projects spelled out in Part Two. For example, different standards of measure (such as square foot, linear foot, and set) are used according to the type of material; and the quantity of building components must be calculated with the measurements you make yourself. A "How To Use" page follows this introduction and explains the format in more detail. Some organization is required when you use the "Details" to draw up plans and make estimates, and a bit of imagination is helpful, too, when it comes to putting the separate elements together. The benefits of this independent approach are the ability to create your own remodeling system, to expand or reduce the size of a given system, and to change certain aspects within that system for reasons of economy, practicality, or personal preference. As a result, you can be the designer of your own room or facility and can make plans in advance according to the commitment of time, effort, and money that is right for you.

HOW TO USE PART THREE

Part Three, "Details," is provided as a quick, efficient reference for estimating the cost of home improvements. It enables you to closely match dollars and cents to your ideas before you commit yourself to a project.

Basic building elements are grouped together to create a complete "system." These systems can then be used to make up a proposed project. For example, the "Three Fixture Bathroom System" lists all the basic elements required to plumb a bathroom. The result is the total cost for the fixtures and installation.

The systems can also be combined to create an estimate for a more complex project. For example, if you need to add new wall finish and flooring as part of the project, you can consult the "Drywall and Thincoat System" and the "Flooring System" pages to calculate the cost per square foot for redoing the interior walls and floor, and then add in the cost of the plumbing fixtures. In this manner, an entire project, matched to your specifications for size and proposed design, can be cost estimated.

Two types of pages are found in the "Details". The most common format is composed of a graphic illustration of the construction process and a list of the building elements that make it up. In many cases, several different options are presented. For example, the "Three Fixture Bathroom System" pages show the cost for systems with a lavatory installed in a vanity or with a wall-hung lavatory.

The second type of page contains the same kind of information in a slightly different format. These pages list the cost for items that do not require a grouping of basic building elements. For example, the "Insulation" page lists the cost per square foot for various types of building insulation.

The illustration below defines terms that appear in this section and will help you use the information presented. The illustration on the facing page shows how this part of the book is organized.

Key to Abbreviations

C.Y. – cubic yard
Ea. – each
L.F. – linear foot
Pr. – pair
Sq. – square (100 square feet of area)
S.F. – square foot
S.Y. – square yard
V.L.F. – vertical linear foot

System Descriptions uniquely define the system being priced. Descriptions range from item-by-item component lists to single entries differentiating systems by size or other relevant characteristics.

Quantity of each component per system unit

Unit of measure for each item. A Key to Abbreviations appears to the left on this page.

Material Cost for each component of the system. In systems comprised of a group of components, the total material cost for one unit of the system appears in a box at the bottom of the column.

System Description	QUAN.	UNIT	LABOR-HOURS	MAT. COST	COST EACH
BATHROOM INSTALLED WITH VANITY					
Water closet, floor mounted, 2 piece, close coupled, white	1.000	Ea.	3.019	154.00	
Rough-in, waste, 4" diameter DWV piping	1.000	Ea.	.828	31.20	
Vent, 2" diameter DWV piping	1.000	Ea.	.955	25.60	
Supply, 1/2" diameter type "L" copper supply piping	1.000	Ea.	.593	10.14	
Lavatory, 20" x 18", P.E. cast iron with accessories, white	1.000	Ea.	2.500	185.00	
Rough-in, supply, 1/2" diameter type "L" copper supply piping	1.000	Ea.	.988	16.90	
Waste, 1-1/2" diameter DWV piping	1.000	Ea.	1.803	50.00	
Bathtub, P.E. cast iron, 5' long with accessories, white	1.000	Ea.	3.636	410.00	
Rough-in, waste, 4" diameter DWV piping	1.000	Ea.	.828	31.20	
Vent, 1-1/2" diameter DWV piping	1.000	Ea.	.593	15.28	
Supply, 1/2" diameter type "L" copper supply piping	1.000	Ea.	.988	16.90	
Piping, supply, 1/2" diameter type "L" copper supply piping	20.000	L.F.	1.975	33.80	
Waste, 4" diameter DWV piping	9.000	L.F.	2.483	93.60	
Vent, 2" diameter DWV piping	6.000	L.F.	1.500	27.12	
Vanity base cabinet, 2 door, 30" wide	1.000	Ea.	1.000	204.00	
Vanity top, plastic laminated square edge	2.670	L.F.	.712	58.74	Contractor's Fee
TOTAL			24.401	1363.48	**3365.06**

Labor-Hours required to install the system. In systems made up of components, the total labor-hours to install one unit of the system appears in a box at the bottom of the column.

Total Cost per unit for each system. This figure includes the price of materials (without sales tax), labor costs, and the contractor's overhead expenses and profit.

PART THREE
TABLE OF CONTENTS

Information in Part Three is organized into the four divisions listed here. The schematic diagram pictured below shows how the parts of a typical house fit into these Part Three divisions.

Note: For Divisions 1-5 see *Exterior Home Improvement Costs.*

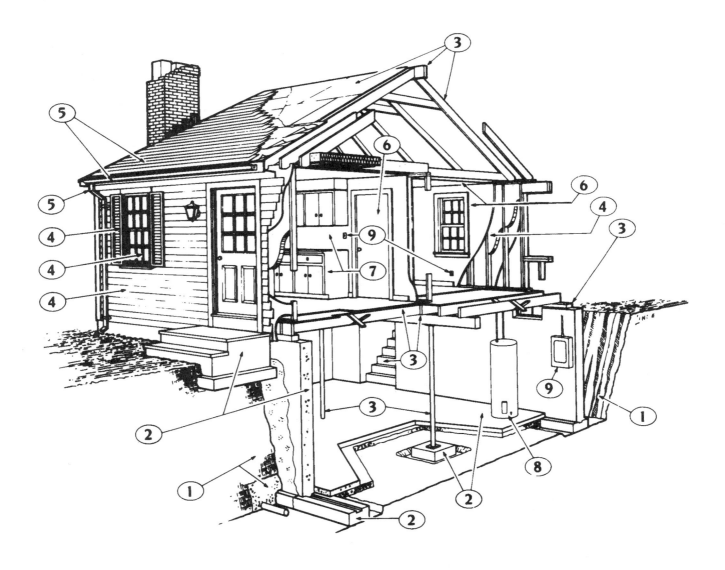

Division 6
INTERIOR FINISHES

Note: For Divisions 1-5, see
Exterior Home Improvement Costs

Drywall & Thincoat Wall Systems

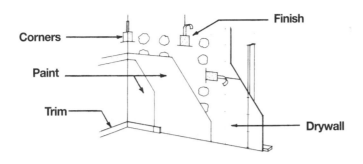

System Description	QUAN.	UNIT	LABOR-HOURS	MAT. COST	COST PER S.F.
1/2" SHEETROCK, TAPED & FINISHED					
Drywall, 1/2" thick, standard	1.000	S.F.	.008	.23	
Finish, taped & finished joints	1.000	S.F.	.008	.04	
Corners, taped & finished, 32 L.F. per 12' x 12' room	.083	L.F.	.001	.01	= 50¢ PSF.
Painting, primer & 2 coats	1.000	S.F.	.011	.15	
Paint trim, to 6" wide, primer + 1 coat enamel	.125	L.F.	.001	.01	Contractor's Fee
Trim, baseboard	.125	L.F.	.005	.18	
TOTAL			.034	.62	**2.62**

Drywall & Thincoat Ceiling Systems

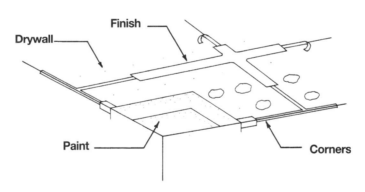

System Description	QUAN.	UNIT	LABOR-HOURS	MAT. COST	COST PER S.F.
1/2" SHEETROCK, TAPED & FINISHED					
Drywall, 1/2" thick, standard	1.000	S.F.	.008	.23	
Finish, taped & finished	1.000	S.F.	.008	.04	
Corners, taped & finished, 12' x 12' room	.333	L.F.	.005	.02	Contractor's Fee
Paint, primer & 2 coats	1.000	S.F.	.011	.15	
TOTAL			.032	.44	**2.21**

Be sure to read pages 172-173 for proper use of this section.

Plaster & Stucco Wall Systems

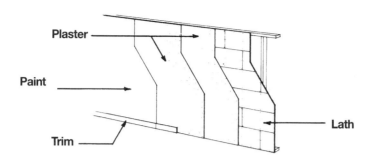

System Description	QUAN.	UNIT	LABOR-HOURS	MAT. COST	COST PER S.F.
PLASTER ON GYPSUM LATH					
Plaster, gypsum or perlite, 2 coats	1.000	S.F.	.053	.38	
Lath, 3/8" gypsum	1.000	S.F.	.010	.43	
Corners, expanded metal, 32 L.F. per 12' x 12' room	.083	L.F.	.002	.01	
Painting, primer & 2 coats	1.000	S.F.	.011	.15	
Paint trim, to 6" wide, primer + 1 coat enamel	.125	L.F.	.001	.01	
Trim, baseboard	.125	L.F.	.005	.18	
					Contractor's Fee
TOTAL			.082	1.16	**5.64**

Plaster & Stucco Ceiling Systems

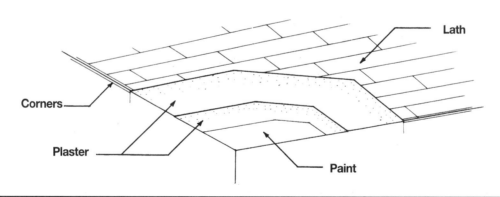

System Description	QUAN.	UNIT	LABOR-HOURS	MAT. COST	COST PER S.F.
PLASTER ON GYPSUM LATH					
Plaster, gypsum or perlite, 2 coats	1.000	S.F.	.061	.38	
Gypsum lath, plain or perforated, nailed, 3/8" thick	1.000	S.F.	.010	.43	
Gypsum lath, ceiling installation adder	1.000	S.F.	.004		
Corners, expanded metal, 12' x 12' room	.330	L.F.	.007	.04	
Painting, primer & 2 coats	1.000	S.F.	.011	.15	**Contractor's Fee**
TOTAL			.093	1.00	**5.86**

Be sure to read pages 172-173 for proper use of this section.

Suspended Ceiling Systems

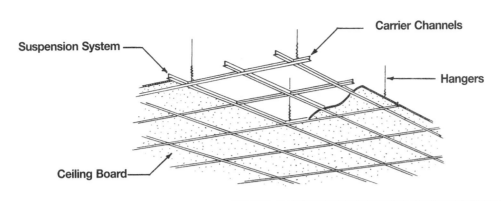

System Description	QUAN.	UNIT	LABOR-HOURS	MAT. COST	COST PER S.F.
2' X 2' GRID, FILM FACED FIBERGLASS, 5/8" THICK					
Suspension system, 2' x 2' grid, T bar	1.000	S.F.	.012	.82	
Ceiling board, film faced fiberglass, 5/8" thick	1.000	S.F.	.013	.57	
Carrier channels, 1-1/2" x 3/4"	1.000	S.F.	.017	.24	Contractor's Fee
Hangers, #12 wire	1.000	S.F.	.002	.03	
TOTAL			.044	1.66	**4.80**
2' X 4' GRID, FILM FACED FIBERGLASS, 5/8" THICK					
Suspension system, 2' x 4' grid, T bar	1.000	S.F.	.010	.71	
Ceiling board, film faced fiberglass, 5/8" thick	1.000	S.F.	.013	.57	
Carrier channels, 1-1/2" x 3/4"	1.000	S.F.	.017	.24	Contractor's Fee
Hangers, #12 wire	1.000	S.F.	.002	.03	
TOTAL			.042	1.55	**4.52**
2' X 2' GRID, MINERAL FIBER, REVEAL EDGE, 1" THICK					
Suspension system, 2' x 2' grid, T bar	1.000	S.F.	.012	.82	
Ceiling board, mineral fiber, reveal edge, 1" thick	1.000	S.F.	.013	1.31	
Carrier channels, 1-1/2" x 3/4"	1.000	S.F.	.017	.24	Contractor's Fee
Hangers, #12 wire	1.000	S.F.	.002	.03	
TOTAL			.044	2.40	**5.96**
2' X 4' GRID, MINERAL FIBER, REVEAL EDGE, 1" THICK					
Suspension system, 2' x 4' grid, T bar	1.000	S.F.	.010	.71	
Ceiling board, mineral fiber, reveal edge, 1" thick	1.000	S.F.	.013	1.31	
Carrier channels, 1-1/2" x 3/4"	1.000	S.F.	.017	.24	Contractor's Fee
Hangers, #12 wire	1.000	S.F.	.002	.03	
TOTAL			.042	2.29	**5.68**

Be sure to read pages 172-173 for proper use of this section.

Interior Door Systems

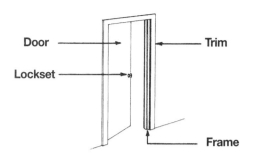

Door — Trim

Lockset —

Frame

System Description	QUAN.	UNIT	LABOR-HOURS	MAT. COST	COST EACH
LAUAN, FLUSH DOOR, HOLLOW CORE					
Door, flush, lauan, hollow core, 2'-8" wide x 6'-8" high	1.000	Ea.	.889	32.50	
Frame, pine, 4-5/8" jamb	17.000	L.F.	.725	96.90	
Trim, stock pine, 11/16" x 2-1/2"	34.000	L.F.	1.133	25.50	
Paint trim, to 6" wide, primer + 1 coat enamel	34.000	L.F.	.340	3.40	
Butt hinges, chrome, 3-1/2" x 3-1/2"	1.500	Pr.		24.60	
Lockset, passage	1.000	Ea.	.500	13.85	
Prime door & frame, oil, brushwork	2.000	Face	1.600	4.52	
Paint door and frame, oil, 2 coats	2.000	Face	2.667	7.28	Contractor's Fee
TOTAL			7.854	208.55	**696.11**
BIRCH, FLUSH DOOR, HOLLOW CORE					
Door, flush, birch, hollow core, 2'-8" wide x 6'-8" high	1.000	Ea.	.889	46.50	
Frame, pine, 4-5/8" jamb	17.000	L.F.	.725	96.90	
Trim, stock pine, 11/16" x 2-1/2"	34.000	L.F.	1.133	25.50	
Butt hinges, chrome, 3-1/2" x 3-1/2"	1.500	Pr.		24.60	
Lockset, passage	1.000	Ea.	.500	13.85	
Prime door & frame, oil, brushwork	2.000	Face	1.600	4.52	
Paint door and frame, oil, 2 coats	2.000	Face	2.667	7.28	Contractor's Fee
TOTAL			7.514	219.15	**697.13**
RAISED PANEL, SOLID, PINE DOOR					
Door, pine, raised panel, 2'-8" wide x 6'-8" high	1.000	Ea.	.889	157.00	
Frame, pine, 4-5/8" jamb	17.000	L.F.	.725	96.90	
Trim, stock pine, 11/16" x 2-1/2"	34.000	L.F.	1.133	25.50	
Butt hinges, bronze, 3-1/2" x 3-1/2"	1.500	Pr.		28.05	
Lockset, passage	1.000	Ea.	.500	13.85	
Prime door & frame, oil, brushwork	2.000		1.600	4.52	
Paint door and frame, oil, 2 coats	2.000		2.667	7.28	Contractor's Fee
TOTAL			7.514	333.10	**874.56**

Be sure to read pages 172-173 for proper use of this section.

Closet Door Systems

Door — Trim

Frame

System Description	QUAN.	UNIT	LABOR-HOURS	MAT. COST	COST EACH
BI-PASSING, FLUSH, LAUAN, HOLLOW CORE, 4'-0" X 6'-8"					
Door, flush, lauan, hollow core, 4'-0" x 6'-8" opening	1.000	Ea.	1.333	174.00	
Frame, pine, 4-5/8" jamb	18.000	L.F.	.768	102.60	
Trim, stock pine, 11/16" x 2-1/2"	36.000	L.F.	1.200	27.00	
Prime door & frame, oil, brushwork	2.000	Face	1.600	4.52	
Paint door and frame, oil, 2 coats	2.000	Face	2.667	7.28	Contractor's Fee
TOTAL			7.568	315.40	**850.21**
BI-PASSING, FLUSH, BIRCH, HOLLOW CORE, 6'-0" X 6'-8"					
Door, flush, birch, hollow core, 6'-0" x 6'-8" opening	1.000	Ea.	1.600	235.00	
Frame, pine, 4-5/8" jamb	19.000	L.F.	.811	108.30	
Trim, stock pine, 11/16" x 2-1/2"	38.000	L.F.	1.267	28.50	
Prime door & frame, oil, brushwork	2.000	Face	2.000	5.65	
Paint door and frame, oil, 2 coats	2.000	Face	3.333	9.10	Contractor's Fee
TOTAL			9.011	386.55	**1027.21**
BI-FOLD, PINE, PANELED, 3'-0" X 6'-8"					
Door, pine, paneled, 3'-0" x 6'-8" opening	1.000	Ea.	1.231	126.00	
Frame, pine, 4-5/8" jamb	17.000	L.F.	.725	96.90	
Trim, stock pine, 11/16" x 2-1/2"	34.000	L.F.	1.133	25.50	
Prime door & frame, oil, brushwork	2.000		1.600	4.52	
Paint door and frame, oil, 2 coats	2.000		2.667	7.28	Contractor's Fee
TOTAL			7.356	260.20	**753.35**
BI-FOLD, PINE, LOUVERED, 6'-0" X 6'-8"					
Door, pine, louvered, 6'-0" x 6'-8" opening	1.000	Ea.	1.600	247.00	
Frame, pine, 4-5/8" jamb	19.000	L.F.	.811	108.30	
Trim, stock pine, 11/16" x 2-1/2"	38.000	L.F.	1.267	28.50	
Prime door & frame, oil, brushwork	2.500		2.000	5.65	
Paint door and frame, oil, 2 coats	2.500		3.333	9.10	Contractor's Fee
TOTAL			9.011	398.55	**1047.40**

Be sure to read pages 172-173 for proper use of this section.

Stairways

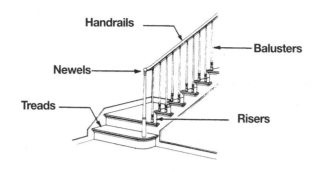

System Description	QUAN.	UNIT	LABOR-HOURS	MAT. COST	COST EACH
7 RISERS, OAK TREADS, BOX STAIRS					
Treads, oak, 9-1/2″ x 1-1/16″ thick	6.000	Ea.	2.667	153.00	
Risers, 3/4″ thick, beech	7.000	Ea.	2.625	133.35	
Balusters, birch, 30″ high	12.000	Ea.	3.429	83.40	
Newels, 3-1/4″ wide	2.000	Ea.	2.286	74.00	
Handrails, oak laminated	7.000	L.F.	.933	50.05	
Stringers, 2″ x 10″, 3 each	21.000	L.F.	.306	8.19	Contractor's Fee
TOTAL			12.246	501.99	**1398.88**
14 RISERS, OAK TREADS, BOX STAIRS					
Treads, oak, 9-1/2″ x 1-1/16″ thick	13.000	Ea.	5.778	331.50	
Risers, 3/4″ thick, beech	14.000	Ea.	5.250	266.70	
Balusters, birch, 30″ high	26.000	Ea.	7.428	180.70	
Newels, 3-1/4″ wide	2.000	Ea.	2.286	74.00	
Handrails, oak, laminated	14.000	L.F.	1.867	100.10	
Stringers, 2″ x 10″, 3 each	42.000	L.F.	5.169	57.54	Contractor's Fee
TOTAL			27.778	1010.54	**2974.81**
14 RISERS, PINE TREADS, BOX STAIRS					
Treads, pine, 9-1/2″ x 3/4″ thick	13.000	Ea.	5.778	197.60	
Risers, 3/4″ thick, pine	14.000	Ea.	5.091	129.36	
Balusters, pine, 30″ high	26.000	Ea.	7.428	117.52	
Newels, 3-1/4″ wide	2.000	Ea.	2.286	74.00	
Handrails, oak, laminated	14.000	L.F.	1.867	100.10	
Stringers, 2″ x 10″, 3 each	42.000	L.F.	5.169	57.54	Contractor's Fee
TOTAL			27.619	676.12	**2444.16**

Be sure to read pages 172-173 for proper use of this section.

Carpet Systems

SYSTEM DESCRIPTION	QUAN.	UNIT	LABOR-HOURS	COST PER S.F. MAT.	COST PER S.F. TOTAL
Carpet, direct glue-down, nylon, level loop, 26 oz.	1.000	S.F.	.018	1.71	3.49
32 oz.	1.000	S.F.	.018	2.41	4.58
40 oz.	1.000	S.F.	.018	3.59	6.41
Nylon, plush, 20 oz.	1.000	S.F.	.018	1.13	2.59
24 oz.	1.000	S.F.	.018	1.20	2.70
30 oz.	1.000	S.F.	.018	1.78	3.60
36 oz.	1.000	S.F.	.018	2.25	4.33
42 oz.	1.000	S.F.	.022	2.35	4.65
48 oz.	1.000	S.F.	.022	3.10	5.82
54 oz.	1.000	S.F.	.022	3.50	6.44
Olefin, 15 oz.	1.000	S.F.	.018	.61	1.78
22 oz.	1.000	S.F.	.018	.72	1.96
Tile, foam backed, needle punch	1.000	S.F.	.014	2.65	4.76
Tufted loop or shag	1.000	S.F.	.014	1.12	2.39
Wool, 36 oz., level loop	1.000	S.F.	.018	7.90	13.11
32 oz., patterned	1.000	S.F.	.020	7.80	13.02
48 oz., patterned	1.000	S.F.	.020	7.95	13.26
Padding, sponge rubber cushion, minimum	1.000	S.F.	.006	.35	.83
Maximum	1.000	S.F.	.006	.95	1.76
Felt, 32 oz. to 56 oz., minimum	1.000	S.F.	.006	.40	.90
Maximum	1.000	S.F.	.006	.74	1.43
Bonded urethane, 3/8" thick, minimum	1.000	S.F.	.006	.43	.95
Maximum	1.000	S.F.	.006	.74	10.67
Prime urethane, 1/4" thick, minimum	1.000	S.F.	.006	.25	.67
Maximum	1.000	S.F.	.006	.46	1.00
Stairs, for stairs, add to above carpet prices	1.000	Riser	.267		12.44
Underlayment plywood, 3/8" thick	1.000	S.F.	.011	.58	1.44
1/2" thick	1.000	S.F.	.011	.70	1.64
5/8" thick	1.000	S.F.	.011	.77	1.78
3/4" thick	1.000	S.F.	.012	.94	2.08
Particle board, 3/8" thick	1.000	S.F.	.011	.42	1.19
1/2" thick	1.000	S.F.	.011	.44	1.24
5/8" thick	1.000	S.F.	.011	.59	1.50
3/4" thick	1.000	S.F.	.012	.64	1.62
Hardboard, 4' x 4', 0.215" thick	1.000	S.F.	.011	.45	1.24

Be sure to read pages 172-173 for proper use of this section.

Flooring Systems

SYSTEM DESCRIPTION	QUAN.	UNIT	LABOR-HOURS	MAT.	TOTAL
				COST PER S.F.	
Resilient flooring, asphalt tile on concrete, 1/8" thick					
Color group B	1.000	S.F.	.020	1.06	2.58
Color group C & D	1.000	S.F.	.020	1.17	2.75
Asphalt tile on wood subfloor, 1/8" thick					
Color group B	1.000	S.F.	.020	1.26	2.89
Color group C & D	1.000	S.F.	.020	1.37	3.06
Vinyl composition tile, 12" x 12", 1/16" thick	1.000	S.F.	.016	.80	1.99
Embossed	1.000	S.F.	.016	.98	2.27
Marbleized	1.000	S.F.	.016	.98	2.27
Plain	1.000	S.F.	.016	1.10	2.46
.080" thick, embossed	1.000	S.F.	.016	1.01	2.32
Marbleized	1.000	S.F.	.016	1.11	2.48
Plain	1.000	S.F.	.016	1.62	3.27
1/8" thick, marbleized	1.000	S.F.	.016	1.08	2.43
Plain	1.000	S.F.	.016	2.05	3.93
Vinyl tile, 12" x 12", .050" thick, minimum	1.000	S.F.	.016	1.75	3.47
Maximum	1.000	S.F.	.016	3.41	6.04
1/8" thick, minimum	1.000	S.F.	.016	2.20	4.17
Maximum	1.000	S.F.	.016	4.93	8.39
1/8" thick, solid colors	1.000	S.F.	.016	3.52	6.21
Florentine pattern	1.000	S.F.	.016	4.05	7.03
Marbleized or travertine pattern	1.000	S.F.	.016	8.30	13.66
Vinyl sheet goods, backed, .070" thick, minimum	1.000	S.F.	.032	2.04	4.66
Maximum	1.000	S.F.	.040	2.76	6.16
.093" thick, minimum	1.000	S.F.	.035	2.20	5.03
Maximum	1.000	S.F.	.040	3.14	6.75
.125" thick, minimum	1.000	S.F.	.035	2.48	5.47
Maximum	1.000	S.F.	.040	3.93	7.97
Wood, oak, finished in place, 25/32" x 2-1/2" clear	1.000	S.F.	.074	3.69	9.11
Select	1.000	S.F.	.074	3.77	9.23
No. 1 common	1.000	S.F.	.074	4.57	10.50
Prefinished, oak, 2-1/2" wide	1.000	S.F.	.047	6.45	12.42
3-1/4" wide	1.000	S.F.	.043	8.25	15.00
Ranch plank, oak, random width	1.000	S.F.	.055	8.00	15.18
Parquet, 5/16" thick, finished in place, oak, minimum	1.000	S.F.	.077	3.68	9.23
Maximum	1.000	S.F.	.107	6.35	14.93
Teak, minimum	1.000	S.F.	.077	5.30	11.73
Maximum	1.000	S.F.	.107	8.70	18.57
Sleepers, treated, 16" O.C., 1" x 2"	1.000	S.F.	.007	.11	.51
1" x 3"	1.000	S.F.	.008	.20	.71
2" x 4"	1.000	S.F.	.011	.55	1.39
2" x 6"	1.000	S.F.	.012	1.23	2.53
Subfloor, plywood, 1/2" thick	1.000	S.F.	.011	.54	1.38
5/8" thick	1.000	S.F.	.012	.62	1.56
3/4" thick	1.000	S.F.	.013	.73	1.79
Ceramic tile, color group 2, 1" x 1"	1.000	S.F.	.087	4.61	10.83
2" x 2" or 2" x 1"	1.000	S.F.	.084	4.37	10.30

Be sure to read pages 172-173 for proper use of this section.

Division 7
SPECIALTIES

Kitchen Systems

Kitchen Systems

Soffit Drywall · Soffit Framing · Top Cabinets · Counter Top · Bottom Cabinets

System Description	QUAN.	UNIT	LABOR-HOURS	MAT. COST	COST PER L.F.
KITCHEN, ECONOMY GRADE					
Top cabinets, economy grade	1.000	L.F.	.171	30.08	
Bottom cabinets, economy grade	1.000	L.F.	.256	45.12	
Counter top, laminated plastic, post formed	1.000	L.F.	.267	9.40	
Blocking, wood, 2" x 4"	1.000	L.F.	.032	.39	
Soffit, framing, wood, 2" x 4"	4.000	L.F.	.071	1.56	
Soffit drywall	2.000	S.F.	.047	.56	Contractor's Fee
Drywall painting	2.000	S.F.	.013	.10	
TOTAL			.857	87.21	**178.60**
AVERAGE GRADE					
Top cabinets, average grade	1.000	L.F.	.213	37.60	
Bottom cabinets, average grade	1.000	L.F.	.320	56.40	
Counter top, laminated plastic, square edge, incl. backsplash	1.000	L.F.	.267	28.50	
Blocking, wood, 2" x 4"	1.000	L.F.	.032	.39	
Soffit framing, wood, 2" x 4"	4.000	L.F.	.071	1.56	
Soffit drywall	2.000	S.F.	.047	.56	Contractor's Fee
Drywall painting	2.000	S.F.	.013	.10	
TOTAL			.963	125.11	**242.78**
CUSTOM GRADE					
Top cabinets, custom grade	1.000	L.F.	.256	101.20	
Bottom cabinets, custom grade	1.000	L.F.	.384	151.80	
Counter top, laminated plastic, square edge, incl. backsplash	1.000	L.F.	.267	28.50	
Blocking, wood, 2" x 4"	1.000	L.F.	.032	.39	
Soffit framing, wood, 2" x 4"	4.000	L.F.	.071	1.56	
Soffit drywall	2.000	S.F.	.047	.56	Contractor's Fee
Drywall painting	2.000	S.F.	.013	.10	
TOTAL			1.070	284.11	**495.13**

Be sure to read pages 172-173 for proper use of this section.

Division 8
MECHANICAL

Two Fixture Lavatory Systems

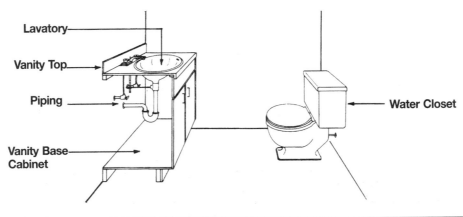

Lavatory

Vanity Top

Piping

Vanity Base Cabinet

Water Closet

System Description	QUAN.	UNIT	LABOR-HOURS	MAT. COST	COST EACH
LAVATORY INSTALLED WITH VANITY, PLUMBING IN 2 WALLS					
Water closet, floor mounted, 2 piece, close coupled, white	1.000	Ea.	3.019	154.00	
Rough-in, vent, 2" diameter DWV piping	1.000	Ea.	.955	25.60	
Waste, 4" diameter DWV piping	1.000	Ea.	.828	31.20	
Supply, 1/2" diameter type "L" copper supply piping	1.000	Ea.	.593	10.14	
Lavatory, 20" x 18", P.E. cast iron white	1.000	Ea.	2.500	185.00	
Rough-in, vent, 1-1/2" diameter DWV piping	1.000	Ea.	.901	25.00	
Waste, 2" diameter DWV piping	1.000	Ea.	.955	25.60	
Supply, 1/2" diameter type "L" copper supply piping	1.000	Ea.	.988	16.90	
Piping, supply, 1/2" diameter type "L" copper supply piping	10.000	L.F.	.988	16.90	
Waste, 4" diameter DWV piping	7.000	L.F.	1.931	72.80	
Vent, 2" diameter DWV piping	12.000	L.F.	2.866	76.80	Contractor's
Vanity base cabinet, 2 door, 30" wide	1.000	Ea.	1.000	204.00	Fee
Vanity top, plastic & laminated, square edge	2.670	L.F.	.712	76.10	**2353.06**
TOTAL			18.236	920.04	
LAVATORY WITH WALL-HUNG LAVATORY, PLUMBING IN 2 WALLS					
Water closet, floor mounted, 2 piece close coupled, white	1.000	Ea.	3.019	154.00	
Rough-in, vent, 2" diameter DWV piping	1.000	Ea.	.955	25.60	
Waste, 4" diameter DWV piping	1.000	Ea.	.828	31.20	
Supply, 1/2" diameter type "L" copper supply piping	1.000	Ea.	.593	10.14	
Lavatory, 20" x 18", P.E. cast iron, wall hung, white	1.000	Ea.	2.000	218.00	
Rough-in, vent, 1-1/2" diameter DWV piping	1.000	Ea.	.901	25.00	
Waste, 2" diameter DWV piping	1.000	Ea.	.955	25.60	
Supply, 1/2" diameter type "L" copper supply piping	1.000	Ea.	.988	16.90	
Piping, supply, 1/2" diameter type "L" copper supply piping	10.000	L.F.	.988	16.90	
Waste, 4" diameter DWV piping	7.000	L.F.	1.931	72.80	
Vent, 2" diameter DWV piping	12.000	L.F.	2.866	76.80	Contractor's
Carrier, steel for studs, no arms	1.000	Ea.	1.143	29.00	Fee
TOTAL			17.167	701.94	**1966.71**

Be sure to read pages 172-173 for proper use of this section.

Three Fixture Bathroom Systems

System Description	QUAN.	UNIT	LABOR-HOURS	MAT. COST	COST EACH
BATHROOM INSTALLED WITH VANITY					
Water closet, floor mounted, 2 piece, close coupled, white	1.000	Ea.	3.019	154.00	
Rough-in, waste, 4" diameter DWV piping	1.000	Ea.	.828	31.20	
Vent, 2" diameter DWV piping	1.000	Ea.	.955	25.60	
Supply, 1/2" diameter type "L" copper supply piping	1.000	Ea.	.593	10.14	
Lavatory, 20" x 18", P.E. cast iron with accessories, white	1.000	Ea.	2.500	185.00	
Rough-in, supply, 1/2" diameter type "L" copper supply piping	1.000	Ea.	.988	16.90	
Waste, 1-1/2" diameter DWV piping	1.000	Ea.	1.803	50.00	
Bathtub, P.E. cast iron, 5' long with accessories, white	1.000	Ea.	3.636	410.00	
Rough-in, waste, 4" diameter DWV piping	1.000	Ea.	.828	31.20	
Vent, 1-1/2" diameter DWV piping	1.000	Ea.	.593	15.28	
Supply, 1/2" diameter type "L" copper supply piping	1.000	Ea.	.988	16.90	
Piping, supply, 1/2" diameter type "L" copper supply piping	20.000	L.F.	1.975	33.80	
Waste, 4" diameter DWV piping	9.000	L.F.	2.483	93.60	
Vent, 2" diameter DWV piping	6.000	L.F.	1.500	27.12	
Vanity base cabinet, 2 door, 30" wide	1.000	Ea.	1.000	204.00	Contractor's
Vanity top, plastic laminated square edge	2.670	L.F.	.712	58.74	Fee
TOTAL			24.401	1363.48	**3365.06**
BATHROOM WITH WALL HUNG LAVATORY					
Water closet, floor mounted, 2 piece, close coupled, white	1.000	Ea.	3.019	154.00	
Rough-in, vent, 2" diameter DWV piping	1.000	Ea.	.955	25.60	
Waste, 4" diameter DWV piping	1.000	Ea.	.828	31.20	
Supply, 1/2" diameter type "L" copper supply piping	1.000	Ea.	.593	10.14	
Lavatory, 20" x 18" P.E. cast iron, wall hung, white	1.000	Ea.	2.000	218.00	
Rough-in, waste, 1-1/2" diameter DWV piping	1.000	Ea.	1.803	50.00	
Supply, 1/2" diameter type "L" copper supply piping	1.000	Ea.	.988	16.90	
Bathtub, P.E. cast iron, 5' long with accessories, white	1.000	Ea.	3.636	410.00	
Rough-in, waste, 4" diameter DWV piping	1.000	Ea.	.828	31.20	
Supply, 1/2" diameter type "L" copper supply piping	1.000	Ea.	.988	16.90	
Vent, 1-1/2" diameter DWV piping	1.000	Ea.	1.482	38.20	
Piping, supply, 1/2" diameter type "L" copper supply piping	20.000	L.F.	1.975	33.80	
Waste, 4" diameter DWV piping	9.000	L.F.	2.483	93.60	
Vent, 2" diameter DWV piping	6.000	L.F.	1.500	27.12	Contractor's
Carrier, steel, for studs, no arms	1.000	Ea.	1.143	29.00	Fee
TOTAL			24.221	1185.66	**3090.59**

Be sure to read pages 172-173 for proper use of this section.

Three Fixture Bathroom Systems

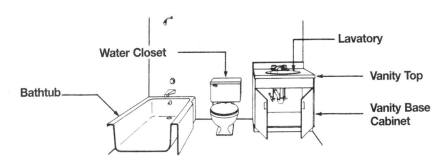

Water Closet

Lavatory

Vanity Top

Bathtub

Vanity Base Cabinet

System Description	QUAN.	UNIT	LABOR-HOURS	MAT. COST	COST EACH
BATHROOM WITH LAVATORY INSTALLED IN VANITY					
Water closet, floor mounted, 2 piece, close coupled, white	1.000	Ea.	3.019	154.00	
Rough-in, waste, 4" diameter DWV piping	1.000	Ea.	.828	31.20	
Vent, 2" diameter DWV piping	1.000	Ea.	.955	25.60	
Supply, 1/2" diameter type "L" copper supply piping	1.000	Ea.	.593	10.14	
Lavatory, 20" x 18", P.E. cast iron with accessories, white	1.000	Ea.	2.500	185.00	
Rough-in, waste, 1-1/2" diameter DWV piping	1.000	Ea.	1.803	50.00	
Supply, 1/2" diameter type "L" copper supply piping	1.000	Ea.	.988	16.90	
Bathtub, P.E. cast iron 5' long with accessories, white	1.000	Ea.	3.636	410.00	
Rough-in, waste, 4" diameter DWV piping	1.000	Ea.	.828	31.20	
Vent, 1-1/2" diameter DWV piping	1.000	Ea.	.593	15.28	
Supply, 1/2" diameter type "L" copper supply piping	1.000	Ea.	.988	16.90	
Piping, supply, 1/2" diameter type "L" copper supply piping	10.000	L.F.	.988	16.90	
Waste, 4" diameter DWV piping	6.000	L.F.	1.655	62.40	
Vent, 2" diameter DWV piping	6.000	L.F.	1.500	27.12	Contractor's Fee
Vanity base cabinet, 2 door, 30" wide	1.000	Ea.	1.000	204.00	
Vanity top, plastic laminated square edge	2.670	L.F.	.712	58.74	**3194.63**
TOTAL			22.586	1315.38	
BATHROOM WITH WALL HUNG LAVATORY					
Water closet, floor mounted, 2 piece, close coupled, white	1.000	Ea.	3.019	154.00	
Rough-in, vent, 2" diameter DWV piping	1.000	Ea.	.955	25.60	
Waste, 4" diameter DWV piping	1.000	Ea.	.828	31.20	
Supply, 1/2" diameter type "L" copper supply piping	1.000	Ea.	.593	10.14	
Lavatory, 20" x 18" P.E. cast iron, wall hung, white	1.000	Ea.	2.000	218.00	
Rough-in, waste, 1-1/2" diameter DWV piping	1.000	Ea.	1.803	50.00	
Supply, 1/2" diameter type "L" copper supply piping	1.000	Ea.	.988	16.90	
Bathtub, P.E. cast iron, 5' long with accessories, white	1.000	Ea.	3.636	410.00	
Rough-in, waste, 4" diameter DWV piping	1.000	Ea.	.828	31.20	
Supply, 1/2" diameter type "L" copper supply piping	1.000	Ea.	.988	16.90	
Vent, 1-1/2" diameter DWV piping	1.000	Ea.	.593	15.28	
Piping, supply, 1/2" diameter type "L" copper supply piping	10.000	L.F.	.988	16.90	
Waste, 4" diameter DWV piping	6.000	L.F.	1.655	62.40	
Vent, 2" diameter DWV piping	6.000	L.F.	1.500	27.12	Contractor's Fee
Carrier, steel, for studs, no arms	1.000	Ea.	1.143	29.00	
TOTAL			21.517	1114.64	**2835.31**

Be sure to read pages 172-173 for proper use of this section.

Three Fixture Bathroom Systems

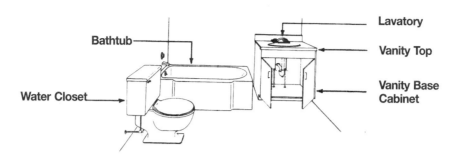

Lavatory

Vanity Top

Vanity Base Cabinet

Bathtub

Water Closet

System Description	QUAN.	UNIT	LABOR-HOURS	MAT. COST	COST EACH
BATHROOM WITH LAVATORY INSTALLED IN VANITY					
Water closet, floor mounted, 2 piece, close coupled, white	1.000	Ea.	3.019	154.00	
Rough-in, vent, 2" diameter DWV piping	1.000	Ea.	.955	25.60	
Waste, 4" diameter DWV piping	1.000	Ea.	.828	31.20	
Supply, 1/2" diameter type "L" copper supply piping	1.000	Ea.	.593	10.14	
Lavatory, 20" x 18", PE cast iron with accessories, white	1.000	Ea.	2.500	185.00	
Rough-in, vent, 1-1/2" diameter DWV piping	1.000	Ea.	1.803	50.00	
Supply, 1/2" diameter type "L" copper supply piping	1.000	Ea.	.988	16.90	
Bathtub, P.E. cast iron, 5' long with accessories, white	1.000	Ea.	3.636	410.00	
Rough-in, waste, 4" diameter DWV piping	1.000	Ea.	.828	31.20	
Supply, 1/2" diameter type "L" copper supply piping	1.000	Ea.	.988	16.90	
Vent, 1-1/2" diameter DWV piping	1.000	Ea.	.593	15.28	
Piping, supply, 1/2" diameter type "L" copper supply piping	32.000	L.F.	3.161	54.08	
Waste, 4" diameter DWV piping	12.000	L.F.	3.310	124.80	
Vent, 2" diameter DWV piping	6.000	L.F.	1.500	27.12	
Vanity base cabinet, 2 door, 30" wide	1.000	Ea.	1.000	204.00	**Contractor's Fee**
Vanity top, plastic laminated square edge	2.670	L.F.	.712	58.74	
TOTAL			26.414	1414.96	**3551.70**
BATHROOM WITH WALL HUNG LAVATORY					
Water closet, floor mounted, 2 piece, close coupled, white	1.000	Ea.	3.019	154.00	
Rough-in, vent, 2" diameter DWV piping	1.000	Ea.	.955	25.60	
Waste, 4" diameter DWV piping	1.000	Ea.	.828	31.20	
Supply, 1/2" diameter type "L" copper supply piping	1.000	Ea.	.593	10.14	
Lavatory, 20" x 18" P.E. cast iron, wall hung, white	1.000	Ea.	2.000	218.00	
Rough-in, waste, 1-1/2" diameter DWV piping	1.000	Ea.	1.803	50.00	
Supply, 1/2" diameter type "L" copper supply piping	1.000	Ea.	.988	16.90	
Bathtub, P.E. cast iron, 5' long with accessories, white	1.000	Ea.	3.636	410.00	
Rough-in, waste, 4" diameter DWV piping	1.000	Ea.	.828	31.20	
Supply, 1/2" diameter type "L" copper supply piping	1.000	Ea.	.988	16.90	
Vent, 1-1/2" diameter DWV piping	1.000	Ea.	.593	15.28	
Piping, supply, 1/2" diameter type "L" copper supply piping	32.000	L.F.	3.161	54.08	
Waste, 4" diameter DWV piping	12.000	L.F.	3.310	124.80	
Vent, 2" diameter DWV piping	6.000	L.F.	1.500	27.12	**Contractor's Fee**
Carrier steel, for studs, no arms	1.000	Ea.	1.143	29.00	
TOTAL			25.345	1214.22	**3192.38**

Be sure to read pages 172-173 for proper use of this section.

Three Fixture Bathroom Systems

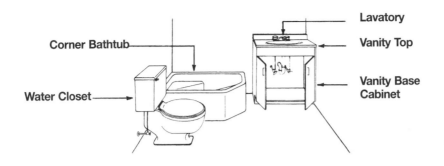

Corner Bathtub

Water Closet

Lavatory

Vanity Top

Vanity Base Cabinet

System Description	QUAN.	UNIT	LABOR-HOURS	MAT. COST	COST EACH
BATHROOM WITH LAVATORY INSTALLED IN VANITY					
Water closet, floor mounted, 2 piece, close coupled, white	1.000	Ea.	3.019	154.00	
Rough-in, vent, 2" diameter DWV piping	1.000	Ea.	.955	25.60	
Waste, 4" diameter DWV piping	1.000	Ea.	.828	31.20	
Supply, 1/2" diameter type "L" copper supply piping	1.000	Ea.	.593	10.14	
Lavatory, 20" x 18", P.E. cast iron with fittings, white	1.000	Ea.	2.500	185.00	
Rough-in, waste, 1-1/2" diameter DWV piping	1.000	Ea.	1.803	50.00	
Supply, 1/2" diameter type "L" copper supply piping	1.000	Ea.	.988	16.90	
Bathtub, P.E. cast iron, corner with fittings, white	1.000	Ea.	3.636	1425.00	
Rough-in, waste, 4" diameter DWV piping	1.000	Ea.	.828	31.20	
Supply, 1/2" diameter type "L" copper supply piping	1.000	Ea.	.988	16.90	
Vent, 1-1/2" diameter DWV piping	1.000	Ea.	.593	15.28	
Piping, supply, 1/2" diameter type "L" copper supply piping	32.000	L.F.	3.161	54.08	
Waste, 4" diameter DWV piping	12.000	L.F.	3.310	124.80	
Vent, 2" diameter DWV piping	6.000	L.F.	1.500	27.12	
Vanity base cabinet, 2 door, 30" wide	1.000	Ea.	1.000	204.00	
Vanity top, plastic laminated, square edge	2.670	L.F.	.712	76.10	
					Contractor's Fee
TOTAL			26.414	2447.32	**5146.85**
BATHROOM WITH WALL HUNG LAVATORY					
Water closet, floor mounted, 2 piece, close coupled, white	1.000	Ea.	3.019	154.00	
Rough-in, vent, 2" diameter DWV piping	1.000	Ea.	.955	25.60	
Waste, 4" diameter DWV piping	1.000	Ea.	.828	31.20	
Supply, 1/2" diameter type "L" copper supply piping	1.000	Ea.	.593	10.14	
Lavatory, 20" x 18", P.E. cast iron, with fittings, white	1.000	Ea.	2.000	218.00	
Rough-in, waste, 1-1/2" diameter DWV piping	1.000	Ea.	1.803	50.00	
Supply, 1/2" diameter type "L" copper supply piping	1.000	Ea.	.988	16.90	
Bathtub, P.E. cast iron, corner, with fittings, white	1.000	Ea.	3.636	1425.00	
Rough-in, waste, 4" diameter DWV piping	1.000	Ea.	.828	31.20	
Supply, 1/2" diameter type "L" copper supply piping	1.000	Ea.	.988	16.90	
Vent, 1-1/2" diameter DWV piping	1.000	Ea.	.593	15.28	
Piping, supply, 1/2" diameter type "L" copper supply piping	32.000	L.F.	3.161	54.08	
Waste, 4" diameter DWV piping	12.000	L.F.	3.310	124.80	
Vent, 2" diameter DWV piping	6.000	L.F.	1.500	27.12	
Carrier, steel, for studs, no arms	1.000	Ea.	1.143	29.00	
					Contractor's Fee
TOTAL			25.345	2229.22	**4760.50**

Be sure to read pages 172-173 for proper use of this section.

Three Fixture Bathroom Systems

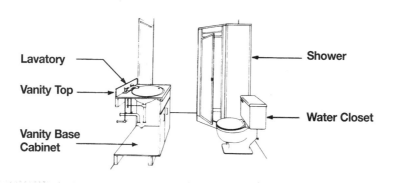

Lavatory

Vanity Top

Vanity Base Cabinet

Shower

Water Closet

System Description	QUAN.	UNIT	LABOR-HOURS	MAT. COST	COST EACH
BATHROOM WITH SHOWER, LAVATORY INSTALLED IN VANITY					
Water closet, floor mounted, 2 piece, close coupled, white	1.000	Ea.	3.019	154.00	
Rough-in, vent, 2″ diameter DWV piping	1.000	Ea.	.955	25.60	
Waste, 4″ diameter DWV piping	1.000	Ea.	.828	31.20	
Supply, 1/2″ diameter type "L" copper supply piping	1.000	Ea.	.593	10.14	
Lavatory, 20″ x 18″ P.E. cast iron with fittings, white	1.000	Ea.	2.500	185.00	
Rough-in, waste, 1-1/2″ diameter DWV piping	1.000	Ea.	1.803	50.00	
Supply, 1/2″ diameter type "L" copper supply piping	1.000	Ea.	.988	16.90	
Shower, steel enameled, stone base, corner, white	1.000	Ea.	8.000	360.00	
Rough-in, vent, 1-1/2″ diameter DWV piping	1.000	Ea.	.225	6.25	
Waste, 2″ diameter DWV piping	1.000	Ea.	1.433	38.40	
Supply, 1/2″ diameter type "L" copper supply piping	1.000	Ea.	1.580	27.04	
Piping, supply, 1/2″ diameter type "L" copper supply piping	36.000	L.F.	4.148	70.98	
Waste, 4″ diameter DWV piping	7.000	L.F.	2.759	104.00	
Vent, 2″ diameter DWV piping	6.000	L.F.	2.250	40.68	
Vanity base 2 door, 30″ wide	1.000	Ea.	1.000	204.00	Contractor's Fee
Vanity top, plastic laminated, square edge	2.170	L.F.	.712	60.08	
TOTAL			32.793	1384.27	**3824.59**
BATHROOM WITH SHOWER, WALL HUNG LAVATORY					
Water closet, floor mounted, close coupled	1.000	Ea.	3.019	154.00	
Rough-in, vent, 2″ diameter DWV piping	1.000	Ea.	.955	25.60	
Waste, 4″ diameter DWV piping	1.000	Ea.	.828	31.20	
Supply, 1/2″ diameter type "L" copper supply piping	1.000	Ea.	.593	10.14	
Lavatory, 20″ x 18″ P.E. cast iron with fittings, white	1.000	Ea.	2.000	218.00	
Rough-in, waste, 1-1/2″ diameter DWV piping	1.000	Ea.	1.803	50.00	
Supply, 1/2″ diameter type "L" copper supply piping	1.000	Ea.	.988	16.90	
Shower, steel enameled, stone base, white	1.000	Ea.	8.000	360.00	
Rough-in, vent, 1-1/2″ diameter DWV piping	1.000	Ea.	.225	6.25	
Waste, 2″ diameter DWV piping	1.000	Ea.	1.433	38.40	
Supply, 1/2″ diameter type "L" copper supply piping	1.000	Ea.	1.580	27.04	
Piping, supply, 1/2″ diameter type "L" copper supply piping	36.000	L.F.	4.148	70.98	
Waste, 4″ diameter DWV piping	7.000	L.F.	2.759	104.00	
Vent, 2″ diameter DWV piping	6.000	L.F.	2.250	40.68	Contractor's Fee
Carrier, steel, for studs, no arms	1.000	Ea.	1.143	29.00	
TOTAL			31.724	1182.19	**3463.11**

Be sure to read pages 172-173 for proper use of this section.

Three Fixture Bathroom Systems

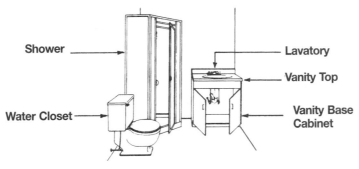

Shower — **Lavatory**

Vanity Top

Water Closet — **Vanity Base Cabinet**

System Description	QUAN.	UNIT	LABOR-HOURS	MAT. COST	COST EACH
BATHROOM WITH LAVATORY INSTALLED IN VANITY					
Water closet, floor mounted, 2 piece, close coupled, white	1.000	Ea.	3.019	154.00	
Rough-in, vent, 2" diameter DWV piping	1.000	Ea.	.955	25.60	
Waste, 4" diameter DWV piping	1.000	Ea.	.828	31.20	
Supply, 1/2" diameter type "L" copper supply piping	1.000	Ea.	.593	10.14	
Lavatory, 20" x 18", P.E. cast iron with fittings, white	1.000	Ea.	2.500	185.00	
Rough-in, waste, 1-1/2" diameter DWV piping	1.000	Ea.	1.803	50.00	
Supply, 1/2" diameter type "L" copper supply piping	1.000	Ea.	.988	16.90	
Shower, steel enameled, stone base, corner, white	1.000	Ea.	8.000	360.00	
Rough-in, vent, 1-1/2" diameter DWV piping	1.000	Ea.	.225	6.25	
Waste, 2" diameter DWV piping	1.000	Ea.	1.433	38.40	
Supply, 1/2" diameter type "L" copper supply piping	1.000	Ea.	1.580	27.04	
Piping, supply, 1/2" diameter type "L" copper supply piping	36.000	L.F.	3.556	60.84	
Waste, 4" diameter DWV piping	7.000	L.F.	1.931	72.80	
Vent, 2" diameter DWV piping	6.000	L.F.	1.500	27.12	
Vanity base, 2 door, 30" wide	1.000	Ea.	1.000	204.00	**Contractor's Fee**
Vanity top, plastic laminated, square edge	2.670	L.F.	.712	58.74	
TOTAL			30.623	1328.03	**3625.96**
BATHROOM, WITH WALL HUNG LAVATORY					
Water closet, floor mounted, 2 piece, close coupled, white	1.000	Ea.	3.019	154.00	
Rough-in, vent, 2" diameter DWV piping	1.000	Ea.	.955	25.60	
Waste, 4" diameter DWV piping	1.000	Ea.	.828	31.20	
Supply, 1/2" diameter type "L" copper supply piping	1.000	Ea.	.593	10.14	
Lavatory, wall hung, 20" x 18" P.E. cast iron with fittings, white	1.000	Ea.	2.000	218.00	
Rough-in, waste, 1-1/2" diameter DWV piping	1.000	Ea.	1.803	50.00	
Supply, 1/2" diameter type "L" copper supply piping	1.000	Ea.	.988	16.90	
Shower, steel enameled, stone base, corner, white	1.000	Ea.	8.000	360.00	
Rough-in, waste, 1-1/2" diameter DWV piping	1.000	Ea.	.225	6.25	
Waste, 2" diameter DWV piping	1.000	Ea.	1.433	38.40	
Supply, 1/2" diameter type "L" copper supply piping	1.000	Ea.	1.580	27.04	
Piping, supply, 1/2" diameter type "L" copper supply piping	36.000	L.F.	3.556	60.84	
Waste, 4" diameter DWV piping	7.000	L.F.	1.931	72.80	
Vent, 2" diameter DWV piping	6.000	L.F.	1.500	27.12	**Contractor's Fee**
Carrier, steel, for studs, no arms	1.000	Ea.	1.143	29.00	
TOTAL			29.554	1127.29	**3266.64**

Be sure to read pages 172-173 for proper use of this section.

Four Fixture Bathroom Systems

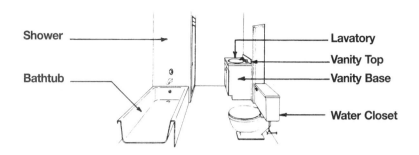

Shower

Bathtub

Lavatory

Vanity Top

Vanity Base

Water Closet

System Description	QUAN.	UNIT	LABOR-HOURS	MAT. COST	COST EACH
BATHROOM WITH LAVATORY INSTALLED IN VANITY					
Water closet, floor mounted, 2 piece, close coupled, white	1.000	Ea.	3.019	154.00	
Rough-in, vent, 2" diameter DWV piping	1.000	Ea.	.955	25.60	
Waste, 4" diameter DWV piping	1.000	Ea.	.828	31.20	
Supply, 1/2" diameter type "L" copper supply piping	1.000	Ea.	.593	10.14	
Lavatory, 20" x 18" P.E. cast iron with fittings, white	1.000	Ea.	2.500	185.00	
Shower, steel, enameled, stone base, corner, white	1.000	Ea.	8.889	825.00	
Rough-in, waste, 1-1/2" diameter DWV piping	2.000	Ea.	4.507	125.00	
Supply, 1/2" diameter type "L" copper supply piping	2.000	Ea.	3.161	54.08	
Bathtub, P.E. cast iron, 5' long with fittings, white	1.000	Ea.	3.636	410.00	
Rough-in, waste, 4" diameter DWV piping	1.000	Ea.	.828	31.20	
Supply, 1/2" diameter type "L" copper supply piping	1.000	Ea.	.988	16.90	
Vent, 1-1/2" diameter DWV piping	1.000	Ea.	.593	15.28	
Piping, supply, 1/2" diameter type "L" copper supply piping	42.000	L.F.	4.148	70.98	
Waste, 4" diameter DWV piping	10.000	L.F.	2.759	104.00	
Vent, 2" diameter DWV piping	13.000	L.F.	3.250	58.76	Contractor's Fee
Vanity base, 2 doors, 30" wide	1.000	Ea.	1.000	204.00	
Vanity top, plastic laminated, square edge	2.670	L.F.	.712	58.74	
TOTAL			42.366	2379.88	**5881.55**
BATHROOM WITH WALL HUNG LAVATORY					
Water closet, floor mounted, 2 piece, close coupled, white	1.000	Ea.	3.019	154.00	
Rough-in, vent, 2" diameter DWV piping	1.000	Ea.	.955	25.60	
Waste, 4" diameter DWV piping	1.000	Ea.	.828	31.20	
Supply, 1/2" diameter type "L" copper supply piping	1.000	Ea.	.593	10.14	
Lavatory, 20" x 18" P.E. cast iron with fittings, white	1.000	Ea.	2.000	218.00	
Shower, steel enameled, stone base, corner, white	1.000	Ea.	8.889	825.00	
Rough-in, waste, 1-1/2" diameter DWV piping	2.000	Ea.	4.507	125.00	
Supply, 1/2" diameter type "L" copper supply piping	2.000	Ea.	3.161	54.08	
Bathtub, P.E. cast iron, 5' long with fittings, white	1.000	Ea.	3.636	410.00	
Rough-in, waste, 4" diameter DWV piping	1.000	Ea.	.828	31.20	
Supply, 1/2" diameter type "L" copper supply piping	1.000	Ea.	.988	16.90	
Vent, 1-1/2" diameter copper DWV piping	1.000	Ea.	.593	15.28	
Piping, supply, 1/2" diameter type "L" copper supply piping	42.000	L.F.	4.148	70.98	
Waste, 4" diameter DWV piping	10.000	L.F.	2.759	104.00	
Vent, 2" diameter DWV piping	13.000	L.F.	3.250	58.76	Contractor's Fee
Carrier, steel, for studs, no arms	1.000	Ea.	1.143	29.00	
TOTAL			41.297	2179.14	**5522.23**

Be sure to read pages 172-173 for proper use of this section.

Four Fixture Bathroom Systems

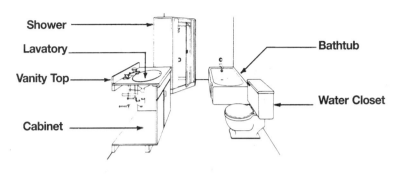

Shower — Lavatory — Vanity Top — Cabinet — Bathtub — Water Closet

System Description	QUAN.	UNIT	LABOR-HOURS	MAT. COST	COST EACH
BATHROOM WITH LAVATORY INSTALLED IN VANITY					
Water closet, floor mounted, 2 piece, close coupled, white	1.000	Ea.	3.019	154.00	
Rough-in, vent, 2" diameter DWV piping	1.000	Ea.	.955	25.60	
Waste, 4" diameter DWV piping	1.000	Ea.	.828	31.20	
Supply, 1/2" diameter type "L" copper supply piping	1.000	Ea.	.593	10.14	
Lavatory, 20" x 18" P.E. cast iron with fittings, white	1.000	Ea.	2.500	185.00	
Shower, steel, enameled, stone base, corner, white	1.000	Ea.	8.889	825.00	
Rough-in, waste, 1-1/2" diameter DWV piping	2.000	Ea.	4.507	125.00	
Supply, 1/2" diameter type "L" copper supply piping	2.000	Ea.	3.161	54.08	
Bathtub, P.E. cast iron, 5' long with fittings, white	1.000	Ea.	3.636	410.00	
Rough-in, waste, 4" diameter DWV piping	1.000	Ea.	.828	31.20	
Supply, 1/2" diameter type "L" copper supply piping	1.000	Ea.	.988	16.90	
Vent, 1-1/2" diameter DWV piping	1.000	Ea.	.593	15.28	
Piping, supply, 1/2" diameter type "L" copper supply piping	42.000	L.F.	4.939	84.50	
Waste, 4" diameter DWV piping	10.000	L.F.	4.138	156.00	
Vent, 2" diameter DWV piping	13.000	L.F.	4.500	81.36	
Vanity base, 2 doors, 30" wide	1.000	Ea.	1.000	204.00	**Contractor's Fee**
Vanity top, plastic laminated, square edge	2.670	L.F.	.712	60.08	
TOTAL			45.786	2469.34	**6194.94**
BATHROOM WITH WALL HUNG LAVATORY					
Water closet, floor mounted, 2 piece, close coupled, white	1.000	Ea.	3.019	154.00	
Rough-in, vent, 2" diameter DWV piping	1.000	Ea.	.955	25.60	
Waste, 4" diameter DWV piping	1.000	Ea.	.828	31.20	
Supply, 1/2" diameter type "L" copper supply piping	1.000	Ea.	.593	10.14	
Lavatory, 20" x 18" P.E. cast iron with fittings, white	1.000	Ea.	2.000	218.00	
Shower, steel enameled, stone base, corner, white	1.000	Ea.	8.889	825.00	
Rough-in, waste, 1-1/2" diameter DWV piping	2.000	Ea.	4.507	125.00	
Supply, 1/2" diameter type "L" copper supply piping	2.000	Ea.	3.161	54.08	
Bathtub, P.E. cast iron, 5' long with fittings, white	1.000	Ea.	3.636	410.00	
Rough-in, waste, 4" diameter DWV piping	1.000	Ea.	.828	31.20	
Supply, 1/2" diameter type "L" copper supply piping	1.000	Ea.	.988	16.90	
Vent, 1-1/2" diameter DWV piping	1.000	Ea.	.593	15.28	
Piping, supply, 1/2" diameter type "L" copper supply piping	42.000	L.F.	4.939	84.50	
Waste, 4" diameter DWV piping	10.000	L.F.	4.138	156.00	
Vent, 2" diameter DWV piping	13.000	L.F.	4.500	81.36	**Contractor's Fee**
Carrier, steel for studs, no arms	1.000	Ea.	1.143	29.00	
TOTAL			44.717	2267.26	**5833.46**

Be sure to read pages 172-173 for proper use of this section.

Five Fixture Bathroom Systems

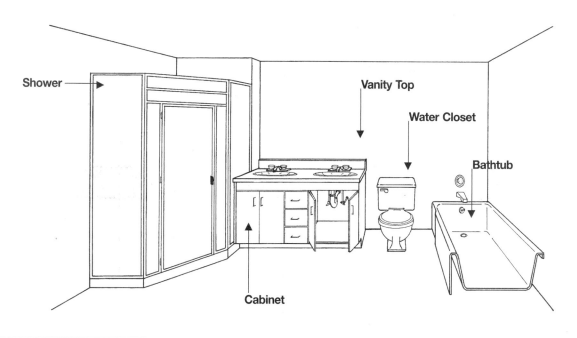

Shower

Vanity Top

Water Closet

Bathtub

Cabinet

System Description	QUAN.	UNIT	LABOR-HOURS	MAT. COST	COST EACH
BATHROOM WITH SHOWER, BATHTUB, LAVATORIES IN VANITY					
Water closet, floor mounted, 1 piece combination, white	1.000	Ea.	3.019	695.00	
Rough-in, vent, 2″ diameter DWV piping	1.000	Ea.	.955	25.60	
Waste, 4″ diameter DWV piping	1.000	Ea.	.828	31.20	
Supply, 1/2″ diameter type "L" copper supply piping	1.000	Ea.	.593	10.14	
Lavatory, 20″ x 16″, vitreous china oval, with fittings, white	2.000	Ea.	5.926	402.00	
Shower, steel enameled, stone base, corner, white	1.000	Ea.	8.889	825.00	
Rough-in, waste, 1-1/2″ diameter DWV piping	3.000	Ea.	5.408	150.00	
Supply, 1/2″ diameter type "L" copper supply piping	3.000	Ea.	2.963	50.70	
Bathtub, P.E. cast iron, 5′ long with fittings, white	1.000	Ea.	3.636	410.00	
Rough-in, waste, 4″ diameter DWV piping	1.000	Ea.	1.103	41.60	
Supply, 1/2″ diameter type "L" copper supply piping	1.000	Ea.	.988	16.90	
Vent, 1-1/2″ diameter copper DWV piping	1.000	Ea.	.593	15.28	
Piping, supply, 1/2″ diameter type "L" copper supply piping	42.000	L.F.	4.148	70.98	
Waste, 4″ diameter DWV piping	10.000	L.F.	2.759	104.00	
Vent, 2″ diameter DWV piping	13.000	L.F.	3.250	58.76	
Vanity base, 2 door, 24″ x 48″	1.000	Ea.	1.400	325.00	
Vanity top, plastic laminated, square edge	4.170	L.F.	1.112	91.74	
TOTAL			47.570	3323.90	Contractor's Fee **7602.67**

Be sure to read pages 172-173 for proper use of this section.

Gas Heating/Cooling Systems

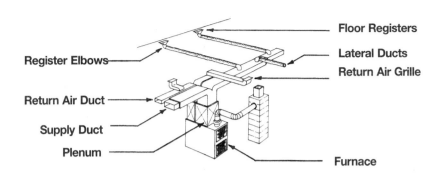

Floor Registers
Lateral Ducts
Return Air Grille
Register Elbows
Return Air Duct
Supply Duct
Plenum
Furnace

System Description	QUAN.	UNIT	LABOR-HOURS	MAT. COST	COST PER SYSTEM
HEATING ONLY, GAS FIRED HOT AIR, ONE ZONE, 1200 S.F. BUILDING					
Furnace, gas, up flow	1.000	Ea.	5.000	720.00	
Intermittent pilot	1.000	Ea.		145.00	
Supply duct, rigid fiberglass	176.000	S.F.	12.068	197.12	
Return duct, sheet metal, galvanized	158.000	Lb.	16.137	132.72	
Lateral ducts, 6" flexible fiberglass	144.000	L.F.	8.862	233.28	
Register, elbows	12.000	Ea.	3.200	318.00	
Floor registers, enameled steel	12.000	Ea.	3.000	220.20	
Floor grille, return air	2.000	Ea.	.727	49.00	
Thermostat	1.000	Ea.	1.000	30.00	
Plenum	1.000	Ea.	1.000	67.50	Contractor's Fee
TOTAL			50.994	2112.82	**5896.36**
HEATING/COOLING, GAS FIRED FORCED AIR, ONE ZONE, 1200 S.F. BUILDING					
Furnace, including plenum, compressor, coil	1.000	Ea.	14.720	3404.00	
Intermittent pilot	1.000	Ea.		145.00	
Supply duct, rigid fiberglass	176.000	S.F.	12.068	197.12	
Return duct, sheet metal, galvanized	158.000	Lb.	16.137	132.72	
Lateral duct, 6" flexible fiberglass	144.000	L.F.	8.862	233.28	
Register elbows	12.000	Ea.	3.200	318.00	
Floor registers, enameled steel	12.000	Ea.	3.000	220.20	
Floor grille return air	2.000	Ea.	.727	49.00	
Thermostat	1.000	Ea.	1.000	30.00	
Refrigeration piping, 25 ft. (pre-charged)	1.000	Ea.		175.00	Contractor's Fee
TOTAL			59.714	4904.32	**10644.02**

Be sure to read pages 172-173 for proper use of this section.

Oil Fired Heating/Cooling Systems

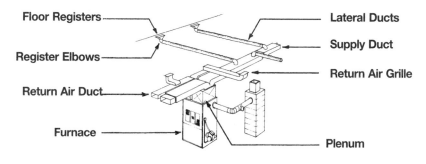

Floor Registers — Lateral Ducts

Register Elbows — Supply Duct

Return Air Duct — Return Air Grille

Furnace — Plenum

System Description	QUAN.	UNIT	LABOR-HOURS	MAT. COST	COST PER SYSTEM
HEATING ONLY, OIL FIRED HOT AIR, ONE ZONE, 1200 S.F. BUILDING					
Furnace, oil fired, atomizing gun type burner	1.000	Ea.	4.571	850.00	
3/8″ diameter copper supply pipe	1.000	Ea.	2.759	39.60	
Shut off valve	1.000	Ea.	.333	7.40	
Oil tank, 275 gallon, on legs	1.000	Ea.	3.200	265.00	
Supply duct, rigid fiberglass	176.000	S.F.	12.068	197.12	
Return duct, sheet metal, galvanized	158.000	Lb.	16.137	132.72	
Lateral ducts, 6″ flexible fiberglass	144.000	L.F.	8.862	233.28	
Register elbows	12.000	Ea.	3.200	318.00	
Floor register, enameled steel	12.000	Ea.	3.000	220.20	
Floor grille, return air	2.000	Ea.	.727	49.00	
Thermostat	1.000	Ea.	1.000	30.00	Contractor's Fee
TOTAL			55.857	2342.32	**6502.18**
HEATING/COOLING, OIL FIRED, FORCED AIR, ONE ZONE, 1200 S.F. BUILDING					
Furnace, including plenum, compressor, coil	1.000	Ea.	16.000	3950.00	
3/8″ diameter copper supply pipe	1.000	Ea.	2.759	39.60	
Shut off valve	1.000	Ea.	.333	7.40	
Oil tank, 275 gallon on legs	1.000	Ea.	3.200	265.00	
Supply duct, rigid fiberglass	176.000	S.F.	12.068	197.12	
Return duct, sheet metal, galvanized	158.000	Lb.	16.137	132.72	
Lateral ducts, 6″ flexible fiberglass	144.000	L.F.	8.862	233.28	
Register elbows	12.000	Ea.	3.200	318.00	
Floor registers, enameled steel	12.000	Ea.	3.000	220.20	
Floor grille, return air	2.000	Ea.	.727	49.00	
Refrigeration piping (precharged)	25.000	L.F.		175.00	Contractor's Fee
TOTAL			66.286	5587.32	**12045.62**

Be sure to read pages 172-173 for proper use of this section.

Hot Water Heating Systems

System Description	QUAN.	UNIT	LABOR-HOURS	MAT. COST	COST EACH
OIL FIRED HOT WATER HEATING SYSTEM, AREA TO 1200 S.F.					
Boiler package, oil fired, 97 MBH, area to 1200 S.F. building	1.000	Ea.	15.000	1400.00	
3/8" diameter copper supply pipe	1.000	Ea.	2.759	39.60	
Shut off valve	1.000	Ea.	.333	7.40	
Oil tank, 275 gallon, with black iron filler pipe	1.000	Ea.	3.200	265.00	
Supply piping, 3/4" copper tubing	176.000	L.F.	18.526	399.52	
Supply fittings, copper 3/4"	36.000	Ea.	15.158	32.04	
Supply valves, 3/4"	2.000	Ea.	.800	99.00	
Baseboard radiation, 3/4"	106.000	L.F.	35.333	361.46	Contractor's Fee
Zone valve	1.000	Ea.	.400	74.50	
TOTAL			91.509	2678.52	**8923.13**
OIL FIRED HOT WATER HEATING SYSTEM, AREA TO 2400 S.F.					
Boiler package, oil fired, 225 MBH, area to 2400 S.F. building	1.000	Ea.	19.704	3125.00	
3/8" diameter copper supply pipe	1.000	Ea.	2.759	39.60	
Shut off valve	1.000	Ea.	.333	7.40	
Oil tank, 550 gallon, with black iron pipe filler pipe	1.000	Ea.	5.926	1250.00	
Supply piping, 3/4" copper tubing	228.000	L.F.	23.999	517.56	
Supply fittings, copper	46.000	Ea.	19.368	40.94	
Supply valves	2.000	Ea.	.800	99.00	
Baseboard radiation	212.000	L.F.	70.666	722.92	
Zone valve	1.000	Ea.	.400	74.50	Contractor's Fee
TOTAL			143.955	5876.92	**16572.99**

Be sure to read pages 172-173 for proper use of this section.

Division 9
ELECTRICAL

Electric Service Systems

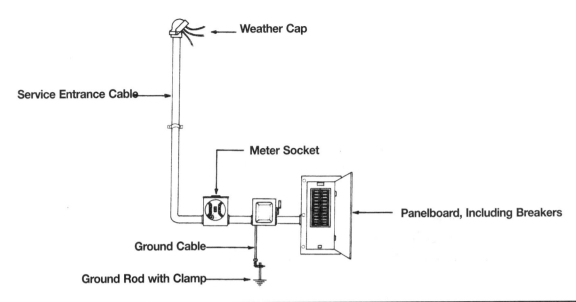

Weather Cap

Service Entrance Cable

Meter Socket

Panelboard, Including Breakers

Ground Cable

Ground Rod with Clamp

System Description	QUAN.	UNIT	LABOR-HOURS	MAT. COST	COST EACH
100 AMP SERVICE					
Weather cap	1.000	Ea.	.667	9.15	
Service entrance cable	10.000	L.F.	.762	23.60	
Meter socket	1.000	Ea.	2.500	32.00	
Ground rod with clamp	1.000	Ea.	1.455	18.05	
Ground cable	5.000	L.F.	.250	6.70	
Panel board, 12 circuit	1.000	Ea.	6.667	182.00	Contractor's Fee
TOTAL			12.301	271.50	**1098.29**
200 AMP SERVICE					
Weather cap	1.000	Ea.	1.000	20.00	
Service entrance cable	10.000	L.F.	1.143	57.50	
Meter socket	1.000	Ea.	4.211	46.00	
Ground rod with clamp	1.000	Ea.	1.818	32.50	
Ground cable	10.000	L.F.	.500	13.40	
3/4" EMT	5.000	L.F.	.308	3.10	
Panel board, 24 circuit	1.000	Ea.	12.308	490.00	Contractor's Fee
TOTAL			21.288	662.50	**2092.06**
400 AMP SERVICE					
Weather cap	1.000	Ea.	2.963	345.00	
Service entrance cable	180.000	L.F.	5.760	275.40	
Meter socket	1.000	Ea.	4.211	46.00	
Ground rod with clamp	1.000	Ea.	2.000	87.00	
Ground cable	20.000	L.F.	.485	19.80	
3/4" greenfield	20.000	L.F.	1.000	9.40	
Current transformer cabinet	1.000	Ea.	6.154	132.00	
Panel board, 42 circuit	1.000	Ea.	33.333	2575.00	Contractor's Fee
TOTAL			55.906	3489.60	**8505.50**

Be sure to read pages 172-173 for proper use of this section.

Electric Perimeter Heating Systems

Thermostat →

Electric Baseboard

System Description	QUAN.	UNIT	LABOR-HOURS	COST EACH	
				MAT.	TOTAL
4' BASEBOARD HEATER					
Electric baseboard heater, 4' long	1.000	Ea.	1.194	54.00	
Thermostat, integral	1.000	Ea.	.500	28.00	
Romex, 12-3 with ground	40.000	L.F.	1.600	14.00	
Panel board breaker, 20 Amp	1.000	Ea.	.300	7.65	Contractor's Fee
TOTAL			3.594	103.65	**360.09**
6' BASEBOARD HEATER					
Electric baseboard heater, 6' long	1.000	Ea.	1.600	73.00	
Thermostat, integral	1.000	Ea.	.500	28.00	
Romex, 12-3 with ground	40.000	L.F.	1.600	14.00	
Panel board breaker, 20 Amp	1.000	Ea.	.400	10.20	Contractor's Fee
TOTAL			4.100	125.20	**420.42**
8' BASEBOARD HEATER					
Electric baseboard heater, 8' long	1.000	Ea.	2.000	91.00	
Thermostat, integral	1.000	Ea.	.500	28.00	
Romex, 12-3 with ground	40.000	L.F.	1.600	14.00	
Panel board breaker, 20 Amp	1.000	Ea.	.500	12.75	Contractor's Fee
TOTAL			4.600	145.75	**480.54**
10' BASEBOARD HEATER					
Electric baseboard heater, 10' long	1.000	Ea.	2.424	150.00	
Thermostat, integral	1.000	Ea.	.500	28.00	
Romex, 12-3 with ground	40.000	L.F.	1.600	14.00	
Panel board breaker, 20 Amp	1.000	Ea.	.750	19.13	Contractor's Fee
TOTAL			5.274	211.13	**618.67**

Be sure to read pages 172-173 for proper use of this section.

Wiring Device Systems

The prices in this system are on a cost each basis and include 20 feet of wire and conduit (as necessary) for each device.

SYSTEM DESCRIPTION	QUAN.	UNIT	LABOR-HOURS	COST EACH MAT.	COST EACH TOTAL
Air conditioning receptacles					
Using non-metallic sheathed cable	1.000	Ea.	.800	13.05	64.24
Using BX cable	1.000	Ea.	.964	23.50	89.78
Using EMT conduit	1.000	Ea.	1.194	26.00	106.41
Disposal wiring					
Using non-metallic sheathed cable	1.000	Ea.	.889	10.45	65.33
Using BX cable	1.000	Ea.	1.067	20.50	90.82
Using EMT conduit	1.000	Ea.	1.333	24.00	110.47
Dryer circuit					
Using non-metallic sheathed cable	1.000	Ea.	1.455	32.00	129.94
Using BX cable	1.000	Ea.	1.739	42.50	161.93
Using EMT conduit	1.000	Ea.	2.162	38.50	179.84
Duplex receptacles					
Using non-metallic sheathed cable	1.000	Ea.	.615	13.05	54.29
Using BX cable	1.000	Ea.	.741	23.50	76.98
Using EMT conduit	1.000	Ea.	.920	26.00	90.77
Exhaust fan wiring					
Using non-metallic sheathed cable	1.000	Ea.	.800	12.70	63.49
Using BX cable	1.000	Ea.	.964	23.50	89.71
Using EMT conduit	1.000	Ea.	1.194	26.00	106.41
Furnace circuit & switch					
Using non-metallic sheathed cable	1.000	Ea.	1.333	19.60	103.47
Using BX cable	1.000	Ea.	1.600	31.00	136.27
Using EMT conduit	1.000	Ea.	2.000	32.50	160.58
Ground fault					
Using non-metallic sheathed cable	1.000	Ea.	1.000	42.00	120.56
Using BX cable	1.000	Ea.	1.212	52.50	148.21
Using EMT conduit	1.000	Ea.	1.481	66.00	184.82
Heater circuits					
Using non-metallic sheathed cable	1.000	Ea.	1.000	12.30	74.81
Using BX cable	1.000	Ea.	1.212	20.00	97.85
Using EMT conduit	1.000	Ea.	1.481	22.50	116.65
Lighting wiring					
Using non-metallic sheathed cable	1.000	Ea.	.500	12.95	47.88
Using BX cable	1.000	Ea.	.602	21.00	65.95
Using EMT conduit	1.000	Ea.	.748	22.00	75.32
Range circuits					
Using non-metallic sheathed cable	1.000	Ea.	2.000	65.50	211.76
Using BX cable	1.000	Ea.	2.424	91.00	274.81
Using EMT conduit	1.000	Ea.	2.963	66.00	265.86
Switches, single pole					
Using non-metallic sheathed cable	1.000	Ea.	.500	12.70	47.14
Using BX cable	1.000	Ea.	.602	23.50	69.80
Using EMT conduit	1.000	Ea.	.748	26.00	81.53
Switches, 3-way					
Using non-metallic sheathed cable	1.000	Ea.	.667	17.05	63.34
Using BX cable	1.000	Ea.	.800	26.50	85.16
Using EMT conduit	1.000	Ea.	1.333	35.00	127.53
Water heater					
Using non-metallic sheathed cable	1.000	Ea.	1.600	19.10	117.63
Using BX cable	1.000	Ea.	1.905	35.00	159.44
Using EMT conduit	1.000	Ea.	2.353	29.00	174.30
Weatherproof receptacle					
Using non-metallic sheathed cable	1.000	Ea.	1.333	104.00	235.22
Using BX cable	1.000	Ea.	1.600	112.00	261.87
Using EMT conduit	1.000	Ea.	2.000	114.00	287.75

Light Fixture Systems

The costs for these fixtures are on an each basis and include installation of the fixture only. See wiring device systems.

DESCRIPTION	QUAN.	UNIT	LABOR-HOURS	COST EACH	
				MAT.	TOTAL
Fluorescent strip, 4' long, 1 light, average	1.000	Ea.	.941	29.00	96.89
Deluxe	1.000	Ea.	1.129	35.00	116.27
2 lights, average	1.000	Ea.	1.000	31.00	103.57
Deluxe	1.000	Ea.	1.200	37.00	124.28
8' long, 1 light, average	1.000	Ea.	1.194	43.50	133.57
Deluxe	1.000	Ea.	1.433	52.00	160.28
2 lights, average	1.000	Ea.	1.290	52.50	153.19
Deluxe	1.000	Ea.	1.548	63.00	183.82
Surface mounted, 4' x 1', economy	1.000	Ea.	.914	62.50	147.97
Average	1.000	Ea.	1.143	78.00	184.96
Deluxe	1.000	Ea.	1.371	93.50	221.95
4' x 2', economy	1.000	Ea.	1.208	80.00	191.19
Average	1.000	Ea.	1.509	100.00	238.99
Deluxe	1.000	Ea.	1.811	120.00	286.78
Recessed, 4'x 1', 2 lamps, economy	1.000	Ea.	1.123	41.00	125.85
Average	1.000	Ea.	1.404	51.00	157.31
Deluxe	1.000	Ea.	1.684	61.00	188.77
4' x 2', 4' lamps, economy	1.000	Ea.	1.362	49.50	152.01
Average	1.000	Ea.	1.702	62.00	190.01
Deluxe	1.000	Ea.	2.043	74.50	228.01
Incandescent, exterior, 150W, single spot	1.000	Ea.	.500	18.60	56.43
Double spot	1.000	Ea.	1.167	75.50	181.82
Recessed, 100W, economy	1.000	Ea.	.800	49.00	120.05
Average	1.000	Ea.	1.000	61.00	150.06
Deluxe	1.000	Ea.	1.200	73.00	180.07
150W, economy	1.000	Ea.	.800	71.00	154.79
Average	1.000	Ea.	1.000	89.00	193.49
Deluxe	1.000	Ea.	1.200	107.00	232.19
Surface mounted, 60W, economy	1.000	Ea.	.800	37.50	102.22
Average	1.000	Ea.	1.000	42.00	120.56
Deluxe	1.000	Ea.	1.194	60.50	159.94
Mercury vapor, recessed, 2' x 2' with 250W DX lamp	1.000	Ea.	2.500	289.00	584.74
2' x 2' with 400W DX lamp	1.000	Ea.	2.759	295.00	606.78
Surface mounted, 2' x 2' with 250W DX lamp	1.000	Ea.	2.963	289.00	613.17
2' x 2' with 400W DX lamp	1.000	Ea.	3.333	325.00	689.09
High bay, single unit, 400W DX lamp	1.000	Ea.	3.478	286.00	634.07
Twin unit, 400W DX lamp	1.000	Ea.	5.000	485.00	1029.30
Low bay, 250W DX lamp	1.000	Ea.	2.500	295.00	592.56
Metal halide, recessed 2' x 2' 250W	1.000	Ea.	2.500	283.00	576.78
2' x 2', 400W	1.000	Ea.	2.759	330.00	660.94
Surface mounted, 2' x 2', 250W	1.000	Ea.	2.963	283.00	605.21
2' x 2', 400W	1.000	Ea.	3.333	335.00	704.45
High bay, single, unit, 400W	1.000	Ea.	3.478	395.00	803.96
Twin unit, 400W	1.000	Ea.	5.000	780.00	1487.09
Low bay, 250W	1.000	Ea.	2.500	375.00	716.53

Be sure to read pages 172-173 for proper use of this section.

ADJUSTING PROJECT COSTS TO YOUR LOCATION

Adjusting Project Costs to Your Location

Costs shown in *Interior Home Improvement Costs* are based on National Averages for materials and installation. To adjust these costs to a specific location, simply multiply the base cost by the factor for that city. The data is arranged alphabetically by state and postal zip code numbers. For a city not listed, use the factor for a nearby city with similar economic characteristics.

STATE	CITY	Residential
ALABAMA		
350-352	Birmingham	.85
354	Tuscaloosa	.80
355	Jasper	.76
356	Decatur	.79
357-358	Huntsville	.81
359	Gadsden	.80
360-361	Montgomery	.82
362	Anniston	.73
363	Dothan	.79
364	Evergreen	.79
365-366	Mobile	.81
367	Selma	.79
368	Phenix City	.82
369	Butler	.79
ALASKA		
995-996	Anchorage	1.25
997	Fairbanks	1.25
998	Juneau	1.24
999	Ketchikan	1.30
ARIZONA		
850,853	Phoenix	.92
852	Mesa/Tempe	.87
855	Globe	.88
856-857	Tucson	.90
859	Show Low	.89
860	Flagstaff	.92
863	Prescott	.90
864	Kingman	.89
865	Chambers	.88
ARKANSAS		
716	Pine Bluff	.80
717	Camden	.70
718	Texarkana	.74
719	Hot Springs	.69
720-722	Little Rock	.81
723	West Memphis	.79
724	Jonesboro	.79
725	Batesville	.75
726	Harrison	.76
727	Fayetteville	.69
728	Russellville	.77
729	Fort Smith	.83
CALIFORNIA		
900-902	Los Angeles	1.08
903-905	Inglewood	1.06
906-908	Long Beach	1.07
910-912	Pasadena	1.07
913-916	Van Nuys	1.09
917-918	Alhambra	1.08
919-921	San Diego	1.10
922	Palm Springs	1.09
923-924	San Bernardino	1.08
925	Riverside	1.11
926-927	Santa Ana	1.09
928	Anaheim	1.10
930	Oxnard	1.13
931	Santa Barbara	1.11
932-933	Bakersfield	1.11

STATE	CITY	Residential
934	San Luis Obispo	1.14
935	Mojave	1.09
936-938	Fresno	1.12
939	Salinas	1.12
940-941	San Francisco	1.21
942,956-958	Sacramento	1.11
943	Palo Alto	1.15
944	San Mateo	1.16
945	Vallejo	1.11
946	Oakland	1.16
947	Berkeley	1.16
948	Richmond	1.14
949	San Rafael	1.26
950	Santa Cruz	1.16
951	San Jose	1.22
952	Stockton	1.13
953	Modesto	1.13
954	Santa Rosa	1.14
955	Eureka	1.10
959	Marysville	1.10
960	Redding	1.11
961	Susanville	1.11
COLORADO		
800-802	Denver	.99
803	Boulder	.88
804	Golden	.97
805	Fort Collins	.98
806	Greeley	.90
807	Fort Morgan	.97
808-809	Colorado Springs	.94
810	Pueblo	.94
811	Alamosa	.89
812	Salida	.89
813	Durango	.88
814	Montrose	.86
815	Grand Junction	.90
816	Glenwood Springs	.95
CONNECTICUT		
060	New Britain	1.04
061	Hartford	1.04
062	Willimantic	1.03
063	New London	1.05
064	Meriden	1.03
065	New Haven	1.04
066	Bridgeport	1.02
067	Waterbury	1.05
068	Norwalk	1.01
069	Stamford	1.04
D.C.		
200-205	Washington	.93
DELAWARE		
197	Newark	1.00
198	Wilmington	1.00
199	Dover	1.00
FLORIDA		
320,322	Jacksonville	.83
321	Daytona Beach	.87
323	Tallahassee	.75

Adjusting Project Costs to Your Location

STATE	CITY	Residential
324	Panama City	.70
325	Pensacola	.84
326,344	Gainesville	.84
327-328,347	Orlando	.86
329	Melbourne	.90
330-332,340	Miami	.83
333	Fort Lauderdale	.83
334,349	West Palm Beach	.86
335-336,346	Tampa	.80
337	St. Petersburg	.81
338	Lakeland	.79
339,341	Fort Myers	.79
342	Sarasota	.78
GEORGIA		
300-303,399	Atlanta	.85
304	Statesboro	.72
305	Gainesville	.76
306	Athens	.77
307	Dalton	.68
308-309	Augusta	.76
310-312	Macon	.81
313-314	Savannah	.80
315	Waycross	.74
316	Valdosta	.76
317	Albany	.77
318-319	Columbus	.78
HAWAII		
967	Hilo	1.27
968	Honolulu	1.27
STATES & POSS.		
969	Guam	1.37
IDAHO		
832	Pocatello	.94
833	Twin Falls	.79
834	Idaho Falls	.83
835	Lewiston	1.09
836-837	Boise	.94
838	Coeur d'Alene	.95
ILLINOIS		
600-603	North Suburban	1.11
604	Joliet	1.11
605	South Suburban	1.10
606	Chicago	1.13
609	Kankakee	1.00
610-611	Rockford	1.05
612	Rock Island	1.06
613	La Salle	1.06
614	Galesburg	1.09
615-616	Peoria	1.09
617	Bloomington	1.05
618-619	Champaign	1.04
620-622	East St. Louis	1.00
623	Quincy	.99
624	Effingham	1.02
625	Decatur	1.01
626-627	Springfield	1.01
628	Centralia	.99
629	Carbondale	.97
INDIANA		
460	Anderson	.95
461-462	Indianapolis	.98

STATE	CITY	Residential
463-464	Gary	1.04
465-466	South Bend	.94
467-468	Fort Wayne	.92
469	Kokomo	.93
470	Lawrenceburg	.93
471	New Albany	.93
472	Columbus	.96
473	Muncie	.94
474	Bloomington	.96
475	Washington	.93
476-477	Evansville	.95
478	Terre Haute	.96
479	Lafayette	.92
IOWA		
500-503,509	Des Moines	.97
504	Mason City	.86
505	Fort Dodge	.84
506-507	Waterloo	.88
508	Creston	.89
510-511	Sioux City	.95
512	Sibley	.80
513	Spencer	.80
514	Carroll	.84
515	Council Bluffs	.96
516	Shenandoah	.82
520	Dubuque	.99
521	Decorah	.88
522-524	Cedar Rapids	1.01
525	Ottumwa	.94
526	Burlington	.92
527-528	Davenport	.98
KANSAS		
660-662	Kansas City	.96
664-666	Topeka	.86
667	Fort Scott	.85
668	Emporia	.81
669	Belleville	.87
670-672	Wichita	.89
673	Independence	.82
674	Salina	.85
675	Hutchinson	.79
676	Hays	.84
677	Colby	.85
678	Dodge City	.84
679	Liberal	.78
KENTUCKY		
400-402	Louisville	.95
403-405	Lexington	.87
406	Frankfort	.92
407-409	Corbin	.78
410	Covington	.98
411-412	Ashland	.96
413-414	Campton	.77
415-416	Pikeville	.82
417-418	Hazard	.76
420	Paducah	.97
421-422	Bowling Green	.96
423	Owensboro	.91
424	Henderson	.94
425-426	Somerset	.75
427	Elizabethtown	.94
LOUISIANA		
700-701	New Orleans	.86

STATE	CITY	Residential
703	Thibodaux	.85
704	Hammond	.84
705	Lafayette	.84
706	Lake Charles	.83
707-708	Baton Rouge	.82
710-711	Shreveport	.81
712	Monroe	.79
713-714	Alexandria	.78
MAINE		
039	Kittery	.86
040-041	Portland	.91
042	Lewiston	.92
043	Augusta	.88
044	Bangor	.93
045	Bath	.89
046	Machias	.87
047	Houlton	.89
048	Rockland	.86
049	Waterville	.86
MARYLAND		
206	Waldorf	.87
207-208	College Park	.90
209	Silver Spring	.89
210-212	Baltimore	.91
214	Annapolis	.89
215	Cumberland	.87
216	Easton	.73
217	Hagerstown	.90
218	Salisbury	.76
219	Elkton	.82
MASSACHUSETTS		
010-011	Springfield	1.04
012	Pittsfield	.99
013	Greenfield	1.02
014	Fitchburg	1.08
015-016	Worcester	1.10
017	Framingham	1.06
018	Lowell	1.08
019	Lawrence	1.09
020-022, 024	Boston	1.14
023	Brockton	1.06
025	Buzzards Bay	1.02
026	Hyannis	1.04
027	New Bedford	1.06
MICHIGAN		
480,483	Royal Oak	1.03
481	Ann Arbor	1.04
482	Detroit	1.07
484-485	Flint	.99
486	Saginaw	.97
487	Bay City	.96
488-489	Lansing	1.01
490	Battle Creek	1.01
491	Kalamazoo	1.00
492	Jackson	.99
493,495	Grand Rapids	.88
494	Muskegon	.95
496	Traverse City	.87
497	Gaylord	.87
498-499	Iron Mountain	.98
MINNESOTA		
550-551	Saint Paul	1.09

STATE	CITY	Residential
553-555	Minneapolis	1.11
556-558	Duluth	1.04
559	Rochester	1.03
560	Mankato	1.00
561	Windom	.90
562	Willmar	.93
563	St. Cloud	1.11
564	Brainerd	1.06
565	Detroit Lakes	.88
566	Bemidji	.91
567	Thief River Falls	.87
MISSISSIPPI		
386	Clarksdale	.70
387	Greenville	.80
388	Tupelo	.71
389	Greenwood	.72
390-392	Jackson	.80
393	Meridian	.76
394	Laurel	.72
395	Biloxi	.84
396	McComb	.72
397	Columbus	.70
MISSOURI		
630-631	St. Louis	1.00
633	Bowling Green	.92
634	Hannibal	.99
635	Kirksville	.86
636	Flat River	.94
637	Cape Girardeau	.93
638	Sikeston	.90
639	Poplar Bluff	.90
640-641	Kansas City	1.04
644-645	St. Joseph	.90
646	Chillicothe	.82
647	Harrisonville	.98
648	Joplin	.84
650-651	Jefferson City	.98
652	Columbia	.99
653	Sedalia	.99
654-655	Rolla	.95
656-658	Springfield	.86
MONTANA		
590-591	Billings	.93
592	Wolf Point	.92
593	Miles City	.91
594	Great Falls	.92
595	Havre	.90
596	Helena	.91
597	Butte	.90
598	Missoula	.89
599	Kalispell	.88
NEBRASKA		
680-681	Omaha	.92
683-685	Lincoln	.88
686	Columbus	.74
687	Norfolk	.84
688	Grand Island	.88
689	Hastings	.82
690	Mccook	.79
691	North Platte	.86
692	Valentine	.77
693	Alliance	.75

Adjusting Project Costs to Your Location

STATE	CITY	Residential
NEVADA		
889-891	Las Vegas	1.05
893	Ely	.93
894-895	Reno	.95
897	Carson City	.96
898	Elko	.90
NEW HAMPSHIRE		
030	Nashua	.94
031	Manchester	.94
032-033	Concord	.93
034	Keene	.78
035	Littleton	.82
036	Charleston	.77
037	Claremont	.76
038	Portsmouth	.93
NEW JERSEY		
070-071	Newark	1.14
072	Elizabeth	1.09
073	Jersey City	1.11
074-075	Paterson	1.12
076	Hackensack	1.10
077	Long Branch	1.10
078	Dover	1.11
079	Summit	1.08
080,083	Vineland	1.11
081	Camden	1.11
082,084	Atlantic City	1.11
085-086	Trenton	1.12
087	Point Pleasant	1.10
088-089	New Brunswick	1.12
NEW MEXICO		
870-872	Albuquerque	.88
873	Gallup	.88
874	Farmington	.88
875	Santa Fe	.88
877	Las Vegas	.88
878	Socorro	.87
879	Truth/Consequences	.87
880	Las Cruces	.84
881	Clovis	.89
882	Roswell	.90
883	Carrizozo	.91
884	Tucumcari	.90
NEW YORK		
100-102	New York	1.35
103	Staten Island	1.31
104	Bronx	1.30
105	Mount Vernon	1.20
106	White Plains	1.19
107	Yonkers	1.22
108	New Rochelle	1.20
109	Suffern	1.14
110	Queens	1.30
111	Long Island City	1.31
112	Brooklyn	1.31
113	Flushing	1.32
114	Jamaica	1.30
115,117,118	Hicksville	1.26
116	Far Rockaway	1.32
119	Riverhead	1.27
120-122	Albany	.97
123	Schenectady	.98
124	Kingston	1.11

STATE	CITY	Residential
125-126	Poughkeepsie	1.13
127	Monticello	1.09
128	Glens Falls	.95
129	Plattsburgh	.95
130-132	Syracuse	.99
133-135	Utica	.91
136	Watertown	.92
137-139	Binghamton	.94
140-142	Buffalo	1.05
143	Niagara Falls	1.07
144-146	Rochester	.99
147	Jamestown	.98
148-149	Elmira	.95
NORTH CAROLINA		
270,272-274	Greensboro	.75
271	Winston-Salem	.74
275-276	Raleigh	.76
277	Durham	.75
278	Rocky Mount	.68
279	Elizabeth City	.70
280	Gastonia	.74
281-282	Charlotte	.74
283	Fayetteville	.75
284	Wilmington	.73
285	Kinston	.67
286	Hickory	.66
287-288	Asheville	.73
289	Murphy	.66
NORTH DAKOTA		
580-581	Fargo	.79
582	Grand Forks	.77
583	Devils Lake	.76
584	Jamestown	.76
585	Bismarck	.80
586	Dickinson	.81
587	Minot	.82
588	Williston	.76
OHIO		
430-432	Columbus	.98
433	Marion	.92
434-436	Toledo	1.02
437-438	Zanesville	.93
439	Steubenville	.97
440	Lorain	1.05
441	Cleveland	1.09
442-443	Akron	1.02
444-445	Youngstown	1.01
446-447	Canton	.97
448-449	Mansfield	.96
450	Hamilton	1.00
451-452	Cincinnati	1.00
453-454	Dayton	.94
455	Springfield	.95
456	Chillicothe	1.02
457	Athens	.92
458	Lima	.96
OKLAHOMA		
730-731	Oklahoma City	.82
734	Ardmore	.83
735	Lawton	.84
736	Clinton	.80
737	Enid	.83
738	Woodward	.82

STATE	CITY	Residential
739	Guymon	.69
740-741	Tulsa	.84
743	Miami	.86
744	Muskogee	.75
745	Mcalester	.76
746	Ponca City	.82
747	Durant	.79
748	Shawnee	.79
749	Poteau	.85
OREGON		
970-972	Portland	1.08
973	Salem	1.06
974	Eugene	1.05
975	Medford	1.05
976	Klamath Falls	1.05
977	Bend	1.06
978	Pendleton	1.03
979	Vale	.98
PENNSYLVANIA		
150-152	Pittsburgh	1.04
153	Washington	1.02
154	Uniontown	1.01
155	Bedford	1.03
156	Greensburg	1.02
157	Indiana	1.05
158	Dubois	1.04
159	Johnstown	1.04
160	Butler	1.01
161	New Castle	1.01
162	Kittanning	1.02
163	Oil City	.91
164-165	Erie	.98
166	Altoona	1.04
167	Bradford	.99
168	State College	.96
169	Wellsboro	.93
170-171	Harrisburg	.98
172	Chambersburg	.96
173-174	York	.97
175-176	Lancaster	.95
177	Williamsport	.91
178	Sunbury	.95
179	Pottsville	.95
180	Lehigh Valley	1.04
181	Allentown	1.01
182	Hazleton	.97
183	Stroudsburg	1.01
184-185	Scranton	.95
186-187	Wilkes-Barre	.93
188	Montrose	.93
189	Doylestown	.94
190-191	Philadelphia	1.13
193	Westchester	1.08
194	Norristown	1.09
195-196	Reading	.97
PUERTO RICO		
009	San Juan	.86
RHODE ISLAND		
028	Newport	1.02
029	Providence	1.02
SOUTH CAROLINA		
290-292	Columbia	.72

STATE	CITY	Residential
293	Spartanburg	.71
294	Charleston	.73
295	Florence	.71
296	Greenville	.70
297	Rock Hill	.64
298	Aiken	.80
299	Beaufort	.68
SOUTH DAKOTA		
570-571	Sioux Falls	.88
572	Watertown	.84
573	Mitchell	.83
574	Aberdeen	.84
575	Pierre	.84
576	Mobridge	.84
577	Rapid City	.85
TENNESSEE		
370-372	Nashville	.86
373-374	Chattanooga	.82
375,380-381	Memphis	.84
376	Johnson City	.80
377-379	Knoxville	.80
382	Mckenzie	.69
383	Jackson	.68
384	Columbia	.76
385	Cookeville	.68
TEXAS		
750	Mckinney	.88
751	Waxahackie	.82
752-753	Dallas	.89
754	Greenville	.78
755	Texarkana	.87
756	Longview	.84
757	Tyler	.91
758	Palestine	.72
759	Lufkin	.76
760-761	Fort Worth	.83
762	Denton	.87
763	Wichita Falls	.80
764	Eastland	.73
765	Temple	.77
766-767	Waco	.81
768	Brownwood	.72
769	San Angelo	.79
770-772	Houston	.87
773	Huntsville	.73
774	Wharton	.75
775	Galveston	.86
776-777	Beaumont	.82
778	Bryan	.81
779	Victoria	.78
780	Laredo	.76
781-782	San Antonio	.82
783-784	Corpus Christi	.80
785	Mc Allen	.78
786-787	Austin	.78
788	Del Rio	.68
789	Giddings	.72
790-791	Amarillo	.81
792	Childress	.75
793-794	Lubbock	.78
795-796	Abilene	.79
797	Midland	.78
798-799,885	El Paso	.79

Adjusting Project Costs to Your Location

STATE	CITY	Residential
UTAH		
840-841	Salt Lake City	.90
842,844	Ogden	.90
843	Logan	.91
845	Price	.81
846-847	Provo	.90
VERMONT		
050	White River Jct.	.73
051	Bellows Falls	.73
052	Bennington	.72
053	Brattleboro	.74
054	Burlington	.85
056	Montpelier	.84
057	Rutland	.87
058	St. Johnsbury	.74
059	Guildhall	.73
VIRGINIA		
220-221	Fairfax	.89
222	Arlington	.89
223	Alexandria	.89
224-225	Fredericksburg	.83
226	Winchester	.78
227	Culpeper	.78
228	Harrisonburg	.75
229	Charlottesville	.83
230-232	Richmond	.86
233-235	Norfolk	.82
236	Newport News	.82
237	Portsmouth	.81
238	Petersburg	.86
239	Farmville	.74
240-241	Roanoke	.76
242	Bristol	.79
243	Pulaski	.73
244	Staunton	.76
245	Lynchburg	.80
246	Grundy	.72
WASHINGTON		
980-981,987	Seattle	1.00
982	Everett	.97
983-984	Tacoma	1.05
985	Olympia	1.05
986	Vancouver	1.11
988	Wenatchee	.94
989	Yakima	1.02
990-992	Spokane	.99
993	Richland	1.00
994	Clarkston	.99
WEST VIRGINIA		
247-248	Bluefield	.89
249	Lewisburg	.90
250-253	Charleston	.93
254	Martinsburg	.75
255-257	Huntington	.93
258-259	Beckley	.90
260	Wheeling	.93
261	Parkersburg	.92
262	Buckhannon	.97
263-264	Clarksburg	.97
265	Morgantown	.97
266	Gassaway	.93
267	Romney	.90
268	Petersburg	.95

STATE	CITY	Residential
WISCONSIN		
530,532	Milwaukee	1.02
531	Kenosha	1.02
534	Racine	1.06
535	Beloit	1.00
537	Madison	1.00
538	Lancaster	.91
539	Portage	.98
540	New Richmond	1.03
541-543	Green Bay	1.00
544	Wausau	.98
545	Rhinelander	.98
546	La Crosse	.98
547	Eau Claire	1.04
548	Superior	1.03
549	Oshkosh	.97
WYOMING		
820	Cheyenne	.86
821	Yellowstone Nat. Pk.	.80
822	Wheatland	.83
823	Rawlins	.81
824	Worland	.78
825	Riverton	.81
826	Casper	.86
827	Newcastle	.79
828	Sheridan	.83
829-831	Rock Springs	.83
CANADIAN FACTORS (reflect Canadian currency)		
ALBERTA		
	Calgary	.99
	Edmonton	.99
BRITISH COLUMBIA		
	Vancouver	1.05
	Victoria	1.04
MANITOBA		
	Winnipeg	.97
NEW BRUNSWICK		
	Moncton	.92
	Saint John	.96
NEWFOUNDLAND		
	St. John's	.94
NOVA SCOTIA		
	Halifax	.96
ONTARIO		
	Hamilton	1.12
	Kitchener	1.05
	London	1.08
	Oshawa	1.09
	Ottawa	1.09
	Sudbury	1.04
	Thunder Bay	1.05
	Toronto	1.12
	Windsor	1.06
PRINCE EDWARD ISLAND		
	Charlottetown	.92

Adjusting Project Costs to Your Location

STATE	CITY	Residential
QUEBEC		
	Chicoutimi	1.02
	Montreal	1.08
	Quebec	1.10
SASKATCHEWAN		
	Regina	.92
	Saskatoon	.91

GLOSSARY

A

Access door or panel

A means of access for the inspection, repair, or service of concealed systems, such as air conditioning equipment.

Anchor bolt, foundation bolt, hold-down bolt

A threaded bolt, usually embedded in a foundation, for securing a sill, framework, or machinery.

Apron

(1) A piece of finished trim placed under a window stool. (2) A slab of concrete extending beyond the entrance to a building, particularly at an entrance for vehicular traffic.

Architectural millwork, custom millwork

Millwork manufactured to meet specifications of a particular job, as distinguished from stock millwork.

Asphalt shingles, composition shingles, strip slates

Roofing felt, saturated with asphalt and coated on the weather side with a harder asphalt and aggregate particles, which has been cut into shingles for application to a sloped roof.

B

Backer

Three studs nailed together in a U-shape, to which a partition is attached.

Baluster, banister

One of a series of short, vertical supporting elements for a handrail or a coping.

Base flashing

In roofing, the flashing supplied by the upturned edges of a watertight membrane.

Batt insulation

Thermal- or sound-insulating material, such as fiberglass or expanded shale, which has been fashioned into a flexible, blanket-like form. It often has a vapor barrier on one side. Batt insulation is manufactured in dimensions which facilitate its installation between the studs or joists of a frame construction.

Bead

Any molding, stop, or caulking used around a glass or panel to hold it in position.

Beam

A large horizontal structure of wood or steel.

Bearing

(1) The section of a structural member, such as a beam or truss, that rests on the supports. (2) Descriptive of any wall that provides support to the floor and/or roof of a building.

Bearing wall

Any wall that supports a vertical load as well as its own weight.

Bed

(1) The mortar into which masonry units are set. (2) Sand or other aggregate on which pipe or conduit is laid in a trench.

B-labeled door

A door carrying a certification from Underwriters' Laboratories that it is of a construction that will pass the standard fire door test for the length of time required for a Class B opening.

Blocking

(1) Small pieces of wood used to secure, join, or reinforce members, or to fill spaces between members. (2) Small wood blocks used for shimming.

Blueprint

Negative image reproduction having white lines on a blue background and made either from an original or from a positive intermediate print.

Board and batten

A method of siding in which the joints between vertically placed boards or plywood are covered by narrow strips of wood.

Board foot

The basic unit of measure for lumber. One board foot is equal to a 1″ board 1′ in width and 1′ in length.

Bonding agent

A substance applied to a suitable substrate to create a bond between it and a succeeding layer.

Box nail

Nail similar to a common nail, but with a smaller diameter shank.

Box sill

A common method of frame construction using a header nailed across the ends of floor joists where they rest on the sill.

Brace

(1) A diagonal tie that interconnects scaffold members. (2) A temporary support for aligning vertical concrete formwork. (3) A horizontal or inclined member used to hold sheeting in place. (4) A hand tool with a handle, crank, and chuck used for turning a bit or auger.

Brick

A solid masonry unit of clay or shale, formed into a rectangular prism while plastic, and then burned or fired in a kiln.

Building paper

A heavy, asphalt-impregnated paper used as a lining and/or vapor barrier between sheathing and an outside wall covering, or as a lining between rough and finish flooring.

Butt hinge

A common form of hinge consisting of two plates, each with one meshing knuckle edge and connected by means of a removable or fixed pin through the knuckles.

Butt joint

(1) A square joint between two members at right angles to each other. The contact surface of the outstanding member is cut square and is flush to the surface of the other member. (2) A joint in which the ends of two members butt each other so only tensile or compressive loads are transferred.

C

Cantilever

Any part of a structure that projects beyond its main support and is balanced on it.

Casement window

A window assembly having at least one casement or vertically hinged sash.

Casing

A piece of wood or metal trim that finishes off the frame of a door or window.

Caulk

(1) To fill a joint, crack, or opening with a sealer material. (2) The filling of joints in bell-and-spigot pipe with lead and oakum.

Cement

Any chemical binder that makes bodies adhere to it or to each other, such as glue, paste, or Portland cement.

Class A, B, C, D, E,

Fire-resistance ratings applied to building components such as doors or windows. The term "class" also refers to the opening into which the door or window will be fitted.

Common brick

Brick not selected for color or texture, and thus useful as filler or backing. Though usually not less durable or of lower quality than face brick, common brick typically costs less. Greater dimensional variations are also permitted.

Common nail

Nail used in framing and rough carpentry having a flat head about twice the diameter of its shank.

Concrete

A composite material that consists essentially of a binding medium within which are embedded particles or fragments of aggregate.

Cornice

The horizontal projection of a roof overhang at the eaves, consisting of lookout, soffit, and fascia.

Crown

The high point of a piece of lumber with a curve in it.

D

d

Abbreviation for penny. Refers only to nail size.

Dampproofing

An application of a water-resisting treatment or material to the surface of a concrete or masonry wall to prevent passage or absorption of water or moisture.

Deck

(1) An uncovered wood platform usually attached to or on the roof of a structure. (2) The flooring of a building. (3) The structural assembly to which a roof covering is applied.

Dressed lumber

Lumber that has been processed through a planing machine for the purpose of attaining a smooth surface and uniformity of size on at least one side or edge.

Drip edge

The edge of a roof that drips into a gutter or into the open.

Drywall

The term commonly applied to interior finish construction using preformed sheets, such as gypsum wallboard, as opposed to using plaster.

E

Eave

The part of a roof that projects beyond its supporting walls.

Estimate

The anticipated cost of materials, labor, services, or any combination of these for a proposed construction project.

F

Fascia, facia

(1) A board used on the outside vertical face of a cornice. (2) The board connecting the top of the siding with the bottom of a soffit. (3) A board nailed across the ends of the rafters at the eaves. (4) The edge beam of a bridge. (5) A flat member or band at the surface of a building.

Finish flooring

The material used to make the wearing surface of a floor, such as hardwood, tile, or terrazzo.

Flashing

A thin, impervious sheet of material placed in construction to prevent water penetration or direct the flow of water. Flashing is used especially at roof hips and valleys, roof penetrations, joints between a roof and a vertical wall, and in masonry walls to direct the flow of water and moisture.

Foundation

The material or materials through which the load of a structure is transmitted to the earth.

Frame

The wood skeleton of a building. Also called "framing."

Framing

(1) Structural timbers assembled into a given construction system. (2) Any construction work involving and incorporating a frame, as around a window or door opening. (3) The unfinished structure, or underlying rough timbers of a building, including walls, roofs, and floors.

Furring

(1) Strips of wood or metal fastened to a wall or other surface to even it, to form an air space, to give appearance of greater thickness, or for the application of an interior finish such as plaster. (2) Lumber 1″ in thickness (nominal) and less than 4″ in width, frequently the product of resawing a wider piece.

G

Gable

The portion of the end of a building that extends from the eaves upward to the peak or ridge of the roof.

Gable roof

A roof shape characterized by two sections of roof of constant slope that meet at a ridge; peaked roof.

Gambrel roof

A roof whose slope on each side is interrupted by an obtuse angle that forms two pitches on each side, the lower slope being steeper than the upper.

Girder

A large principal beam of steel, reinforced concrete, wood, or combination of these, used to support other structural members at isolated points along its length.

Glue laminated, glu-lam

The result of a process in which individual pieces of lumber or veneer are bonded together with adhesives to make a single piece in which the grain of all the constituent pieces is parallel.

Grade

(1) A designation of quality, especially of lumber and plywood. (2) Ground level. (3) The slope of the ground on a building site.

Grain

The direction of fibers in wood.

Ground

(1) A strip of wood that is fixed in a wall of concrete or masonry to provide a place for attaching wood trim or furring strips. (2) A screed, strip of wood, or bead of metal fastened around an opening in a wall and acting as a thickness guide for plastering or as a fastener for trim.

Grout

(1) An hydrous mortar whose consistency allows it to be placed or pumped into small joints or cavities, as between pieces of ceramic clay, slate, and floor tile. (2) Various mortar mixes used in foundation work to fill voids in soils, usually through successive injections through drilled holes.

Gutter

(1) A shallow channel of wood, metal, or PVC positioned just below and following along the eaves of a building for the purpose of collecting and diverting water from a roof. (2) In electrical wiring, the rectangular space allowed around the interior of an electrical panel for the installation of feeder and branch wiring conductors.

Gypsum

A naturally occurring, soft, whitish mineral (hydrous calcium sulfate) that, after processing, is used as a retarding agent in Portland cement and as the primary ingredient in plaster, gypsum board, and related products.

H

Header

(1) A rectangular masonry unit laid across the thickness of a wall, so as to expose its end(s). (2) A lintel. (3) A member extending horizontally between two joists to support tailpieces. (4) In piping, a chamber, pipe, or conduit having several openings through which it collects or distributes material from other pipes or conduits. (5) The wood surrounding an area of asphaltic concrete paving.

Hip

(1) The exterior inclining angle created by the junction of the sides of adjacent sloping roofs, excluding the ridge angle. (2) The rafter at this angle. (3) In a truss, the joint at which the upper chord meets an inclined end post.

Hip roof

A roof shape characterized by four or more sections of constant slope, all of which run from a uniform eave height to the ridge.

J

Jamb

An exposed upright member on each side of a window frame, door frame, or door lining. In a window, these jambs outside the frame are called "reveals."

Joint

(1) The point, area, position, or condition at which two or more things are jointed. (2) The space, however small, where two surfaces meet. (3) The mortar-filled space between adjacent masonry units. (4) The place where seperate but adjacent timbers are

connected, as by nails or screws, or by mortises and tenons, glue, etc.

Joist

A piece of lumber 2″ or 4″ thick and 6″ or more wide, used horizontally as a support for a ceiling or floor. Also, such a support made from steel, aluminum, or other material.

K

Keyway

A recess or groove in one lift or placement of concrete that is filled with concrete of the next lift, giving shear strength to the joint. Also called a "key."

Kneewall

A short wall under a slope, usually in attic space.

L

Labor-hour

A unit describing the work performed by one person in one hour.

Lally column

A trade name for a pipe column 3″ to 6″ in diameter, sometimes filled with concrete.

Layout

A design scheme or plan showing the proposed arrangement of objects and spaces within and outside a structure.

Level

(1) A term used to describe any horizontal surface that has all points at the same elevation and thus does not tilt or slope. (2) In surveying, an instrument that measures heights from an established reference. (3) A spirit level, consisting of small tubes of liquid with bubbles in each. The small tubes are positioned in a length of wood or metal that is hand held and, by observing the position of the bubbles, used to find and check level surfaces.

Lintel

A horizontal supporting member, installed above an opening such as a window or a door, that serves to carry the weight of the wall above it.

M

Main beam

A structural beam that transmits its load directly to columns, rather than to another beam.

Mansard roof

A type of roof with two slopes on each of four sides, the lower slope much steeper than the upper and ending at a constant eave height.

Millwork

All the building products made of wood that are produced in a planing mill such as moldings, door and window frames, doors, windows, blinds, and stairs. Millwork does not include flooring, ceilings, and siding.

Mortar

A plastic mixture used in masonry construction that can be troweled and hardens in place.

N

Nominal dimension

The size designation for most lumber, plywood, and other panel products.

Nominal size

The rounded-off, simplified dimensional name given to lumber.

Nonbearing wall

A dividing wall that supports none of the structure above it.

Nosing

The rounded front edge of a stair tread that extends over the riser.

O

On center (O.C.)

Layout spacing designation that refers to distance from the center of one framing member to another.

Outlet

The point in an electrical wiring circuit at which the current is supplied to an appliance or device.

P

Paint

(1) A mixture of a solid pigment in a liquid vehicle that dries to a protective and decorative coating. (2) The resultant dry coating.

Panel door

A door constructed with panels, usually shaped to a pattern, installed between the stiles and rails that form the outside frame of the door.

Paneling

The material used to cover an interior wall. Paneling may be made from a 4/4 sheet milled to a pattern and may be either hardwood or softwood plywood, often prefinished or overlaid with a decorative finish, or hardboard, also usually prefinished.

Particle board

A generic term used to describe panel products made from discrete particles of wood or other ligno-cellulosic material rather than from fibers. The wood particles are mixed with resins and formed into a solid board under heat and pressure.

Partition

An interior wall that divides a building into rooms or areas, usually nonload-bearing.

Pitch

The angle or inclination of a roof, which varies according to the climate and roofing materials used.

Plaster

A cementitious material or combination of cementitious material and aggregate that, when mixed with a suitable amount of water, forms a plastic mass or paste.

Plumb

Straight up and down, perfectly vertical.

Plumbing system

The water supply and distribution pipes; plumbing fixtures and traps; soil, waste, and vent pipes; building drains and sewers; and respective devices and appurtenances within a building.

Plywood

A flat panel made up of a number of thin sheets, or veneers, of wood, in which the grain direction of each ply, or layer, is at right angles to the one adjacent to it. The veneer sheets are united under pressure by a bonding agent.

Polystyrene foam

A low-cost, foamed plastic weighing about 1 lb. per cu. ft., with good insulating properties; it is resistant to grease.

Polyurethane

Reaction product of an isocyanate with any of a wide variety of other compounds containing an active hydrogen group. Polyurethane is used to formulate tough, abrasion-resistant coatings.

Polyvinyl chloride (PVC)

A synthetic resin prepared by the polymerization of vinyl chloride, used in the manufacture of nonmetallic waterstops for concrete, floor coverings, pipe and fittings.

Post-and-beam framing

Framing in which the horizontal members are supported by a distinct column, as opposed to a wall.

Prehung door

A packaged unit consisting of a finished door on a frame with all necessary hardware and trim.

Purlin

In roofs, horizontal member supporting the common rafters.

Q

Quotation

A price quoted by a contractor, subcontractor, material supplier, or vendor to furnish materials, labor, or both.

R

Rail

A horizontal member supported by vertical posts.

Resilient flooring

A durable floor covering that has the ability to resume its original shape, such as linoleum.

Retaining wall

(1) A structure used to sustain the pressure of the earth behind it. (2) Any wall subjected to lateral pressure other than wind pressure.

Rib

One of a number of parallel structural members backing sheathing.

Rise and run

The angle of inclination or slope of a member or structure, expressed as the ratio of the vertical rise to the horizontal run.

Riser

A vertical member between two stair treads.

Rough flooring

Any materials used to construct an unfinished floor.

Rough opening (R.O.)

Any opening formed by the framing members to accommodate doors or windows.

Run

(1) In a roof with a ridge, the horizontal distance between the edge of the rafter plate (building line) and the center line of the ridgeboard. (2) In a stairway, the horizontal distance between the top and bottom risers plus the width of one tread.

"R" value

A measure of a material's resistance to heat flow given a thickness of material. The term is the reciprocal of the "U" value. The higher the "R" value, the more effective the particular insulation.

S

Sheathing

(1) The material forming the contact face of forms. Also called "lagging" or "sheeting."(2) Plywood, waferboard, oriented strand board, or lumber used to close up side walls, floors, or roofs preparatory to the installation of finish materials on the surface.

Shed dormer

A dormer window having vertical framing projecting from a sloping roof, and an eave line parallel to the eave line of the principal roof.

Shim

A thin piece of material, often tapered (such as a wood shingle) inserted between building materials for the purpose of straightening or making their surfaces flush at a joint.

Shingle

A roof-covering unit made of asphalt, wood, slate, cement, or other material cut into stock sizes and applied on sloping roofs in an overlapping pattern.

Shoe

Any piece of timber, metal, or stone receiving the lower end of virtually any member.

Sill

(1) The horizontal member forming the bottom of a window or exterior door frame. (2) As applied to general construction, the lowest member of the frame of the structure, resting on the foundation and supporting the frame.

Sized lumber

Lumber uniformly manufactured to net surfaced sizes. Sized lumber may be rough, surfaced, or partly surfaced on one or more faces.

Slab

A flat, horizontal (or nearly so) molded layer of plain or reinforced concrete, usually of uniform but sometimes of variable thickness, either on the ground or supported by beams, columns, walls, or other framework.

Sleeper

Lumber laid on a concrete floor as a nailing base for wood flooring.

Slope

The pitch of a roof, expressed as inches of rise per 12" of run.

Soffit

The underside of a projection, such as a cornice.

Span

(1) The distance between supports of a member. (2) The measure of distance between two supporting members.

Spread footing

A generally rectangular prism of concrete, larger in lateral dimensions than the column or wall it supports, that distributes the load of a column or wall to the subgrade.

Stair

(1) A single step. (2) A series of steps or flights of steps connected by landings, used for passage from one level to another.

Stepped footing

A wall footing with horizontal steps to accommodate a sloping grade or bearing stratum.

Stock lumber

Lumber cut to standard sizes and readily available from suppliers.

Strapping

(1) Flexible metal bands used to bind units for ease of handling and storage. (2) Another name for *furring*.

Stringer

(1) A secondary flexural member parallel to the longitudinal axis of a bridge or other structure. (2) A horizontal timber used to support joists or other cross members.

Structural lumber

Any lumber with nominal dimensions of 2″ or more in thickness and 4″ or more in width that is intended for use where working stresses are required.

Stud

(1) A vertical member of appropriate size (2 x 4 to 4 x 10 in. or 50 x 100 to 100 x 250 mm) and spacing (16″ to 30″ or 400 to 750 mm) to support sheathing or concrete forms. (2) A framing member, usually cut to a precise length at the mill, designed to be used in framing building walls with little or no trimming before it is set in place.

Subfloor sheathing

The rough floor, usually plywood, laid across floor joists and under finish flooring.

T

Thermal barrier

An element of low conductivity placed between two conductive materials to limit heat flow, for use in metal windows or curtain walls that are to be used in cold climates.

Threshold

A shaped strip on the floor between the jambs of a door, used to separate different types of flooring or to provide weather protection at an exterior door.

Toe

(1) Any projection from the base of a construction or object to give it increased bearing and stability. (2) That part of the base of a retaining wall that projects beyond the face, away from the retained material.

Toenailing

To drive a nail at an angle to join two pieces of wood.

Tongue and groove

Lumber machined to have a groove on one side and a protruding tongue on the other so that pieces will fit snugly together, with the tongue of one fitting into the groove of the other.

Top plate

A member on top of a stud wall on which joists rest to support an additional floor or form a ceiling.

Truss

A structural component composed of a combination of members, usually in a triangular arrangement, to form a rigid framework; often used to support a roof.

V

Vapor barrier

Material used to prevent the passage of vapor or moisture into a structure or another material, thus preventing condensation within them.

Veneer

(1) A masonry facing that is attached to the backup but not so bonded as to act with it under load. (2) Wood peeled, sawn, or sliced into sheets of a given constant thickness and combined with glue to produce plywood.

Ventilation

A natural or mechanical process by which air is introduced to or removed from a space, with or without heating, cooling, or purification treatment.

W

Waterstop

A thin sheet of metal, rubber, plastic, or other material inserted across a joint to obstruct the seeping of water through the joint.

Wood preservative

Any chemical preservative for wood, applied by washing on or pressure-impregnating. Products used include creosote, sodium fluoride, copper sulfate, and tar or pitch.

Z

Z-bar

A Z-shaped member that is used as a main runner in some types of acoustical ceiling.

Zoning permit

A permit issued by appropriate government officials authorizing land to be used for a specific purpose.

ABBREVIATIONS

A	Area	kW	Kilowatt
Ab	Above	L or Ldr	Leader
Abs	Absolute	L or Lth	Length
AFF	Above finished floor	Lav	Lavatory
Al	Aluminum	Lb	Pound
Avg	Average	LF	Linear feet
BCF	Backfill	LH	Labor-Hours
BOCA	Building Officials and Code	Mat	Material
	Administrators	Max	Maximum
Br	Branch	Mfr	Manufacturer
BTU	British thermal unit	Min	Minimum (or minute)
C	Centigrade	MS	Milled steel
°C	Degrees centigrade	NTS	Not to scale
C to C	Center to Center	Oz	Ounce
CA	Compressed air	P&T	Pressure and temperature
CF	Cubic Feet	Pb	Lead
Cfm	Cubic feet per minute	PG	Pressure gauge
CI	Cubic Inches	pH	Hydrogen concentration
Circ.	Circulator/Circulation	PIV	Post indicator valve
CL el	Centerline elevation	PO	Plugged outlet
Clg	Ceiling	Ppm	Parts per million
CTE	Connection to existing	Press	Pressure
Cu	Copper	PSI	Pounds per square inch
CW	Cold water	PVC	Polyvinylchloride
CY	Cubic yard	Qt	Quart
D	Drain	Qty	Quantity
Deg or °	Degrees	R	Hydraulic radius
Dn	Down	Rad	Radius
Dp	Deep	RCP	Reinforced concrete pipe
Dwg	Drawing	RD	Rate of demand (or roof
Elev.	Elevation		drain)
Exc	Excavation	Red	Reducer
F	Fahrenheit	RT	Running trap
°F	Degrees fahrenheit	RV	Relief valve
FAI	Fresh air intake	S	Soil
FF	Finish floor	S&W	Soil and waste
FG	Finish grade	SA	Shock absorber
Fig	Figure	San	Sanitary
Fixt	Fixture	Sb	Antimony
Flr	Floor	Sc	Sillcock
Ga	Gauge	Sec	Second
Gal	Gallon (231 CI)	SF	Square foot
Galv.	Galvanized	Shwr	Shower
Gas	Gallons	SI	Square inches
H	Hydrogen or handicapped	Spec	Specification
HClg	Hung ceiling	Std	Standard
Hd	Head	T	Temperature (or time)
HP	Horsepower	Therm	Thermometer
Hr	Hour	V	Vent
HT	House trap	Vac	Vacuum
Htr	Heater	Vel	Velocity
HW	Hot water	Vol	Volume
In	Inch	Wgt	Weight
Jt	Joint		

INDEX

REFERENCE BOOKS

Exterior Home Improvement Costs 8th Edition

Quick estimates for 64 projects, including:

- Room Additions
- Garages
- Dormers, Roofs, Skylights
- Patios & Decks
- Painting & Siding
- Walls, Fences
- Landscaping, Driveways & Porches
- New! Price Comparison Sheets for Material and Equipment

Includes tips for working with an architect, evaluating cost versus resale value, and procuring permits.

$19.95 per copy
Over 270 pages, illustrated, softcover
Catalog No. 67309D   ISBN 0-87629-657-6

Means Illustrated Construction Dictionary 3rd Edition

Your comprehensive guide to understanding the words, terms, phrases, concepts, slang, abbreviations, symbols, and acronyms of today's construction. Includes 19,000 construction words, terms, phrases, symbols, weights, measures, and equivalents.

Long regarded as the Industry's finest, the *Means Illustrated Construction Dictionary* is now even better. With the addition of over 1000 new terms and hundreds of new illustrations, it is the clear choice for the most comprehensive and current information.

The companion CD-ROM that comes with this new edition adds many extra features: larger graphics, expanded definitions, and links to both CSI MasterFormat numbers and production information.

$99.95 per copy
Over 800 pages, illustrated, hardcover
Catalog No. 67292A   ISBN 0-87629-538-3

Residential & Light Commercial Construction Standards

A Unique Collection of Industry Standards That Define Quality in Construction—For Contractors & Subcontractors, Owners, Developers, Architects & Engineers, Attorneys & Insurance Personnel

Compiled from the nation's major building codes, and from scores of publications and reports from professional institutes and other authorities, this one-of-a-kind resource enables you to:

- Set a standard for subcontractors and employees
- Protect yourself against defect claims
- Resolve disputes
- Overview installation methods
- Answer client questions

$59.95 per copy
Over 500 pages, illustrated, softcover
Catalog No. 67322   ISBN 0-87629-499-9

Builder's Essentials
Plan Reading & Material Takeoff

A complete course in reading and interpreting building plans—and performing quantity takeoffs to professional standards.

Organized by CSI division, this book shows and explains, in clear language and with over 160 illustrations, typical working drawings encountered by contractors in residential and light commercial construction. The author describes not only how all common features are represented, but how to translate that information into a materials list. Each chapter uses plans, details and tables, and a summary checklist.

$35.95 per copy
Over 420 pages, illustrated, softcover
Catalog No. 67307   ISBN 0-87629-348-8

Contractor's Pricing Guide:
Residential Detailed Costs 2002

Every aspect of residential construction, from overhead costs to residential lighting and wiring, is in here. All the detail you need to accurately estimate the costs of your work with or without markups— labor-hours required, typical crews and equipment are included as well. When you need a detailed estimate, this publication has all the costs to help you come up with a complete, on-the-money price you can rely on to win profitable work.

$36.95 per copy
Over 300 pages, with charts and tables, 8-1/2 x 11
Catalog No. 60332   ISBN 0-87629-648-7

Contractor's Pricing Guide:
Residential Repair & Remodeling Costs 2002

This book provides total unit price costs for every aspect of the most common repair & remodeling projects. Organized in the order of construction by component and activity, it includes demolition and installation, cleaning, painting, and more.

With simplified estimating methods; clear, concise descriptions; and technical specifications for each component, the book is a valuable tool for contractors who want to speed up their estimating time, while making sure their costs are on target.

$36.95 per copy
Over 250 pages, illustrated, 8-1/2 x 11
Catalog No. 60342   ISBN 0-87629-655-X

MATERIAL SHOPPING LIST

Use this form to develop a shopping list of materials needed to complete your home improvement project. Materials have been grouped under headings that will make it easier for you to find items at your lumberyard or home center. Use the blank lines under each category for other items you may select.

The shaded section of the form is provided to help you track multiple prices, if you do comparison shopping. Simply enter the name of each store at the top of the column, and put the prices obtained from each one in the spaces provided. Circle the best price for each item.

This form can become a Project Estimate by performing the following steps:

1. Enter the quantity and the cost of each item in the appropriate spaces in the **Project Estimate** section. Multiply **Quantity** times **Cost** to determine the **Total Cost** for each item.

2. Add all the numbers in the Total Cost column *on each page* to determine **Page Totals**. Enter each Page Total in the space provided at the bottom of the page.

3. On the last page, add all Page Totals to determine the **Project Subtotal**.

4. Multiply the Project Subtotal by your local sales tax (if necessary) to determine the **Project Total**. (For example, if the sales tax is 5%, multiply the Project Subtotal by 1.05)

ITEM DESCRIPTION	Unit	PRICE COMPARISON			PROJECT ESTIMATE			
		Store 1	Store 2	Store 3	Quantity	× Cost Used	=	Total Cost
Building Materials								
3/8" Gypsum Board								
1/2" Gypsum Board								
1/2" Gypsum Board, Water Resistant								
Cement Board								
Joint Compound								
Drywall Accessories								
Suspended Ceiling System								
Main Runner								
4' Tee								
2' Tee								
2' x 2' Ceiling Tile								
Hanger Wire								
Paneling								
Mouldings								
Baseboard, pine								
Chair Rail								
Crown Moulding								
1 x 3								
Door Casing								
Fireplace Mantel								

Page Total _____

ITEM DESCRIPTION	Unit	PRICE COMPARISON			PROJECT ESTIMATE		
		Store 1	Store 2	Store 3	Quantity ×	Cost Used =	Total Cost
Window Casing							
Shelving							
Wood							
Plastic Laminate Covered Wood							
Stair Parts							
Treads							
Risers							
Starting Strip							
Handrail							
Skirt Board							
Newel Post							
Balusters							
Railing							
Circular Stairs—Metal							
Disappearing Attic Stairs							
Insulation							
Fiberglass Batt, Kraft Faced, 3-1/2″							
Fiberglass Batt, Kraft Faced, 6″							
Fiberglass Batt, Foil Faced, 6″							
Fiberglass Batt, Unfaced, 6″							
Rigid Molded Beadboard, 1/2″							
Polyethylene Vapor Barrier							
Asphalt Felt Building Paper							
Lumber							
Framing Lumber							
1 x 4							
2 x 3							
2 x 4							
2 x 6							
2 x 8							
2 x 10							
2 x 12							
4 x 4							
2 x 4 Pressure Treated							
Plywood							
1/4″ AC Birch							
3/4″ AC Birch							
Cedar Closet Lining							
3/16″ Hardboard Underlay							
5/8″ Underlayment							

Page Total _____

ITEM DESCRIPTION	Unit	PRICE COMPARISON			PROJECT ESTIMATE		
		Store 1	Store 2	Store 3	Quantity ×	Cost Used =	Total Cost
1/2" Interior							
5/8" CDX							
Doors & Windows							
Interior Door							
Flush							
6-Panel							
Hollow Core, Pre-hung							
Closet Door							
Bi-Fold							
Bi-Pass							
Hardware and Fasteners							
Nails							
Screws							
Hinges							
Door Hardware							
Lock Set (Door knob with key lock)							
Privacy Set (Door knob with privacy lock)							
Passage Set (Door knob without lock)							
Closet System—Vinyl-coated Wire							
Drawer Pulls							
Plumbing							
Fixtures							
Water Closet							
Bathtub							
Shower Receptor							
Bidet							
Vanity Sink							
Kitchen Sink							
Fittings							
Shower Fittings							
Bathtub Fittings							
Vanity Sink Fitting							
Kitchen Sink Fitting							
Rough Plumbing							
Supply Pipe							

Page Total _____

ITEM DESCRIPTION	Unit	PRICE COMPARISON			PROJECT ESTIMATE		
		Store 1	Store 2	Store 3	Quantity ×	Cost Used =	Total Cost
Drain Pipe							
Vent Pipe							
Traps							
Closet Flange and Ring							
Paint and Decor							
Paint							
Stain							
Urethane							
Drop Cloth							
Brushes							
Sandpaper							
Electrical							
Switches							
Light Switch							
Dimmer Switch							
Receptacles							
Common							
GFI							
Telephone							
Boxes							
Wire							
Baseboard Heat							
Appliances							
Range							
Cooktop							
Microwave							
Refrigerator							
Range Hood							
Dishwasher							
Lighting and Ceiling Fans							
Lights							
Undercabinet Light							
Pendant Light							
Recessed Light							
Vanity Light Strip							
Hall Light							
Track Light Strip and Fixtures							
Chandelier							

Page Total _____

ITEM DESCRIPTION	Unit	PRICE COMPARISON			PROJECT ESTIMATE		
		Store 1	Store 2	Store 3	Quantity ×	Cost Used =	Total Cost
Vents and Fans							
Whole House Fan							
Vent Exhaust Fan							
Vent/Heat/Light Ceiling Fixture							
Floor and Wall Coverings							
Wallpaper							
Vinyl Wall Covering							
Wallpaper Paste/Adhesive							
Carpet							
Carpet Pad							
Wood Flooring							
Wood Parquet							
Unfinished Wood Strip							
Prefinished Wood Strip							
Floating Floor							
Tile for Floors and Walls							
Quarry Tile							
Ceramic Tile							
Marble Tile							
Tile Base							
Resilient Flooring							
Sheet Vinyl							
Vinyl Tile							
Adhesive							
Grout							
Cleaner							
Latex Underlayment/Floor Leveler							
Kitchen/Bath Furnishings							
Cabinets							
Base Cabinet							
Wall Cabinet							
Sink Base							
Vanity Base							
Island Base Unit							
Bookcase							
Valance Board							
Countertop							
Plastic Laminate							
Solid Surface							

Page Total _____

ITEM DESCRIPTION	Unit	PRICE COMPARISON			PROJECT ESTIMATE		
		Store 1	Store 2	Store 3	Quantity ×	Cost Used =	Total Cost
Ceramic Tile							
Granite							
Butcher Block							
Marble							
Cultured Marble							
Medicine Cabinet							
Bathroom Accessories							
Towel Bar							
Toilet Tissue Dispenser							
Shower Curtain Rod							
Closet Pole							
Tools/Equipment							
Eye Protection							
Dust Mask							
Fire Extinguisher							
Hard Hat							
Work Boots							
Gloves							
Rubbish Barrels							
Plastic Film/Drop Cloths							
Rental Equipment							
Saw Blades							
Flat Bar, Cat's Paw							
Drill Bits							
Extension Cords							
Drop Light							
Saw Horses							

PAGE TOTAL _____

TOTAL FROM PREVIOUS PAGES _____

PROJECT SUBTOTAL _____

TAX @ _____ %

PROJECT TOTAL _____

NOTES

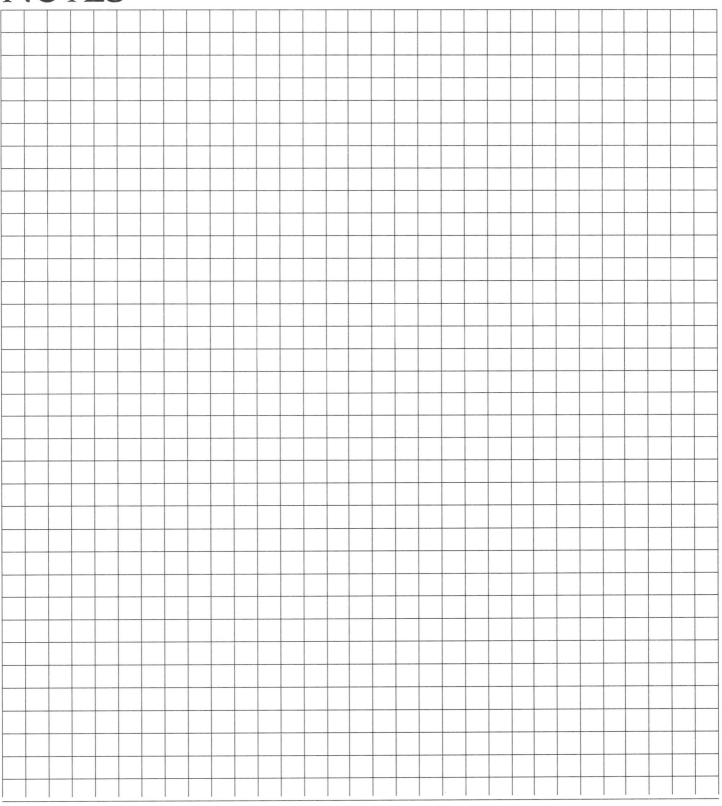

NOTES

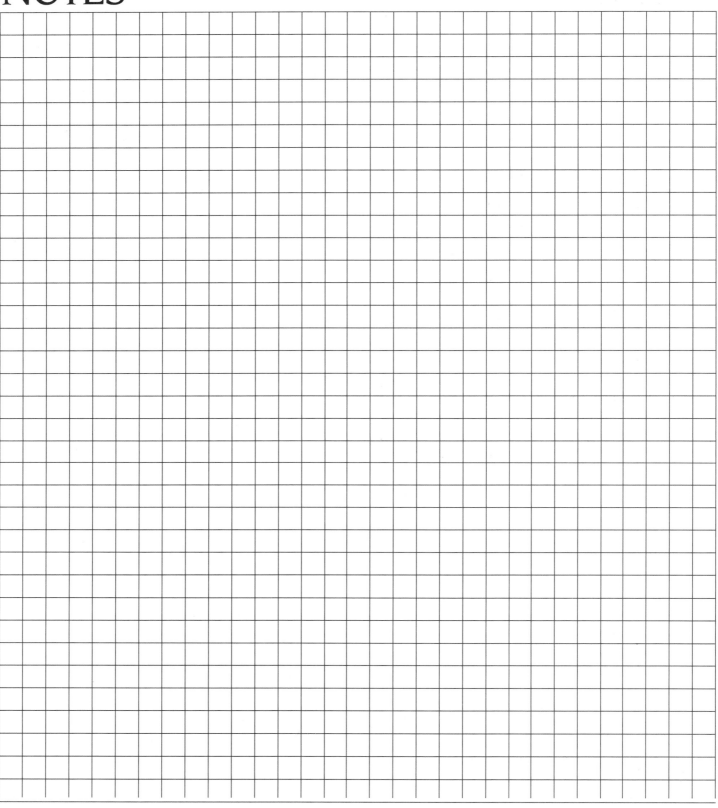

NOTES

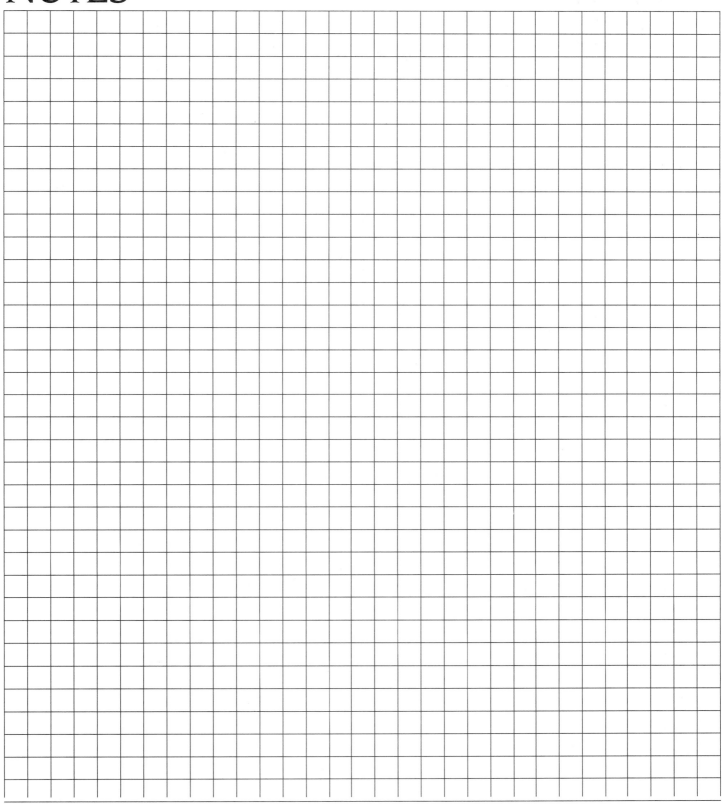

NOTES

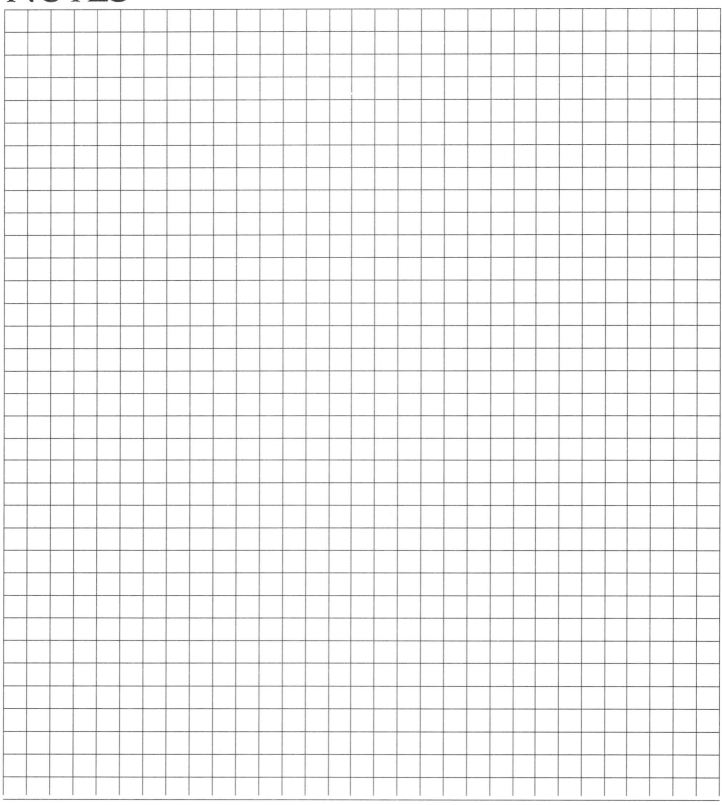

NOTES

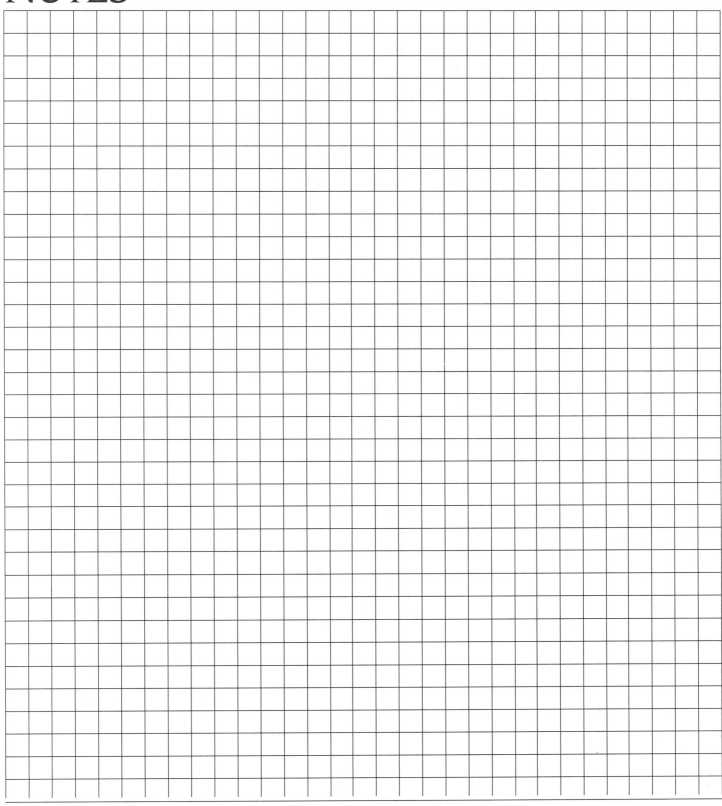

NOTES

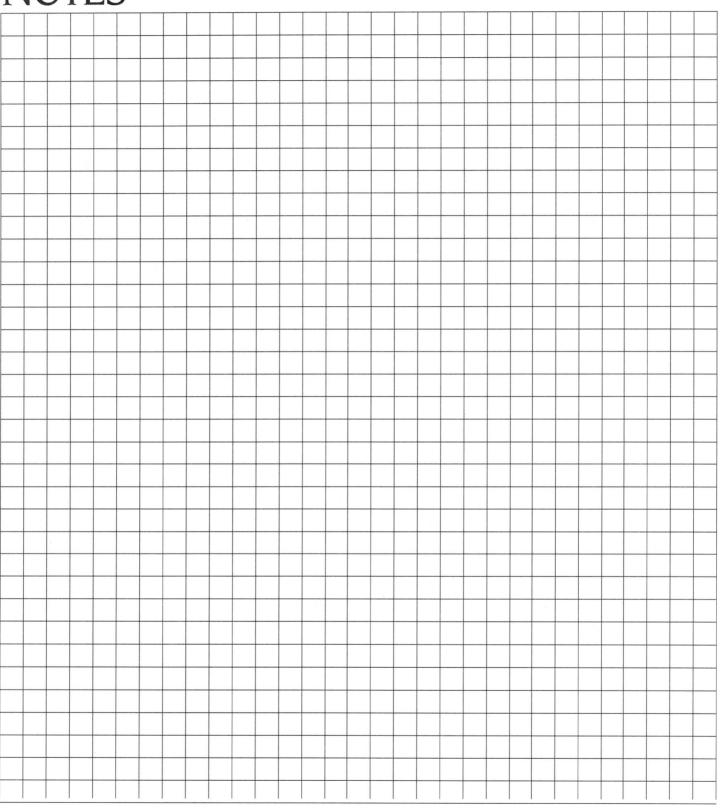

NOTES

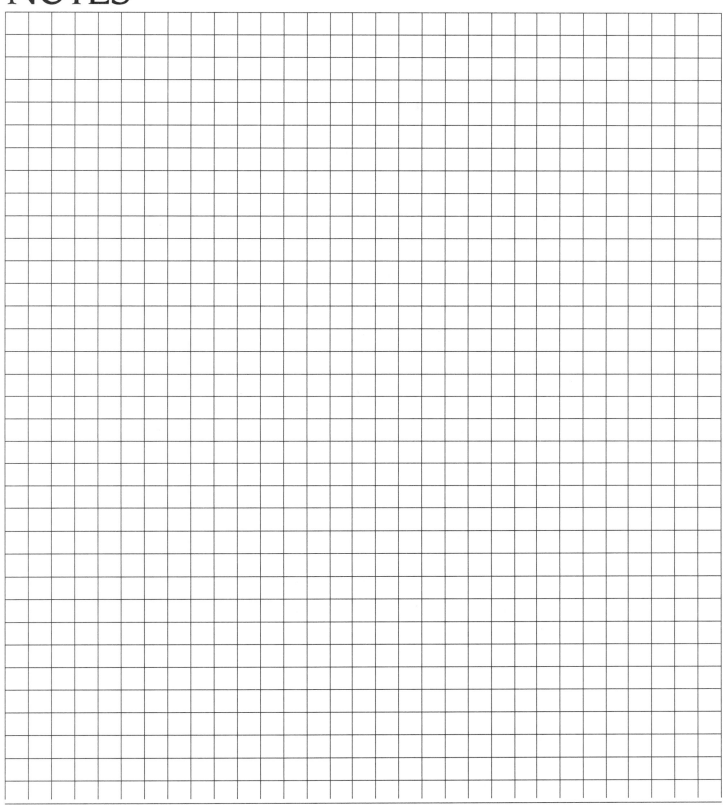

NOTES

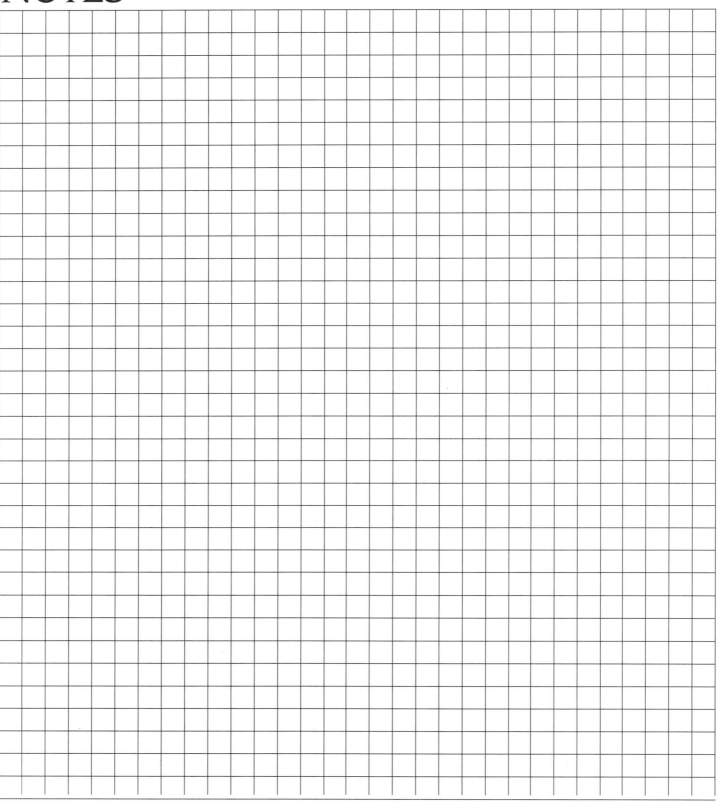

NOTES

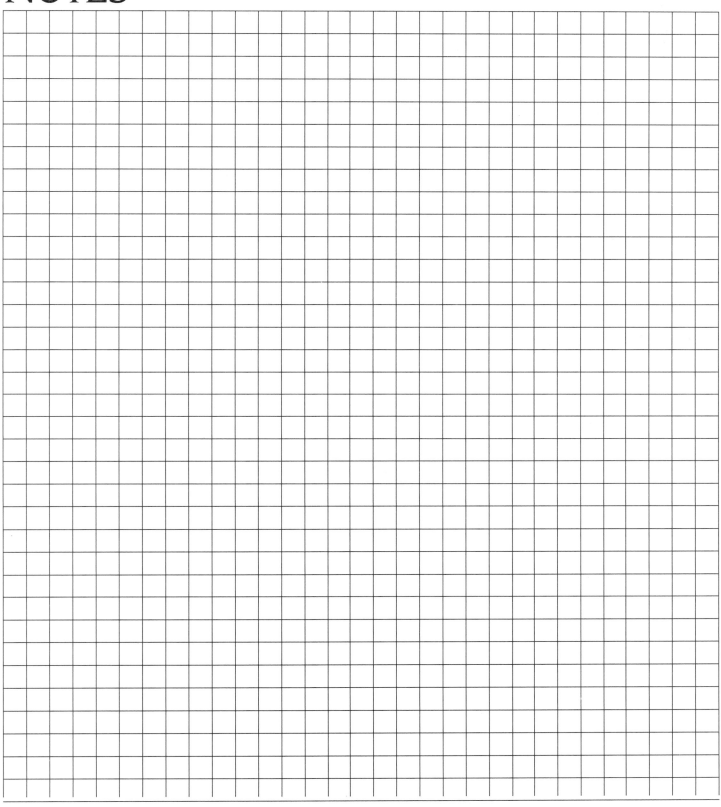

NOTES

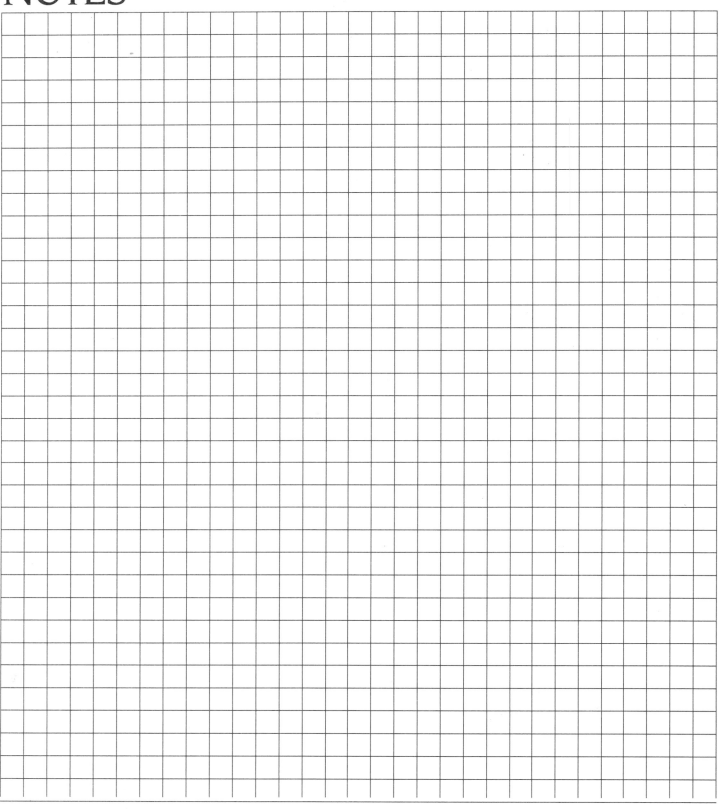